"This is a very useful resource for graduate students anu ....
ested in one of the most challenging puzzles in the theory of probability."

Hykel Hosni, *University of Milan, Italy*

"This is essential reading for anyone seriously interested in Bertrand's chord paradox or the broader epistemic issue of the status of the principle of indifference (and the maximum entropy principle)."

Darrell P. Rowbottom, *Lingnan University, Hong Kong*

Bertrand's Paradox and the Principle of Indifference

Events between which we have no epistemic reason to discriminate have equal epistemic probabilities. Bertrand's chord paradox, however, appears to show this to be false, and thereby poses a general threat to probabilities for continuum sized state spaces. Articulating the nature of such spaces involves some deep mathematics and that is perhaps why the recent literature on Bertrand's paradox has been almost entirely from mathematicians and physicists, who have often deployed elegant mathematics of considerable sophistication. At the same time, the philosophy of probability has been left out. In particular, left out entirely are the philosophical ground of the principle of indifference, the nature of the principle itself, the stringent constraint this places on the mathematical representation of the principle needed for its application to continuum sized event spaces, and what these entail for rigour in developing the paradox itself. This book puts the philosophy and its entailments back in and in so doing casts a new light on the paradox, giving original analyses of the paradox, its possible solutions, the source of the paradox, the philosophical errors we make in attempting to solve it and what the paradox proves for the philosophy of probability. The book finishes with the author's proposed solution—a solution in the spirit of Bertrand's, indeed—in which an epistemic principle more general than the principle of indifference offers a principled restriction of the domain of the principle of indifference.

Bertrand's Paradox and the Principle of Indifference will appeal to scholars and advanced students working in the philosophy of mathematics, epistemology, philosophy of science, probability theory and mathematical physics.

Nicholas Shackel is Professor of Philosophy, Cardiff University and Distinguished Research Fellow at the Oxford Uehiro Centre for Practical Ethics, Oxford University. His research is mainly on paradoxes and rationality. He has published numerous articles in books and leading journals including *Journal of Philosophy*, *Mind*, and *Philosophy and Phenomenological Research*.

Routledge Studies in the Philosophy of Mathematics and Physics
Edited by Elaine Landry, University of California, Davis, USA and Dean Rickles, University of Sydney, Australia

A Minimalist Ontology of the Natural World
Michael Esfeld and Dirk-André Deckert

Naturalizing Logico-Mathematical Knowledge
Approaches from Philosophy, Psychology and Cognitive Science
Edited by Sorin Bangu

The Emergence of Spacetime in String Theory
Tiziana Vistarini

Origins and Varieties of Logicism
On the Logico-Philosophical Foundations of Mathematics
Edited by Francesca Boccuni, and Andrea Sereni

The Foundations of Spacetime Physics
Philosophical Perspectives
Edited by Antonio Vassallo

Scientific Understanding and Representation
Modeling in the Physical Sciences
Edited by Kareem Khalifa, Insa Lawler, and Elay Shech

The Non-Fundamentality of Spacetime
General Relativity, Quantum Gravity, and Metaphysics
Kian Salimkhani

Bertrand's Paradox and the Principle of Indifference
Nicholas Shackel

For more information about this series, please visit: https://www.routledge.com/Routledge-Studies-in-the-Philosophy-of-Mathematics-and-Physics/book-series/PMP

Bertrand's Paradox and the Principle of Indifference

Nicholas Shackel

Routledge
Taylor & Francis Group

NEW YORK AND LONDON

First published 2024
by Routledge
605 Third Avenue, New York, NY 10158

and by Routledge
4 Park Square, Milton Park, Abingdon, Oxon, OX14 4RN

Routledge is an imprint of the Taylor & Francis Group, an informa business

Library of Congress Cataloging-in-Publication Data
Names: Shackel, Nicholas, 1955- author.
Title: Bertrand's paradox and the principle of indifference / Nicholas Shackel.
Description: New York, NY : Routledge, 2024. | Series: Routledge studies in the philosophy of math and physics | Includes bibliographical references and index.
Identifiers: LCCN 2023032608 (print) | LCCN 2023032609 (ebook) | ISBN 9781032597935 (hardback) | ISBN 9781032597980 (paperback) | ISBN 9781003456308 (ebook)
Subjects: LCSH: Probabilities. | Paradox. | Logic, Symbolic and mathematical. | Bertrand, Joseph, 1822-1900.
Classification: LCC QA273.4 .S53 2024 (print) | LCC QA273.4 (ebook) | DDC 165--dc23/eng/20231026
LC record available at https://lccn.loc.gov/2023032608
LC ebook record available at https://lccn.loc.gov/2023032609

ISBN: 978-1-032-59793-5 (hbk)
ISBN: 978-1-032-59798-0 (pbk)
ISBN: 978-1-003-45630-8 (ebk)

DOI: 10.4324/9781003456308

Typeset in Sabon
by SPi Technologies India Pvt Ltd (Straive)

For my Daughter
Eleanor Louise Grant

Contents

Preface

The mathematical theory of probability is a segment of measure theory and, on its own, neither tells us what probabilities are, nor determines their quantities, nor justifies their normative role. A philosophical theory of probability adopts some portion of the mathematical theory and adds the missing ingredients, thereby giving a metaphysical interpretation of the mathematical theory with determinable quantities and a grounding of their normative role in defining rational degrees of belief and methodologies of enquiry. Current philosophical theories that defend epistemic probability continue to rely, at least some of the time, on the principle of indifference to provide the second and third ingredients.

Bertrand formulated his paradoxes to undermine anything but a finitist classical theory of probability by showing that the principle of indifference produces inconsistent probabilities when applied to infinite possibilities. Contemporary theories that use the principle continue to face the threat posed by the paradoxes and for that reason philosophers, physicists and mathematicians have continued to offer solutions. Here I offer the first fully comprehensive and mathematically rigorous exposition and analysis of the paradox.

The normative ground and structure of the principle of indifference have been almost entirely neglected in the literature on the paradoxes. Here I rectify that neglect. What is presupposed by and what is defined by the principle are disentangled, on which basis its determination of quantities and normative roles is properly articulated. This leads to a comprehensive definition of the principle including a definition of its inputs and outputs, both in its most general form and in its generation of subordinate principles with narrower domains of application. This formulation of the principle avoids some of the simplest paradoxes due to a feature I call its generality, which feature plays a significant role throughout the book. I introduce elements of measure theory in order to articulate the principle's application to infinite state spaces. A brief analysis makes clear the need to distinguish two entirely different roles for measures in that application. I then develop mathematical expressions of the principle with a precision and rigor lacking elsewhere and prove the mathematical principle for sets as a principle subordinate to the principle of indifference. The rectification is continued as needed at various places in the book and along the way I prove novel mathematical subordinate principles for what I call meta-indifference and symmetric indifference.

I give Bertrand's four paradoxes for infinite state spaces in their first full translation into English (so far as I know), of which the paradox renownedly known as Bertrand's paradox, namely, the chord paradox, is our topic. I identify a number of significant frailties in Bertrand's procedure, each of which might justify rejecting the paradox. An extensive and mathematically rigorous reformulation of his cases shows that two can be well founded without any reliance on Bertrand's own flawed procedures or any vulnerability to his frailties, and shows the third to be irreparably flawed. I then turn to discussing exactly how the paradox threatens the principle, articulating an explicit argument on Bertrand's behalf (he gave none) from his chord paradox to the falsity of the principle of indifference. The analysis needed to formulate the argument leads to categorizing the four strategies to solve the paradox.

Over a number of chapters I examine the many instances of claimed solutions in the literature, among which I include the most recent based on radical mathematical proposals, solutions that have been neither analysed nor rebutted previously. I also develop solutions not in the literature, from permissivism, from a proposed criterion of identity for the principle of indifference and finally from the most general ground available for a solution, the hope from symmetry.

I have found the solutions discussed herein interesting and usually admirable. The mathematical resources deployed, especially those I have called radical, are valuable, elegant and ingenious additions to our thinking about probability. In part because of the nature of this mathematics the book is significantly interdisciplinary. My discussion requires substantial analysis and development of the modes and methods of these mathematical resources and thereby is not only relevant to analysis of the solutions but also adds, I think, to our understanding of their deployment in the philosophy of probability.

Returning once again to measure theory allows us to unearth the root of the failure to solve Bertrand's paradox. The way that informative measures are developed argues for a distinction between state spaces that have a natural measure on which to base a probability measure and state spaces that don't. Bertrand's original four paradoxes fall into two groups on either side of the naturalness distinction, which distinction explains why the chord paradox is rightly, renownedly, known as Bertrand's paradox.

In the penultimate chapter I review the temptations that lead us into erroneous solutions in our struggle with the paradox. The book concludes by first discussing whether our ultimate understanding should be that the paradox proves the principle false or whether its paradoxicality remains. I then propose my own solution. The negation of the principle of indifference means that events between which we have no epistemic reason to discriminate can have distinct epistemic probabilities, but that is impossible. And yet the paradox still stands, as has been shown both by the failures of the other solutions and by the unearthed root. Consequently, there is only one answer left: that the paradox is a proof that not all events have probabilities. I now need a principled restriction of the domain of the principle of indifference. From the proportioning ground used earlier I derive a principle more general than the principle of indifference and prove it true from some further assumptions about the nature of epistemic reasons. This then gives us a principled restriction of the domain of the principle of indifference.

Material from Shackel, N. 2007. Bertrand's Paradox and the Principle of Indifference. *Philosophy of Science*, 74, 150–75 reproduced with permission of Cambridge University Press. Copyright 2007 by the Philosophy of Science Association. All rights reserved.

Material from Rowbottom, D. P. & Shackel, N. 2010. Bangu's Random Thoughts on Bertrand's Paradox. *Analysis*, 70, 689–92 appears by permission of Oxford University Press and the Analysis Trust.

Material from Shackel, N. & Rowbottom, D. P. 2020. Bertrand's Paradox and the Maximum Entropy Principle. *Philosophy and Phenomenological Research*, 101, 505–23 appears by permission of John Wiley and Sons. © 2019 Philosophy and Phenomenological Research, LLC.

Material from Shackel, N. 2008. Paradoxes of Probability. In *Handbook of Probability Theory with Applications*. Ed. Rudas, T. Thousand Oaks: Sage. 49–66 appears by permission of Sage Publications.

Material from these articles appears with varying amounts of rewriting and development. The *Analysis* paper is the basis for §7.2 in Chapter 7. Of the *Philosophy of Science* paper: a principle and a theorem first stated there (but not proved) appear in Chapters 2 and 3 respectively; roughly a page of the set up analysis of that paper is developed in greater depth in about three pages in Chapter 4; the material on Marinoff, including some material on meta-indifference, contributes to Chapter 5; the material on

Jaynes contributes to §6.2 and 6.5 in Chapter 6 and to some remarks in the one-page introduction to that chapter. The *Philosophy and Phenomeno-logical Research* paper is the basis for Chapter 8. §16.4 draws on material in my chapter about probability paradoxes in Shackel 2008.

Figures 17, 19 and 21 originally appeared as Figures 1–3 in Rizza, D. 2018. A Study of Mathematical Determination through Bertrand's Paradox. *Philosophia Mathematica*, 26, 375–95 and are reproduced here by permission of Oxford University Press.

Figure 26 is reprinted from Figure 2 in Aerts, D. & Sassoli de Bianchi, M. 2014. Solving the Hard Problem of Bertrand's Paradox. *Journal of Mathematical Physics*, 55, 083503 with the permission of AIP Publishing.

I must express my gratitude to the Mind Association for awarding me their Major Research Fellowship for 2021–22, without which this book would not have been written; to Roger Crisp, Krister Bykvist and Brad Hooker for their encouragement and support; to the two anonymous reviewers for Routledge and to my editors, Andrew Weckenmann, Rosaleah Stammler and Andrea Harris, who have been unfailingly helpful.

1 The Principle of Indifference

1.1 Introduction

Our common talk has always acknowledged the uncertainty in our lives and has articulated aspects of it by speaking of good and bad luck, of things being more or less likely: in short, by speaking of chance and chances. Chance is the realm of uncertainty and chances are in one sense inhabitants of that realm and in another sense the feature had by those inhabitants that constitute a comparative ordering among them. Gamblers had the intuition that talk of chances, at least in their domain, was of a quantitative property and this led to the mathematical theory of probability as we now know it, in which theory probabilities are not only ordinals but cardinals as well.

In this book I assume that a certain care must be taken in distinguishing philosophical and mathematical theories of probability. The philosopher of probability is concerned with what probability is, just as the physicist is concerned with what the magnetic field is, and so is concerned with its metaphysical nature and its epistemic roles. The mathematical theory of probability is a segment of measure theory and, on its own, neither tells us what probabilities are nor explains how they are related to normative epistemic theory. Nor is there agreement that measure-theoretic probability is an adequate axiomatization of quantitative probability.

A philosophical theory of probability adopts some portion of the mathematical theory and adds the missing ingredients, thereby giving an interpretation of the mathematical theory with determinable quantities whose relation to epistemic requirements, such as proportioning belief to evidence and defining rational methodologies of enquiry, is explained. So the relation of probability itself to the mathematical theory is assumed to be that of the quantitative aspects of the former being representable by the latter because a region of the realm of uncertainty and the chances of that region can be a model of a segment of the mathematical theory, preferably in the technical sense of a model in formal logic.

DOI: 10.4324/9781003456308-1

In part because of the history, and in part because in formal logic what we call models of a formal theory we also call interpretations of that theory, philosophical theories of probability have been spoken of as interpretations of the mathematical theory of probability. In so doing we do not want to find ourselves incidentally committed to identifying probability itself with part or all of the mathematical theory of probability, any more than we want to identify the magnetic field with its representation by a vector field, since that identification is a contentious philosophical claim in its own right. We do not intend to get the metaphysics of probability the wrong way round just because we are calling the philosophical theories 'interpretations'.

There are good reasons for thinking the representation of probabilities by numbers cannot be a matter of identity. The mathematical theory of probability is by definition about normal measures[1] but there is nothing that stops probability being represented by other finite measures. This point does philosophical work. For example, Joyce shows how to derive the laws of probability from truth accuracy defined by a quadratic scoring rule (2009). Howson (2014b) faults Joyce on the grounds that Joyce's derivation of a normalized measure requires a coding of truth such that $1 = \text{code}(T) > \text{code}(\bot) = 0$ when that coding is mere convention. Howson's critique assumes that representing probabilities by a normal measure is necessary. In reply Joyce shows that 'a [measure] satisfies the scale invariant laws of probability if and only if all of its linear transforms satisfy them as well' (2015:418). Consequently, since measures got from his derivation vary linearly with the coding, they still represent probability even if not normal. So Howson's assumption

> is yet another instance of confusing the artefacts of a representation with genuine facts about the quantity being represented. Normality is a mere convention.
>
> (2015:417)

Normality is a convention of convenience in the *representation* of probability, whereas it is a *defining property* of the normal measures of measure theory that are called probability measures.

Another distinction between probability and its representation is that certainty is represented by the number 1, and yet in any continuum sized event space with a normal measure, an event can have measure 1 without being the whole space, so it is not certain. Similarly, the event that makes up the rest of the space has measure 0 and yet it is *not* certain *not* to

---

1 Normal measures are finite measures whose measure of the entire space is 1. See next chapter.

happen because it is not the empty event. So 0 and 1 are flawed representations of the extremal probabilities. Hence we have (at least) two reasons why probability itself is not a normal measure-theoretic mathematical entity. I discuss further the context and relevance of alternative axiomatizations of quantitative probability at the end of the next chapter.

So whilst positions in the philosophy of probability, such as the classical, logical, Bayesian, frequentist and propensity positions, may be called interpretations of the mathematical theory, what they really are is competing theories about what probability itself is and how it has an epistemic role. As such, they are philosophical theories of probability (rather than heuristic or methodological theories) because they are metaphysical and epistemic theories. For this reason when, for brevity, I speak of a *probabilism* I mean a philosophical theory of probability.

The philosophy of probability has developed simultaneously with the mathematical theory of probability and has for this reason concerned itself largely with probability as a precise quantitative property. As such, probability has the virtue of going beyond being more or less likely and allows of scalar comparisons such as being half as likely or twice as likely. Keynes took a step away in doubting that probabilities are necessarily totally ordered, let alone allowing of scalar comparison, but might be only partially ordered.[2] More recently the theories of imprecise probabilities return us in a different way to probabilities which do not always allow of scalar comparison.

In this book I assume that probability is a precise, totally ordered, scalar comparable, measurable (at least in principle) quantitative property, which probabilities can therefore be represented by cardinalities in the reals, even in the reals extended with infinitesimals. I assume that the density and completeness properties of the reals do not end up misrepresenting probabilities. What I shall address in these terms is complex enough already without attempting to extend the work to imprecise probabilities and for this reason I leave the work on imprecise probabilities to others. That being said, in the chapter on permissivism and the final chapter I shall say more about imprecise probabilities.

For simplicity and smoothness of exposition, the primary bearers of probabilities will be called possibilities or events, but no specific commitment to the metaphysical status of the primary bearers is thereby intended. Rather, these terms are to be understood as interpretable by whatever kind of object a probabilism takes to be the primary bearers of probability. There are some mismatches between measure-theoretic probability and

---

2 For Keynes' scepticism about the extent to which probabilities could be represented numerically see 1921/1973:ch.3. von Kries was an earlier sceptic who influenced Keynes: see Fioretti 2001.

some kinds of bearers (see Hájek and Hitchcock 2016:§1.8) but I will not be addressing such questions beyond the indications in Chapter 2 of some kinds of bearers that can match. Sometimes the bearers of probabilities I will speak of are secondary bearers, being the names of possibilities or events. The details of this will be made clear in Chapter 2.

1.2 The Ascertainability Criterion

The mathematical theory of probability does not, beyond the convention of normality, identify specific numerical probabilities of events. How that identification is achieved is one part of a philosophical theory of probability. Salmon calls the necessity for such identification the Ascertainability criterion: 'that there be some method by which, in principle at least, we can ascertain values of probabilities' (1967:64). The numerical probabilities, once identified, should satisfy Bishop Butler's criterion: 'Probability is the very Guide of Life' (Butler 1736:iij), called by Salmon the Applicability criterion (1967:64). Keynes holds that 'the probable is the hypothesis on which it is rational for us to act' (1921/1973). To be a guide to life means that they have normative import, that they are about what ought or ought not to be believed and are so such as to be the normatively correct cognitive input to discerning what ought or ought not to be done.[3] Evidently, they won't be much use in guiding life if it turns out that the method ascertains multiple values for the same event. So there are both practical and theoretical reasons why the method of ascertaining probability should produce a unique number rather than several distinct numbers.

The requirement for a method satisfying these criteria is not satisfied by a bag of heuristic tricks nor may its field of application be restricted by arbitrary convention or unprincipled convenience. We seek an epistemically principled method sufficient to determine consistent probabilities over various domains of uncertainty. Since it is what we *don't* know that produces the uncertainty, I shall say that what determines the probabilities is our state of ignorance. I take a state of ignorance to be an epistemic state that includes both what we do and don't know about a given domain, which state is determined by our possessed epistemic reasons.[4] This I express as

> The *ignorance requirement*: ([*subjective version*] for anyone,) for any state of ignorance over a domain of events, there exists a single

---

3 Cf. Hájek's list of applicabilities: frequencies, rational beliefs, rational decisions, ampliative inferences, science. (Hájek 2019).
4 This is compatible with prior work, see Jeffreys 1998 Chapter 3, Jaynes 2003 Chapter 12 and Norton 2008:45.

probability measure over that domain determined by an epistemically principled method ([*objective version*] for anyone).

I have used here the word 'single' rather than the natural 'unique' in order to avoid importing the special meaning that the latter has come to have in the debate over permissivism. The debate over permissivism is about what that singleness is relative to and uniqueness there means that singleness is relative to evidence. I shall carry forward this particular regimentation only when developing material that will be adverted to in the chapter on permissivism and otherwise revert to 'unique'. A theory or method that produces more than one probability measure is inconsistent because there is at least one event for which it gives more than one probability.

The domains of quantification are up for principled restriction, of course. The difference between objective and subjective versions is in the order of the universal personal quantifiers with respect to the existential probability measure quantifier, as indicated. The subjective version has the existential quantifier in the scope of the universal quantifier and vice versa for the objective version.

The method should also satisfy the

non-presumptuous requirement: the epistemically principled method, in some identifiable way, *respects* the state of ignorance rather than treats it arbitrarily.

Exactly what it is to respect the state of ignorance is up for philosophical argument, hence the requirement for respect in some identifiable way. Any specific claim to be the correct method would have to state why it was respectful. What lies behind this is the assumption that probability is related directly (for ontically subjective theories) or indirectly (for ontically objective theories) to rational belief and so lack of respect would entrain irrational bias in belief.

1.3 Origins

Classical probabilism is defined by its foundational principle, the principle of indifference, which identifies the base quantities and justifies probability's normative role. The principle of indifference is the name Keynes gave to the principle first set out by the 17th-century originators of probability theory and named the principle of insufficient reason by Bernoulli (1713). Laplace's definition of classical probabilism gives it thus:

The theory of chance consists in reducing all the events of the same kind to a certain number of cases equally possible, that is to say, to such as we may be equally undecided about in regard to their existence, and in determining the number of cases favourable to the event whose probability is sought. The ratio of this number to that of all the cases possible is the measure of this probability.[5]

(1814/1995:6–7)

Since Bertrand's paradox is the topic of this book, I note his near quotation of Laplace:

The probability of an event is the ratio of the number of favourable cases to the total number of possible cases. A condition is implied: all cases must be equally possible.

(1889:2 my translation here and throughout the book)

Without some such principle, classical probabilism is unable to satisfy Salmon's Ascertainability criterion.

Hajek regards Laplace's definition as inadequate

for what is it for an agent to *be* "*equally uncertain*" about a set of cases, other than assigning them equal probability? The notion of "equally possible" cases faces the charge of either being a category mistake (for possibility does not come in degrees), or circular (for what is meant is really 'equally probable').

(Hájek 2019: section 3.1, his emphasis on what in my translation is 'equally undecided'. From hereon emphasis is in the original unless I say otherwise.)

Hajek directs his point against Laplace but it is better taken against Bertrand's definition, except to the extent Bertrand is relying on Laplace. The point, however, is uncharitable. There is an obvious and better understanding of Laplace.

Equal possibility is had by those events over which we are 'equally undecided', for which we must attend to the import of Laplace's remarks that

probability is relative... to... ignorance [and]... to knowledge'.

(1814/1995:6)

---

5 Laplace's view on partitions for equal probability is not that there is a single canonical partition, but that it is a matter of judgement. This may make him closer to subjective Bayesians than he may at first appear, depending on whether or not he holds that rational judgement on the same reasons produces a canonical partition.

when there is no reason to believe that one of these cases must occur rather than the others, that makes them, for us, equally possible.

(Laplace 1812/1886:181, my translation)

So being equally possible is not a solecism about possibilities but is about their relation to epistemic reasons, and being equally undecided is not about a mere psychological state but about a warranted psychological state. Both are therefore a matter of equality of normative epistemic status. This is the identifiable way in which the principle respects ignorance.

Hence classical probabilism takes the needed step into epistemology to define equal possibility, thereby giving the theory its normative import. For example, the mathematical theory of probability alone tells me nothing of the probability of heads, but because I am equally undecided normatively over which side of the coin to expect, heads has the probability of one half and that is what I ought to believe.

Logical probabilism also identifies quantities using the principle of indifference. Keynes' version make more explicit the normative basis of the equality of probabilities:

if there is no known reason for predicating of our subject one rather than another of several alternatives, then relative to such knowledge... these alternatives have an equal probability.

(1921/1973:45)

Bayesians of various stripes have also a commitment to the principle of indifference, and since the history of their commitment is included in Chapter 15 I shall not address it here. I have placed in the chapter appendix versions of the principle in the literature that are not given elsewhere in the book.

1.4 The Principle of Indifference

The, or at least a, motivation behind taking the principle of indifference to constitute base probabilities is the epistemic requirement to believe in proportion to our possessed epistemic reasons, which I shall call the *Proportioning Ground*. This requirement has been spoken of in terms of sufficient and insufficient reason: that your credal commitments should discriminate between possibilities in ways for which you have sufficient reason and not discriminate in ways for which you lack sufficient reason. The last conjunct we might reasonably call the most general Principle of Insufficient Reason.

I am going to depart from this way of speaking, since what does the work is the matter of the various balances of epistemic reasons over the possibilities and that balance is prior to the question of meeting a threshold of sufficiency. I can subsume both sufficiency and insufficiency under

> *The Principle of Proportionate Reason*: your credal commitments should discriminate between possibilities in proportion to their balances of epistemic reasons.

The order this entails depends on the nature of the order given by balances of reasons: it need not entail a total order. Indeed, it might even define no order at all but only whether pairs of possibilities have an equal or an unequal balance of reasons. As we shall see shortly, that need not be a bad thing, and it is at least a sharper distinction than that between sufficiency and insufficiency of reason.

I shall say that we have *equal reason* for some possibilities just in case the balance of reasons for each is equal, which equality includes when we have no reasons. The latter is misleading in the following way. Consider possibilities of which we know no more than there are, for example, two of them. Although we may have no further information or evidence, and for that reason people speak of such a situation as one in which we have no reasons, that is not, strictly speaking, true. If we had no reason *at all* for either (as opposed to no reason for one over the other) then each is an epistemic impossibility, not a possibility.[6] So talk of having no reasons for possibilities is loose talk.

Equal reason generalizes over the varieties of normative epistemic equality appealed to in the many and various formulations of indifference principles, covering conditions such as epistemic indifference, epistemic neutrality, evidential equality, evidential symmetry, null states of background information and so on, to taste. Nor in so generalizing am I claiming these conditions are equivalent.[7] Our concerns do not turn on disputes over the epistemic equality used. When it suits the context, I shall also use the terms 'equal ignorance' or 'equal epistemic status' to express this generality. When discussing other authors I discuss them in terms of their own vocabulary, assuming that the link to my own is clear enough.

Since in taking the balance we have countenanced all the reasons, whatever they are, that bear on each possibility, possibilities having equal reason are therefore possibilities between which we lack any reason to discriminate. When we take credal commitments to be necessarily

---

6   Here I am disagreeing with the ancient Greeks: 'if the reasons in favour of two outcomes are equally balanced, neither should occur' (Zabell 2016b:317). In other words, equal reason implies impossibility whereas I say only zero reason implies impossibility.

7   For example, whether epistemic reasons are confined to evidence is contested.

rationally related to probabilities[8] and we apply the Principle of Proportionate Reason it issues in:

> *The Principle of Indifference* (POI): necessarily, probabilities of possibilities between which we lack any reason to discriminate epistemically are equal.[9]

If successful, this satisfies the ignorance requirement in a non-presumptuous way.[10]

One of the promises of this principle is that if we can successfully define the quantities of probabilities in this way, this of itself extends the coarse epistemic resource of only equal or unequal balances of reasons for possibilities. As is shown in the next chapter, the mathematical extension of the resultant equal probabilities gives us a total order over the whole domain of possibilities.

When it comes to applying the principle, I will have to define the mathematical representation of what I shall call the *structure of possibilities*, and of their having equal reason, in such a way that I can then derive the numerical probabilities. This is done in the next chapter, which is devoted to deriving the mathematical principle of indifference. For the time being, a simple example gives the general idea. The sides of a die are represented by the set of numbers $\{1, 2, 3, 4, 5, 6\}$ and the equal reason for each side is represented by the cardinality of the singleton sets of those numbers being the same for each. This then suffices to define the probability of any particular possibility in terms of the ratio of the cardinalities of subsets of $\{1, 2, 3, 4, 5, 6\}$.

I have been careful *not* to say simply that all the possibilities have one and the same probability. Stating the principle like that can fall into immediate difficulty. Take the book paradox: it is red or not red, gives

---

8 Standard relations are such as the rational strength of a belief being proportionate to, even identical to, the probability, and an even closer relation holds for those identifying probability itself with rational degrees of belief. I need not take a position on what this rational relation is.

9 To make the logical form absolutely clear: Necessarily, for all possibilities x, y, if we have no epistemic reason to discriminate between x and y then Probability (x) = Probability (y).

10 I note in passing that this could be derived from the following principle: *The Principle of Unequal Reason* (POUR): necessarily, the probability of a possibility for which you have at least as much epistemic reason as another is at least as great as the probability of the other. If you have no reason to discriminate between a pair of possibilities then you have at least as much epistemic reason for each as for the other, hence by POUR each of their probabilities is at least as great as the other and therefore they are equal. POI, however, may yet be more fundamental since it requires less of the relation of balances of reasons, namely, only that they be equal or unequal, rather than the partial order required by POUR.

probability of red a half; it is red or yellow or blue, gives probability of red a third. Attending to our possessed reasons and the consequent equal reason, however, groups red, yellow and blue together and not-red, not-yellow and not-blue, and the principle says the probabilities of the members of each group are the same but it does not say that the probability of red and not-red are the same.

Keynes discusses this and his particular way of analysing the problem leads him to a similar conclusion: 'our knowledge of the form and meaning of the alternatives may be a relevant part of the evidence' (1921/1973:66). If there is some implausibility in calling this knowledge evidence, I do not have to call it evidence since I do not have to say that epistemic reasons are all and only evidences.

Some conclude that we must avoid this problem by stating the principle in terms of 'basic' possibilities, such as in terms of a mutually exclusive and jointly exhaustive set of possible outcomes (see examples in the chapter appendix). This is in fact what Keynes cornered himself into doing, possibly because of his view of propositional functions (see 1921/1973:60–65). von Kries also took a similar approach in requiring 'suppositions to... correspond to elementary possibilities [ursprünglich spielräume] (von Kries 1886:36 my translation).

Certainly it is true that getting to the actual numerical probabilities from here does indeed depend on the structure of the possibilities. But it is not part of the principle to sort out the structure of possibilities. That is an input and the principle cannot be blamed for garbled output if it is handed garbled input.

There are three further reasons for not hobbling the principle in this way. The first is that it is simply not needed in order to avoid the book paradox, as was just shown. The second reason is that, whilst for many cases a principle confined to defining the same equal probability for all the basic possibilities suffices to define a probability measure, there are cases where the basic possibilities do not all have equal reason. Worse, for a continuum of possibilities such a principle cannot define a probability measure. My principle can—as is demonstrated in the next chapter. It is the third reason, the feature I shall call the *generality* of my principle, that explains the avoidance and the power.

The principle of indifference above allows that there may be distinct sets of possibilities, even distinct sets of basic possibilities, where belonging to a set is defined by each pair in the set having equal reason. Call such a set an equal reason set. In the book paradox we had two: a set of basic possibilities {red, yellow, blue} and a set of non-basic possibilities {not red, not yellow, not blue}. For such equal reason sets, the principle has two important features. First, it entails that possibilities in the same set have the same probability and this equality of probability is not confined to basic possibilities. Second, the principle does not entail that members of distinct equal reason sets have the same probability. It is the latter feature that blocks the book paradox and the former feature that makes the principle capable of defining probability measures for continua.

In offering the Proportioning Ground I do not intend to reject other ways of normatively grounding the principle. The point is to capture in somewhat general terms the kind of grounding given by theories of probability in which the principle plays a central rather than peripheral role in at least part of the theory. Such theories are epistemic, in the sense meant by Gillies when he says

> Interpretations of probability will be divided into (1) *epistemological...* and (2) *objective*. The difference is this. Epistemological interpretations of probability take probability to be concerned with the knowledge or belief of human beings. On this approach probability measures degree of knowledge, degree of rational belief, degree of belief or something of this sort. Clearly the logical, subjective and intersubjective interpretations are all epistemological. Objective interpretations of probability by contrast take probability to be a feature of the objective material world, which has nothing to do with human knowledge or belief. Clearly the frequency and propensity interpretations are objective.
>
> (2000:2)

Carnap earlier draws essentially the same distinction with his probability$_1$ (epistemic) and probability$_2$ (objective) (1950:ch.2) and Hume draws it more succinctly:

> Probability is of two kinds, either when the object is really in itself uncertain, and to be determin'd by chance; or when, tho' the object be already certain, yet'tis uncertain to our judgment, which finds a number of proofs on each side of the question.
>
> (1739/1978: II.3.9)

So in this sense, objective probabilities are ontically objective because they are mind independent and epistemic probabilities are ontically subjective because they are mind dependent, in at least the extended sense of being about the cognitive relation of minds to the world. I'm not sure whether classical and logical probabilists would accept that probabilities are, for that reason, ontologically subjective, at least in the sense in which Bayesians standardly take them to be so in identifying probabilities with degrees of belief.

I do not propose to resolve this question of ontology but will accept Gillies' distinction as given. For our purposes, frequency and propensity theories do not take the principle to central: they may take it as a bridge principle to do with rational belief or as having only an heuristic role in coming to know probabilities or reject it altogether. It is consequently not clear to what extent the paradoxes are really a problem for such theories. By contrast, for epistemic theories of probability it is central and the paradoxes are therefore a threat.

The principle is quite unrestricted: it is supposed to apply to any possibilities among which we have no reason to discriminate and to allow that to be sufficient to determine the equal probabilities, where such probabilities are the rational probabilities for these possibilities given the reasons possessed. It implies the method of obtaining numerical probabilities: they are got from ratios of suitable measures of sizes of the possibilities.

1.5 Epistemic Reason

In formulating the principle I have taken the facts about epistemic reason, such as what the epistemic reasons are, what the balance of reasons for and against any particular proposition is, and so on, to be basic. The natural assumption is that only theoretical reasons, in the sense of reasons having to do with the truth or falsity of a proposition rather than having to do with any practical benefit that may obtain from believing or disbelieving, can be epistemic reasons. What I have to say goes most smoothly on that assumption. I do not, however, need to take a position on whether there are any practical reasons that are epistemic reasons and so I do not.

Epistemic reason and epistemic reasons are, in my opinion, ineliminable. Since arguments against reasons must consist in the offering of reasons against them, I cannot but regard such arguments as self-refuting. I suppose they could be offered as a reductio, namely, proving the conditional that if there are reasons then there are none, which in classical logic entails that there are none. Yet I cannot understand a claim that I should accept that conclusion unless an argument of that pattern is itself a reason to accept its conclusion.

More importantly, there are formal epistemologists who reject epistemic reason, or who regard it as something to be defined only by logically prior probabilities. Those, like myself, who argue that probability cannot represent theoretical reason, may grant that the obscurities of epistemic reasons require further work, but may nevertheless take the Moorean pledge: I am more sure that there are epistemic reasons for and against propositions than I am sure of the arguments against them. The point of the principle of indifference is to satisfy Salmon's Ascertainability criterion by taking us from a state of epistemic reason to numerical probabilities. If you reject epistemic reason you reject the principle and if you define it by probabilities you do not need the principle. In either case, you have no reason to trouble yourself with the principle and the paradoxes. I shall say a little more about this later when speaking of reversed principles and consilience. In the meantime, epistemic reason and epistemic reasons are therefore a central commitment of the book just because they are a fundamental assumption of the principle of indifference.

1.6 Transitivity

It is uncontroversial that equal epistemic status is reflexive and symmetric and it is commonly wanted to be transitive. For example, 'we want equipossibility to be an equivalence relation' (Bartha and Johns 2001:116). Whether it is transitive has been controverted. Lando (2021:346), in a discussion of one of Bertrand's paradoxes, denies that White's account of 'evidential symmetry' given in terms of 'no more reason' (2009:162) is transitive. Lando's argument relies on defining evidential symmetry by imprecise probabilities. In my opinion this kind of reversal of definition amounts to an abandonment of the principle of indifference, for the reasons I give in §1.9 about inputs and outputs for the principle. Setting that aside, if Lando is right, then 'symmetry' is the wrong word for White to use, because any relation fit to be called a symmetry is transitive.

Fitelson, in an example reported by Novack, puts pressure on transitivity using a sorites case:

> imagine the police are asking you about who the robber is. They ask you to make pairwise comparisons between a man with a full head of hair and his clone with one hair plucked. Then between the one-hair lacking clone, and a two-hair-lacking clone, and so on. Eventually they'll ask you to compare the original, bushy-haired man with a completely bald clone.
>
> (2010:664)

A problem with this kind of case is that you will see the direction it is going: in this case, although for each pair you were unable to discriminate between them, the new man had less hair than the old man. So at some point, if you knew the robber was not bald, although your perceptual evidence will no longer discriminate the men in front of you, knowing the direction means you will have less reason for the new man than the old man. We can restore the difficulty if you don't know which one is the new man. However, the restoration is incomplete. You will have equal reason to believe each is the new man: you will no longer have equal reason to think the new man and the old man, whoever they are, is the robber.

A further issue is that the contradiction Fitelson intends (through transitivity giving equal reason for bushy-haired man and completely bald clone) requires that the equal reason for each pair stays in place throughout. But once you know the direction, you now have a reason that means that it is no longer the case that the earlier pairs have equal reason. So your possessed reasons no longer preserve the earlier equal epistemic status for each pair. Consequently, whilst up to a certain point you may through transitivity have equal reason between the first and most recent man, spotting the direction will now mean you no longer have equal reason between them because you no longer have equal reason between any of the pairs, *even despite* your inability to distinguish between the new and old man in each pair.

It may be true that some objections to the principle of indifference can be blocked by the failure of equal epistemic status to be transitive or by transitivity of equal epistemic status failing to imply transitivity of equal rational credal status (see Novack 2010 for an illuminating discussion). On the other hand, if equal epistemic status entails equal probability, which is transitive, there are obviously going to be problems if equal epistemic status is not itself transitive. Indeed, the converse principle, that equal probability entails equal epistemic status, is defensible when we recall that we are talking of epistemic probability rather than objective probability.[11] In that case equal epistemic status has to be transitive. It would also be a methodological virtue if the principle settles further questions of equal epistemic status by equality of probability in cases where the answer was not at all obvious to us, and this virtue requires transitivity. I shall give an example of this in §1.7.

I am not going to carry this discussion of transitivity further. Insofar as equal epistemic status rests on symmetry, as it often does in the literature I shall examine, transitivity is guaranteed. Consequently, transitivity of equal epistemic status is not an issue I need to resolve for the work of this book. I will, however, point out the consequences if it is generally transitive during the chapter on symmetry. In the final chapter I will discuss some theories that can or do deny transitivity, although the issue of transitivity will not there be our topic.

1.7 Possessed reasons

The principle may appear to be covertly idealising our access to our possessed reasons. Apparently I lack reason to discriminate between the possibilities of, tomorrow, a meteor hitting my house and someone riding a

---

11 During his prolonged critique of the principle of indifference, the converse principle is the one Keynes has confidence in: 'the principle certainly remains as a *negative* criterion: two propositions cannot be equally probable, so long as there is any ground for discriminating between them' (1921/1973:55).

unicycle along their roof ridge in New Zealand, but whether they should have the same probability is unclear. Whether I really lack reason in such a case is no easy thing to trace, since my knowledge of astronomy and peculiar people may in fact discriminate between them in a way hard to determine. And then there is the question of whether equal reason is itself vague. These are the kinds of problems that led Keynes to scepticism about the extent of possibilities with probabilities and about whether probabilities are necessarily totally ordered (1921/1973:ch.3).

One way round this is to assume there are unexpressed and sometimes hard to determine domain restrictions on the principle's application to possibilities. For any particular possibility, there is a restriction to the domain over which equal ignorance determines equal probability. In that way, strange questions of equal ignorance, such as my meteor/unicycle example, might be settled as much by the equality of probabilities determined by each in its own domain as directly by the principle of indifference. Indeed, such possibilities are part of the epistemic power promised by a satisfactory probabilism. Such domain restrictions would not undermine generality of the principle provided every possibility belonged to such a domain. Notorious worries about domain restriction are the possibility of vague borders and other problems over what determines such domains uniquely. These, however, are not worries for the principle of indifference alone and I note that the cases that will concern us have obvious, simple and severe restrictions of domain, if they are needed.

That being said, I think it is better instead to distinguish metaphysical and methodological roles of the principle of indifference. Metaphysically, it determines what the rational probabilities are, given the possessed reasons, and does so irrespective of whether our *access* to our possessed reasons is obscure. Methodologically, our use of the principle in order to know the rational probabilities is constrained by our access to our possessed reasons. In this way I can place the worry about domains as belonging to the imperfections of methodology.

1.8 Ignorance cannot imply knowledge

With these points in place I think we can reject a standard objection to the principle of indifference, namely, that ignorance cannot imply knowledge, in this case, that equal ignorance cannot imply knowledge of equal probabilities. The objection confuses the metaphysical job of the principle in defining probabilities with our methodological use of the principle in coming to know probabilities. The metaphysical job does not imply knowledge of the probabilities, it only implies their equality of quantity for possibilities of equal epistemic status. The methodological use is not a matter of ignorance implying knowledge, but of *knowledge* of equal ignorance implying knowledge. It is *knowing* that we have no reason to discriminate between the possibilities that allows us to infer equal probabilities.

1.9 Inputs and outputs

Grounded as previously described, by which I mean grounded in a normative epistemic principle or requirement such as the Proportioning Ground, it is no job *of the principle of indifference* to determine what the possibilities are or what it is to lack reason to discriminate between them. These two factors are logically prior to the job of the principle. The job of the principle depends on the prior determination of those two factors, and when given them, the principle's job is to assign probabilities, and it does so by assigning equal probability to the possibilities of which we are equally ignorant. The prior determination of those two factors and their relation to elements of the mathematical theory is a necessary part of any philosophical theory advancing the principle. The only things the principle defines are these equalities of probabilities.[12]

Consequently, any putative expression of the principle of indifference which attempts also to *define* what the possibilities are or what equal ignorance is, is faulty. For example, if equal ignorance is *defined* by equal probability the definition is in danger of being circular, or alternatively, if we already have probabilities by which to define equal ignorance we never needed the principle in the first place to satisfy the Ascertainability criterion. I will call such reversals of the order of definition—in other words, reversals that take what I called the converse principle above to be the *defining* principle—*reversed principles of indifference* and we will see some instances in §1.11.

De Finetti objected to the principle of indifference in part because it confused defining and evaluating probability (de Finetti 1970). I think it is now evident that he was incorrect: it has no concern with defining what probability itself is but is concerned only with satisfying Salmon's Ascertainability criterion by defining the equality of probabilities under a certain condition which, as we shall see in the next chapter, leads to defining their numerical evaluation.

1.10 Subordinate principles

The principle as formulated in §1.4 is what I shall call *the* principle of indifference. Anything else that could be called *a* principle of indifference

12 Cf. 'POI.... takes an epistemic input ('having no more reason...') to deliver an epistemic output (equal credence)' (White 2009:168).

must express essentially the same idea of assigning equal probability to the possibilities of which we are equally ignorant. When it amounts to applying the principle to some specific domain of possibilities, perhaps characterising that domain in special ways that makes evident how it is amenable to equal ignorance, it inherits its normative ground from the principle and I shall call such principles subordinate principles.

For example, Cristoforo initially gives equal ignorance in terms of 'no more reason to believe' (de Cristofaro 2008:332). Having earlier remarked that

> The correct version of Bayes' formula shows that, not only the data, but also the... [experimental] design is one part of the evidence and it may affect the a priori probabilities.
>
> (2008:331)

he then interprets 'no more reason' to include this point and gives his subordinate principle

> new formulation of the Principle of Indifference: Given the set of all admissible hypotheses H, let h denote any one element of the partition of H and let d denote the projected design, then we are allowed to assign the same probability to every h if (i) prior information is considered to be irrelevant, and (ii) there is no discriminating treatment to any h caused by d (that is, d is to be impartial).
>
> (Cristofaro 2008:332)

He then gives a mathematical condition on what it is for the design to be impartial.

A more subtle instance of a subordinate principle is Norton's

> Principle of the Invariance of Ignorance (PII). An epistemic state of ignorance is invariant under a transformation that relates symmetric descriptions.
>
> (Norton 2008:48)

What Norton calls symmetric descriptions are renamings constrained by conditions he gives (Norton 2008:49). Norton's PII is entailed by the conjunction of the principle of indifference with either Paris' Renaming principle (1999:79 and §2.4) or a symmetry principle such as van Fraassen's Symmetry Requirement (1989:236 and §6.2). I prove this entailment formally in the Equal Epistemic Status Group theorem 48 and sequent theorems, on which basis I define the Principle of Symmetric Indifference in the chapter on symmetry.

1.11 Consilient derivations

The principle of indifference derived from the Principle of Proportionate Reason took credal commitments to be necessarily rationally related to probability. Some authors offer what they call principles of indifference that do not define probabilities but instead concern themselves with credal commitments that might be ordered but are not necessarily rationally related to probabilities (e.g. see Norton 2008; Novack 2010; Eva 2019). From my point of view, they are a distinct branch off the Principle of Proportionate Reason and I do not concern myself with them until the final chapter. Until then, I am only concerned with commitments that are rationally related to probabilities and hence are concerned only with the principle and its subordinates as defined above.

I am also concerned with reversed principles. As I defined them, reversed principles take the converse of the principle of indifference—in brief, that equal probabilities entail equal epistemic status—to be the defining principle. Their inputs are possibilities and probabilities and their output is equal epistemic status. Reversed principles and what follows from them are otiose in the sense that they rely on some other source of probabilities to satisfy Salmon's Ascertainability criterion; they therefore lack the philosophical significance had by the principle of indifference for probabilisms that advance it as part of their satisfaction of that role.

For example, using 'chance' for objective probability as opposed to epistemic probability, we have the reversed principle

> *Principle of Equal Chances*: necessarily, possibilities of known equal chances are possibilities between which you lack any reason to discriminate epistemically.[13]

Assuming that epistemic probabilities should equal known chances and respond via equal reason would grant us the principle of indifference as before, and then from the pair we would derive

> *Equal Chances have Equal Probabilities*: necessarily, possibilities of known equal chances have equal epistemic probabilities.

This pattern is not unique to starting from chance but applies to any reversed principle. A problem is that here, epistemic reason looks like an idle cog, and, indeed, it is. A further problem is that being an idle cog here, the

13 See Strevens 1998 for an argument that knowledge of symmetries can give us knowledge of chances. The paper is overtly hostile to the principle of indifference as an explanation but does not say whether the author rejects or is merely not discussing epistemic probability.

question of explaining the normativity of probability arises. Although I will not go into the arguments here, any attempt to derive normativity directly from probability faces the problems with normativity Putnam highlighted for naturalized epistemology in general, as well as the notorious problems with which non-naturalists confront naturalists in metaethics.[14]

Let us turn to a couple of examples of authors deriving what they call a principle of indifference from formal principles of probability. First I consider the derivation from the principal principle by Hawthorne *et al.*

> Now suppose that E is a non-defeater and XE contains no information relevant to F or that renders F relevant to A. By Conditions 1 and 2, neither EF nor $E(A \leftrightarrow F)$ are defeaters. Hence by Proposition 2, $P(F|XE) = 0.5$. But this is a version of the principle of indifference, since it says that given a suitable lack of information about F, one should believe F and $\neg F$ to exactly the same degree, under evidence XE. Indeed, since XE contains no information relevant to F, $P(F|XE) = P(F)$ so $P(F) = 0.5$ too. Thus the principle of indifference also holds for unconditional initial credences.
>
> (Hawthorne *et al.* 2017:125)

Conditions 1 and 2 of their argument have been controverted and defended (Pettigrew 2017; Titelbaum and Hart 2020; Landes *et al.* 2021) but for the sake of our concerns I assume their derivation is sound.

This is a plausible, but typically swift, skate over exactly how epistemic reason is playing its role and whether what has been shown is really the principle. Yes, one account of equal epistemic status is lack of information. But is that the role being played here? The way lack of information is getting into the antecedent is that it is a condition on the truth of the principal principle but not itself a normative condition. So when they show 'if we have a suitable lack of information about F, $P(F) = 0.5$' follows from the principal principle they have not shown that equal reason implies equal probability. The normative demand of equal probability is *not* got from the lack of information but is instead inherited from the normative demand of the principal principle. Consequently, Hawthorne *et al.* have not derived the principle of indifference from the principal principle. What they *have* shown is that the principal principle requires and produces the same probability as the principle of indifference. That much suffices for their wider conclusion:

> The above considerations suggest that the Bayesian epistemologist should either embrace both principles in line with objective Bayesianism

---

14 E.g. Hume 1739/1978: III,1.1, Moore 1903 and a recent defence of Moore in Shackel 2021, and Parfit 2011a: part 1 and Parfit 2011b: part 6.

(see, for example, Williamson [2010]), or deny both principles in line with radical subjectivism. Either way, the principal principle, as a half-way house between radically subjective Bayesianism and objective Bayesianism, becomes an unstable position.

(Hawthorne *et al.* 2017:126)

No doubt that is why they do not concern themselves with the questions I have just raised. If their argument is sound, what they in fact show is a consilience between the principal principle and the principle of indifference and a commitment to agreeing with the output of the latter by the former.

Pettigrew derives what he calls the principle of indifference from the requirement of accuracy:

> PoI Suppose $\mathscr{F}$ is a finite, rank-complete set of propositions. If an agent has an initial credence function c_0 defined on $\mathscr{F}$, then rationality requires that c_0 is the uniform distribution on $\mathscr{F}$.
>
> (2016a:164)

What this does show, granting his various premisses, is that accuracy requires and produces the same probability as the principle of indifference, thereby demonstrating its consilience with his general theory.

When we ask our questions about normativity and the order of explanation, we find the principle on which Pettigrew bases his derivation is risk aversion:

> we assume that rationality requires an agent not to risk greater than necessary inaccuracy.
>
> (2016a:161)

Risk aversion is formally defined by Maximin (2016a:161), which is

> the principle of extreme epistemic conservatism.... [with which]... we have reached normative bedrock.
>
> (2016a:166)

In broad, the credence function you ought to have is the one that is the least bad, least inaccurate, assessed over all the worlds. Williamson also offers an argument from risk aversion, where being least bad is a practical rather than theoretical matter (2018).

Pettigrew's objection to White's (2009) evidentialist justification of the principle, and Pettigrew's wider project, makes it clear that what he has derived is a reversed principle.

The problem with the Argument from Evidential Support is this. It must assume that, for every body of evidence, and every pair of propositions, there is... a fact of the matter about whether the evidence provides any reason to favour one over the other. But what warrant is there for thinking that there is such a fact that can do the job required of it? According to the most promising account of evidential support... viz., Bayesian confirmation theory—there is no such fact.... facts about evidential support are entirely determined by facts about rational initial credences and rules for rational updating upon receipt of a body of evidence.... If facts about equal evidential support are in fact determined by the rational principles governing credences... and facts about a particular rational principle—namely, PoI$_{Ev}$—depend on facts about equal evidential support, that is circular.... Thus... [we are owed]... a notion of equal evidential support that is prior to the notion of rational credence.

(Pettigrew 2016a:156)

He might reject my Proportioning Ground for the same reason. No such account is owed, however, since I take facts about epistemic reason, including facts about equal or unequal reasons, to be basic and determinative of rational credal commitments. What we have here is really an expression of the skepticism many formal epistemologists have about epistemic reasons in general. Against that view and for arguments that probability is in fact incapable of representing theoretical reason, see Dancy 2018:ch.4 and Shackel MS-b. Against the view that traditional epistemology must give up whatever formal epistemologists fail to model, see Weatherson 2007.

What matters more for us is that he is defining equal reason by equal probability. That this is exactly what he intends is shown when he says

the Argument from Accuracy presented in this paper [is] part of a larger project.... to show that we can establish important evidential norms by appealing only to the good of accuracy, and thereby reduce the virtue of responding appropriately to the evidence to the virtue of accuracy. The Principle of Indifference is one such norm—it says how an agent should respond in the absence of evidence.

(Pettigrew 2016b:42)

The aim to reduce evidential norms to the good of accuracy means their normativity must be inherited from the good of accuracy and so this is how the principle of indifference must acquire it. What he actually proves is that accuracy implies a uniform initial credence function and merely *calls* that theorem the principle of indifference. From his earlier remarks about facts of equal evidential support being determined by facts about credences, we have that if credences of propositions are equal then they have equal evidential support, and hence more generally what he has actually proved is

> *Pettigrew's reversed principle of indifference*: if credences of propositions are equal then we have no reason to discriminate between them.

The normativity of equal reason has been defined from the good of accuracy via the equality of credence. His denial of any other source of epistemic status than the good of accuracy makes this the only definition of their epistemic status and so *from* his reversed principle we could now offer a derivation of the principle of indifference. However, what answers to Salmon's Ascertainability criterion are the probabilities from accuracy and there is nothing left for epistemic reason and the principle of indifference to do: as in the previously described chance case, both are idle cogs.

Having shown how these other derivations related to the principle of indifference, we can see that the formal interests of their authors mean they do not attend much to the relation to epistemic normativity nor to the order of explanation. They are not by any means alone in this and having exhibited these two examples I shall not spend time showing how to reconcile those two concerns for other authors. In general, formal derivations of the principle of indifference, assuming they are sound, at least achieve a commitment to consilience with the principle of indifference, namely, that in cases to which the principle applies they are committed to agreeing with it. Consequently, a difference in formal derivation, or over the detail of normative grounding or the order of explanation, will not make a difference in the features of the Principle of Indifference for Sets given in the next chapter that do the crucial work in what follows, nor, therefore, to the challenge posed by Bertrand's paradox.

1.12 Note on mathematics

One of the problems for philosophical theories with significant formal inputs is that the important philosophical moves can be incidentally buried under, or even deliberately concealed behind the mathematics. I have tried to avoid such interments and concealments.

In many cases it has been useful to give definitions or theorems with names that can be used as grammatical phrases, which are then used as such, but strictly in the sense given by its definition or its theorem. For the sake of the reader that cares only about the philosophy and is willing to take the mathematical inputs to the philosophy as established, I have tried to write this in a way that most of the mathematics can be skipped without significant loss of philosophical understanding. In aid of this, where theorems and lemmas are deployed, the names, preambles and postambles aim to set up and draw out what matters philosophically. Once the basic definitions and axioms of measure theory are understood, the proofs are not especially sophisticated and for the sake of those interested I have erred on the side of spelling things out.

Appendix

I collect here statements of principles of indifference in the literature that are similar in various ways to various degrees, which on occasion leave the normative relation only weakly expressed. Some are covertly circular.

> When several hypotheses are presented to our mind, which we believe to be mutually exclusive and exhaustive, but about which we know nothing further, we distribute our belief equally amongst them.
>
> (Donkin 1851:358)

> Princip der Spielräume: Spielräume (a range of possibilities) that are irreducible [ursprünglich], without logical preference [keinerlei logische Bevorzugung] and comparable [vergleichbar] are equally probable.
>
> (von Kries 1886—this is not a translation but a summary, relying in part on Fioretti 2001 and Zabell 2016a)

> if the evidence does not contain anything that would favour either of two or more possible events, in other words, if our knowledge situation is symmetrical with respect to these events, then they have equal probabilities relative to the evidence.
>
> (Carnap 1955:318)

> if there is no evidence leading one to believe that one event from an exhaustive set of mutually exclusive events is more likely to occur than another, then the events should be judged equally probable.
>
> (Luce and Raiffa 1957/1985:284)

> The assignment of equal weights to the state descriptions may be regarded as an application of the principle of indifference to the fundamental possibilities of our model universe.
>
> (Salmon 1967:71)

> two possibilities are equiprobable if and only if there is no ground for choosing between them.
>
> (Kyburg 1970:31)

principle of nonsufficient reason... assume alternatives to be equiprobable in the absence of known reasons to expect the contrary.... principle of indifference... assume alternatives to be equiprobable when there is a balance of evidence in favour of each alternative.

(Fine 1973:167)

The first indifference principle for assigning probabilities is to assume a uniform distribution in the absence of reasons to the contrary.... The second indifference principle is to assume statistical independence, in the absence of reasons to the contrary.

(van Fraassen 1989:299)

in the absence of any known reason to assign two events differing probabilities, they ought to be assigned the same probability.

(Strevens 1998:231)

Each member of a set of propositions should be assigned the same probability... in the absence of any reason to assign them different probabilities.

(Castell 1998:387)

the elements of a set of outcomes are equally possible if we have no reason to prefer one of them to any other.

(Kyburg and Teng 2001:43)

In the absence of any known reason to assign two outcomes different probabilities, they ought to be assigned the same probability.

(Bartha and Johns 2001:109)

We take the uniform distribution... to be one presentation... of our evidence.

(Bovens and Hartmann 2003:119)

we should assign equal probability to any mutually exclusive and jointly exhaustive set of possible outcomes, iff we have insufficient reason to consider any one of these outcomes more or less likely than any other.

(Mikkelson 2004:137)

evidence which gives us no reason to think that any one of a number of mutually exclusive possibilities... is more probable than any other will give those possibilities equal epistemic probabilities.

(Mellor 2005:29)

equal parts of the possibility space should receive equal probabilities relative to a null state of background information.

(Howson and Urbach 2006:266)

If we are indifferent among several outcomes, that is, if we have no grounds for preferring one over any other, then we should assign equal belief to each.

(Norton 2008:47)

The equivocation norm says that degrees of belief should... be equivocal.

(Williamson 2010:49)

This first version says that if there is no evidence at all then one should believe each basic possibility to the same extent.... The second version... says that if the evidence treats each basic possibility symmetrically, then one should believe each such possibility to the same extent.... The third version... says that if it is compatible with the evidence to believe each basic possibility to the same extent, then one should do so.

(Williamson 2018:561–2)

Principle of Indifference: Let $X = \{x_1, x_2, ..., x_n\}$ be a partition of the set W of possible worlds into n mutually exclusive and jointly exhaustive possibilities. In the absence of any relevant evidence pertaining to which cell of the partition is the true one, a rational agent should assign an equal initial credence of $1/n$ to each cell.

(Eva 2019:390)

2 The Principle of Indifference for Sets

2.1 Introduction

I now turn to defining a subordinate principle of indifference that is suitable for our purposes. It must suffice to define probabilities for continua of events. This turns out to involve rather more analysis than might be expected, but with that analysis in place we will find that a principle of indifference for sets suffices quite generally.

2.2 Mathematical probability theory

The mathematical theory of probability does not identify numerical probabilities beyond the convention of normality: it tells us only what follows from the assumption that certain possibilities have certain numerical probabilities. The principle of indifference supplies such an assumption. Classically, it was applied to a base set of mutually exclusive and jointly exhaustive 'atomic' events among which there was no reason to discriminate. These were then assigned equal probabilities summing to 1. For example, since there are six sides to a die, only one of which can be on top at any one time, the set of atomic events are the six distinct possibilities for which face is on top, each of which is assigned the probability of 1/6. Extending this to the case of infinite sets of events is more complicated. Given a countable infinite set of mutually exclusive and jointly exhaustive atomic events, there isn't a way of assigning equiprobability to members of that set that will sum to 1, but for an uncountable infinite set, there is. The sets of events that we will be concerned with are continuum sized, and therefore uncountable.

For most of my analysis I need only the most abstract features of the standard measure theoretic formulation of probability. In a later chapter I will need to go into some fundamentals of measure theory. For now, however, I shall articulate the basics of probability as a normal measure sufficient to derive a subordinate principle of indifference for continua. I start by defining a measure.

DOI: 10.4324/9781003456308-2

Definitions

$\mathbb{P}(\Omega)$ is the power set of a set Ω.

A *family* of sets is a set of sets.[15]

A *difference* of sets, $A, B \in \mathbb{P}(\Omega)$, $A - B = \{\omega \in \Omega: \omega \in A \text{ and } \omega \notin B\}$

A *sequence* is a countably infinite ordered set, $S = \{S_1, S_2, \ldots\}$.

Given a function $f: X \to Y$, if $S \subset X$, $f(S) = \{y \in Y: y = f(x), x \in S\}$ and $f|_S$ is f restricted to S.

If $\mathcal{F}$ is a family of sets, the union of all sets in $\mathcal{F}$ is $\bigcup \mathcal{F}$ and the intersection is $\bigcap \mathcal{F}$.

The *extended reals* is the set $\mathbb{R} \cup \{-\infty, \infty\}$ on which the usual algebra between $\{-\infty, \infty\}$ and $\mathbb{R}$ is applied (e.g. see Halmos 1974:2). We designate the extended non-negative reals, $[0, \infty]$, by $\mathbb{R}^+$.

$\mathcal{M}$, a family of subsets of a set Ω, is a *σ-algebra* on Ω iff

1. $\mathcal{M}$ is a subset of the power set of Ω and contains the empty set, $\varnothing$.
2. If S is in $\mathcal{M}$ then the complement of S, $\Omega - S$, is in $\mathcal{M}$.
3. If S is a sequence of elements in $\mathcal{M}$ then the union of S, $\bigcup S$, is in $\mathcal{M}$.

The latter two imply, by de Morgan's law, that $\bigcap S$ is also in $\mathcal{M}$.

$(\Omega, \mathcal{M})$ is a *measurable space* iff $\mathcal{M}$ is a σ-algebra on a set Ω.

μ is a *measure on, of* or *for* a set Ω, a σ-algebra $\mathcal{M}$, and a measurable space $(\Omega, \mathcal{M})$ iff

4. μ is a non-negative set function to the extended non-negative reals $\mu: \mathcal{M} \to \mathbb{R}^+$
5. $\mu(\varnothing) = 0$
6. μ is countably additive, that is if S is a sequence of disjoint elements in $\mathcal{M}$ then

$$\mu\left(\bigcup S\right) = \sum_{n \in N} \mu(S_n)$$

μ is a *finite measure* iff $\mu(\Omega) < \infty$ and a *normal measure* iff $\mu(\Omega) = 1$.

A *measure space*, $(\Omega, \mathcal{M}, \mu)$, is a triple where $\mathcal{M}$ is a σ-algebra on Ω, μ is a measure on $\mathcal{M}$ and the elements of $\mathcal{M}$ are the μ-measurable subsets of Ω.

A *probability space*, (Ω, Σ, P), is a measure space on the set Ω where Σ is a σ-algebra on Ω and P is a *probability measure*, which last by definition is a normal measure. The elements of Σ are the items that have probabilities given by P.

A function that is a probability measure satisfies Kolmogorov's original axiomatization of probability (e.g. see Capinski and Kopp 2004:46).

The standard measures for spaces of real numbers, $\mathbb{R}^n$, are *Lebesgue measures*.

---

15 We use the word 'family' for a set of sets because it helps keep track of which level we are talking about.

For the contextually salient $\mathbb{R}^n$, $\mathscr{L}$ is the set of *Lebesgue Measurable* subsets of $\mathbb{R}^n$ and λ is the *Lebesgue measure*. When we need to keep track of which $\mathbb{R}^n$ we are in, subscripts are used, for example $\mathscr{L}_2$, λ_2.

Abuse of notation: In general, for the sake of brevity we cease to make any distinction between elements in Ω and their singleton sets. So for any measurable space, $(\Omega, \mathscr{M})$, since $\omega \in \Omega$ iff $\{\omega\} \in \mathscr{M}$ we abuse our notation from hereon by using 'ω' for both the element in Ω and the singleton in $\mathscr{M}$ as is convenient. Likewise, if $m \in \mathscr{M}$, where m is a singleton set, 'm' might also be used for an element in Ω.

2.3 Naming and renaming

In general the way we apply the mathematical theory of probability to events of interest is to name those events by mathematical objects. Different treatments of a problem using different parameterisations can be done by distinct namings or by taking one naming and defining a renaming. Renamings are useful shortcuts that are often made use of inexplicitly, since in many contexts they are obvious. We, however, will at times need to disentangle some such procedures and so I now give them rigorous formal definitions.

Events are atomic or compounded of atomic events, where atomic events are mutually exclusive and jointly exhaustive.[16] We assume the mereology of compound events is such that compound events correspond with sets of atomic events of which they are compounded. The compound event of e or e' corresponds to $e \cup e'$ and the compound event of e and e' corresponds with $e \cap e'$. We therefore treat the set of all the random events of interest, $\mathscr{E}$, as a family of sets of atomic events, and we assume it is a σ-algebra on the set of atomic events, E. We call $(E, \mathscr{E})$ the *event space*. We name the events by a function into and onto a measurable space and renaming by replacing the names from one measurable space by those in another.[17]

Definition: Given an event space $(E, \mathscr{E})$ and a measurable space $(\Omega, \mathscr{M})$, a *naming* is a bijection $n : \mathscr{E} \to \mathscr{M}$ and we call $(\Omega, \mathscr{M})$ a *name space*.

Definition: Given a naming, $n : \mathscr{E} \to \mathscr{M}$, and a measurable space $(\Omega', \mathscr{M}')$, a *renaming* is a bijection $r : \mathscr{M} \to \mathscr{M}'$.

Notation: When we need to make use of various namings and renamings between various spaces we use notations such as $n : (E, \mathscr{E}) \to (\Omega, \mathscr{M})$, $m : (E', \mathscr{E}') \to (\Omega', \mathscr{M}')$, $r : (\Omega, \mathscr{M}) \to (\Omega', \mathscr{M}')$, $q : (\Omega, \Sigma, P) \to (\Omega', \Sigma', P')$ to keep track of the distinct bijections between the distinct σ-algebras.

---

16 Note that we are not hereby committed to atomic events corresponding to Carnap's state descriptions rather than structure descriptions.
17 What Norton 2008:49 calls symmetric descriptions are effectively renamings of a restricted variety.

With these definitions in place, when we do not need to keep track of the distinction, for the sake of brevity I will frequently identify events with their names and vice versa.

The exact relation of our event space to events *qua* metaphysical species can be given further spelling out for particular cases but I will not do so. In many cases the correspondence will be straightforward because atomic events will correspond with metaphysically individual events, such as a particular side of a die landing up. In general, however, atomic events of our event space do not correspond to metaphysically individual events. Rather, the atomic events will consist of relevantly maximal sums of such individual events. If we have two dice, for example, there will be no pair of atomic events corresponding to the pair of the individual events of the red die landing 5 and the blue die landing 1. Instead, one of the atomic events will be the mereological sum of the red die landing 5 and the blue die landing 1. Whether there are negations or disjunctions, properly so called, of metaphysically individual events can be disputed (e.g. Hájek and Hitchcock 2016:17) but our event space contains all such negations and disjunctions due to the properties of σ-algebras. A natural interpretation of the set of atomic events is, therefore, a partition of the possible worlds, and on some views, the cells of such a partition are propositions.

Lemma 1

Under namings and renamings (*a*) Every event receives a name, (*b*) names are unique, that is no distinct events receive the same name, (*c*) every member of $\mathcal{M}$ names an event, (*d*) every event is measurable in the sense that its name is measurable and (*e*) renamings preserve measurability.

Proof: Let $n: \mathcal{E} \to \mathcal{M}$ be a naming and $r: \mathcal{M} \to \mathcal{M}'$ a renaming. They are bijections so $r \circ n: \mathcal{E} \to \mathcal{M}'$ is a bijection and therefore a naming, whence proofs for n suffice for r. (*a*) Namings are functions with domains the set of events. (*b*) Namings are injections so if $n(e) = n(d)$ then $e = d$. (*c*) Namings are surjections so if $m \in \mathcal{M}'$ then $n^{-1}(m) \in \mathcal{E}$. (*d*) Each event has a name belonging to the σ-algebra of a measurable space. (*e*) Preserving measurability means for all $m \in \mathcal{M}, m' \in \mathcal{M}'$

$$m \in \mathcal{M} \text{ iff } r(m) \in \mathcal{M}'$$
$$m' \in \mathcal{M}' \text{ iff } r^{-1}(m) \in \mathcal{M}$$

which follows because r is a bijection. QED

Lemma 2

Function compositions of namings, renamings and their inverses produce namings, renamings and their inverses and likewise for compositions of restricted namings, restricted renamings and their inverses.

Proof: Compositions of bijections and their inverses are possible for any pair of bijections or their inverses where the codomain of the first is the domain of the second and the composition is a bijection. QED

Remark: For example, given namings $n: \mathcal{E} \to \mathcal{M}$, $m: \mathcal{E} \to \mathcal{M}'$ and renaming $r: \mathcal{M} \to \mathcal{M}'$, $mon^{-1}: \mathcal{M} \to \mathcal{M}'$ is a renaming, $ron: \mathcal{E} \to \mathcal{M}'$ is a naming, $mon^{-1}or^{-1}: \mathcal{M}' \to \mathcal{M}'$ is a renaming etc.

Lemma 3

(a) Given an event space $(E, \mathcal{E})$, any bijection $f{:}E \to \Omega$, where $\mathcal{M}$ is defined by $\mathcal{M} = \{m \subset \Omega : m = f(e), e \in \mathcal{E}\}$ defines a naming $n : \mathcal{E} \to \mathcal{M}$ by $n(e) = f(e)$.

(b) Given a naming, $n : \mathcal{E} \to \mathcal{M}$, any bijection $f{:}\Omega \to \Omega'$ defines a renaming $r : \mathcal{M} \to \mathcal{M}'$, $r(m) = f(m)$ for which $\mathcal{M}'$ is defined by $\mathcal{M}' = \{m' \subset \Omega' : m' = f(m), m \in \mathcal{M}\}$.

Proof: We prove both parts together. The codomains $\mathcal{M}$ and $\mathcal{M}'$ are subsets of the power sets of Ω and Ω' respectively. Any bijection between two sets defines a bijection between their power sets, hence n and r are bijections. We need to check that $\mathcal{M}$ in (a) and $\mathcal{M}'$ in (b) are σ-algebras. In each case they are the codomain of a bijection from a σ-algebra, so we prove the general result that applies to both.

Let $\mathcal{E}$ be a σ-algebra on E and $g : \mathcal{E} \to \mathcal{M}$ a bijection. $g(\varnothing) = \varnothing \in \mathcal{M}$. If $M \in \mathcal{M}'$ then there exists $e \in \mathcal{E}$ such that $e = g(m)$. Hence

$$\Omega - m = g(E) - g(e) = g(E - e) \in \mathcal{M} \text{ because } \mathcal{E} - e \in \mathcal{E}$$

So $\mathcal{M}$ is closed under complements. Suppose $M = \{m_1, m_2, ...\}$ is a sequence in $\mathcal{M}$. Then there exists a sequence $S = \{e_1, e_2, ...\}$ in $\mathcal{E}$ such that $m_n = g(e_n)$ for all n. $\bigcup S$ is in $\mathcal{E}$, so $g(\bigcup S)$ is in $\mathcal{M}$ and

$$g\left(\bigcup S\right) = \bigcup_{n\in\mathbb{N}} g(e_n) = \bigcup_{n\in\mathbb{N}} m_n = \bigcup M$$

So $\bigcup M$ is in $\mathcal{M}$ and $\mathcal{M}$ is closed under countable unions. Hence $\mathcal{M}$ is a σ-algebra and $(\Omega, \mathcal{M})$ is a measurable space. QED

Remark: In these cases we may abuse our notation by calling the naming or renaming by the same name as the bijection between the base spaces.

2.4 The renaming principle

Paris codified a number of defensible common sense principles, truths about probability that any theory of probability must respect, among which is the

RENAMING PRINCIPLE. Changing the names we call things should not change the probabilities we assign to them.

(Paris 1999:79)

Plainly this is an instance of Juliet's point:

> What's in a name? That which we call a rose, by any other name would smell as sweet.
>
> (Shakespeare 1750:2.2)

Names don't change properties of the named and so don't change probabilities of events. In short, changing names should *preserve probabilities*.

There are some familiar background issues about intensional contexts (e.g. I may not believe that Romeo loves the only daughter of Capulet even though I believe he loves Juliet) that may require a more careful characterisation of Paris's principle. I shall ignore them as they are not the source of the difficulties we shall encounter.

Evidently this principle is a desirable constraint on our formal renamings. A formally defined renaming between probability spaces preserves probabilities iff it is a probability isomorphism, so applying Paris's principle gives:

Renaming Theorem 4

A renaming between probability spaces, $r:(\Omega, \Sigma, P) \rightarrow (\Omega', \Sigma', P')$ satisfies Paris's Renaming principle iff it is a probability isomorphism.

Proof: r is probability isomorphism between probability spaces iff it is a bijection that preserves both measurability and the probability measure, that is for all $\sigma \in \Sigma$ and $\sigma' \in \Sigma'$

preserves measurability
(a) $\sigma \in \Sigma$ iff $r(\sigma) \in \Sigma'$
(b) $\sigma' \in \Sigma'$ iff $r^{-1}(\sigma') \in \Sigma$

preserves the probability measure
(c) $P(\sigma) = P'(r(\sigma))$
(d) $P'(\sigma') = P(r^{-1}(\sigma'))$.

LTR: Lemma 1 (e) gives us (a) and (b). (c) Suppose $\sigma \in \Sigma$ names the event e. Then $r(\sigma)$ is its new name. Since r satisfies Paris's Renaming principle, it does not change the probability of e, and hence $P(\sigma) = P'(r(\sigma))$. (d) Suppose $\sigma' \in \Sigma'$. Then there exists $\sigma \in \Sigma$ such that $\sigma' = r(\sigma)$ so $r^{-1}(\sigma') = \sigma$. Let event e be the event named by σ. Then since r satisfies Paris's renaming principle we have $P(\sigma) = P'(r(\sigma))$ and so $P'(\sigma') = P'(r(\sigma)) = P(\sigma) = P(r^{-1}(\sigma'))$.

RTL: If a renaming is a probability isomorphism, by definition it preserves the probability measure, hence it preserves probabilities, whence it satisfies Paris's Renaming principle. QED

2.5 Probability measures and the principle of indifference

We now turn to linking the principle of indifference to the probability measures that represent the probabilities defined by the principle. Our abuse of notation that identifies atomic events and their singleton sets is carried over for probabilities of atomic events, where we use '$P(x)$' to mean $P(\{x\})$.

To rehearse this for the example of a coin toss, we name heads by 1, tails by 0, when

$$\Omega = \{0,1\}$$

$$\Sigma = \{\emptyset, \{0\}, \{1\}, \{0,1\}\}$$

$$P : \Sigma \mapsto [0,1]$$

$$P(\sigma) = \begin{cases} 0 & \text{if } \sigma = \emptyset \\ \dfrac{1}{2} & \text{if } \sigma = \{0\} \text{ or } \{1\} \\ 1 & \text{if } \sigma = \{0,1\} \end{cases}$$

We applied the POI in giving heads and tails equal probabilities.

If we have countably many events, so that Ω is countable and therefore orderable as a sequence, $\{\omega_n\}$,

$$\Omega = \bigcup_{n \in N} \{\omega_n\}$$

so

$$P(\Omega) = \sum_{n \in N} P(\omega_n) = 1$$

This requires that the sequence $\{P(\omega_n)\}$ converges to 0. If it does so by all its members being 0 then $P(\Omega) = 0 \neq 1$. So if countably many events have equal epistemic status then the principle of indifference assigns to them all a positive value. But for $c > 0$

$$P(\Omega) = \sum_{n \in N} P(\omega_n) = \sum_{n \in N} c = \infty$$

So for countably many events of equal epistemic status, we cannot apply the principle of indifference and get a probability measure.

If we have continuum many events, Ω is uncountable. Assuming $\omega \in \Omega$ have equal epistemic status, the POI assigns them equal probabilities. If those probabilities are non-zero, as we just saw, any compound event with countably many constituent atomic events would have infinite probability, so they are all zero. That does not pose the problem it did for the countable events since there is no countable sequence S in Ω such that $\Omega = \bigcup S$ and hence we don't end up with $P(\Omega) = 0$. Unfortunately, this much fails to specify a unique probability measure on Ω, the reason being that almost every probability measure on an uncountable set assigns probability zero to all the atomic events. This is a manifestation of the theorem that countable subsets of a continuum sized measure space with a finite measure have measure zero.[18]

This means that the apparently straightforward application of the principle of indifference to an uncountable set of events, under which we give equal probability of zero to each of the atomic events (i.e. zero to each singleton set in $\mathcal{E}$), whilst fine so far as it goes, does not go far enough. Since equal probabilities over the atomic events fails to pick out a particular probability measure, equal ignorance over atomic events does not suffice for the POI to pick out a particular measure. This is precisely why we need the feature I called the *generality* of the principle of indifference as formulated in the last chapter. It is that feature, and only that feature, that allows equal epistemic status to apply not only to members of the subset of singleton sets in $\mathcal{E}$ but also, and independently, to members of other subsets of $\mathcal{E}$. So we need

> *More equal ignorance*: equal ignorance over compound events as well as atomic events.

Over which compound events, though?

Let us consider an example in which we think we are equally ignorant of where on a cricket square a treasure is to be found buried. Naturally, cricket squares are not generally square, since the English do not make a fetish out of reason: we, however, will address the exception with sides 22 yards long. We don't just think that we are equally ignorant of each point in the square. We are equally ignorant of squares within that square that have the same area, circles which have the same area, squares and circles of the same area, peculiar patches of the same shape that have the same area, peculiar patches of distinct shapes that have the same area. In other

---

18 This is generally true due to the way in which measures are usually defined on continua (e.g. methods I, II and III in Bruckner *et al.* 2008: chapter 3). For the Lebesgue measure version of this theorem see Weir 1973:18. For exceptions, the counting measure assigns infinite measure to countable sets, but is not a finite measure. It is true that one can gerrymander a measure to have countably many non-zero provided, for reasons already seen, the sum of the probabilities of the events constituting that compound event converges to less than 1 (less than because otherwise this is just a covert countable event space). But that still leaves uncountably many events with probability zero and infinitely many compatible probability measures over those.

words, we are equally ignorant over any and all patches having the same area: we are equally ignorant of all the 1 square yard patches, the 2 square yard patches, indeed, of all the A square yard patches in the square. And in telling us to assign events of equal ignorance the same probability means the POI tells us to assign patches of equal area the same probability. We can do this quite simply by assigning to any patch the probability

$$P(\text{Patch}) = \frac{\text{area of patch}}{\text{area of cricket square}}$$

What we have done here is followed out the wider implications of which sets of events whose members we have no reason to discriminate between. From this it followed that equal ignorance over compound events of the event space was a matter of equal ignorance over compound events of equal area. So those are the possibilities and the equal ignorance that give us the 'more equal ignorance' to be the input to the POI. And then the output of the POI was to assign to those possibilities the probability proportional to their area, which is easily done by dividing their area by the area of the whole. In so doing, we have implicitly defined a uniform probability density function over the cricket square, fitting with our intuition about the case. Let $S = [0, 22] \times [0, 22]$, λ be the Lebesgue measure on $\mathbb{R}^2$ and $f{:}S \rightarrow \mathbb{R}$ the probability density function that satisfies the equation

$$P(\text{Patch}) = \int_{\text{Patch}} f \, d\lambda = \frac{\text{area of patch}}{\text{area of cricket square}} = \frac{\lambda(\text{Patch})}{\lambda(S)}$$

$$\Leftrightarrow \quad f \text{ is the constant, } k, \text{ where } \int_S k \, d\lambda = 1 \quad \text{i.e. } k = \frac{1}{484}$$

Hence f is the uniform density function, $f(x,y) = 1/484$.

The critical step that made applying the POI result in a determinate probability measure was (1) attending to how equal ignorance in the case applied to compound events and (2) for the compound events over which we were equally ignorant to be definable in terms of a measure space *and* (3) for the *equal ignorance* over compound events to be *represented by a measure* on that measure space, under which measure compound events with equal epistemic status have equal measure. The important thing here is that the measure of item (3) is an *input* to the principle, *not* an output!

Of course, when we are trying to determine probabilities, we also are trying to determine an output measure of the event space, namely, a probability measure. Recalling that the possibilities and equal ignorance are an input for the principle and the probabilities are an output, where the input is logically prior to the output, we cannot simply conflate these two measures, that it to say, we cannot conflate the measure of item (3) with the probability measure. Even if the measure in item (3) happened to be a normal measure, its role is distinct. In typical cases like the cricket square, the obviousness of the equal ignorance including equal ignorance over patches of equal area tends to obscure the subtle distinction of these measures and of their difference in logical priority.

We needed a measure of the event space as an input for the POI before we could use the POI to determine the probability measure as an output. Furthermore, it is now evident that this need of the POI for a prior input measure before it can determine the posterior output measure is not confined to continuum state spaces. When we applied it to the dice we were using the counting measure (Bruckner *et al.* 2008:73) as the input when taking the cardinality of each of {1}, {2}, {3}, {4}, {5}, {6} to represent mathematically our equal ignorance over the possibilities consisting of the six sides of the dice that might land up.[19]

The reason we didn't need more equal ignorance in the latter case but do need it for continua is this: to define a specific measure over a measurable space (Ω, Σ), it suffices to have a measure over a sub-basis of the σ-algebra.[20] For any σ-algebra, Σ, over Ω, if Ω is finite it is a sub-basis of Σ, but if Ω is a continuum it is not. Consequently, equal ignorance over atomic events suffices for the dice but not for the cricket square.

So in using the POI we cannot determine the probability measure out of the equal ignorance and possibilities alone: we must be able to represent mathematically those possibilities and that equal ignorance by a measure, where that measure represents possibilities of equal ignorance by their measures being equal. In other words, we need a logically prior input measure to get the logically posterior output probability measure:

Need-a-Measure-to-Get-a-Measure principle: Inputs to the POI include the representation of equal ignorance of events by a measure over the space representing the atomic and compound events, under which measure events of equal epistemic status have equal measure, whereby the equal ignorance over those events is represented by their equality of measure.

---

19 We do this routinely without fully realising we do it. For example, see Lando exploiting this in his LAPD and Multiple Beauties examples by using two different measures for inputs to the POI (Lando 2021:343 and 350).

20 A sub-basis of a σ-algebra is any set-system that generates the σ-algebra using the operations that define a σ-algebra, i.e. complementation and countable union (Jost 2015:105). In Chapter 13 we will see that once we add in the Carathéodory extension theorems, we can weaken this further. The need for the input measure can be satisfied if we have what is called a premeasure (of a certain kind) on a semialgebra of the events. The minimal requirement is a pre-measure on a π-system of events.

Technically: Given an event space $(E, \mathscr{E})$ and a naming $n:(E, \mathscr{E}) \to (\Omega, \mathscr{M})$ to a naming space $(\Omega, \mathscr{M})$, an input needed by the POI is a non-zero finite measure, μ, on Ω such that we are equally ignorant over events d, $e \in \mathscr{E}$ iff $\mu(d) = \mu(e)$.[21]

This principle is the source of the well known problem of the priors.

2.6 The Full Principle of Indifference for Sets

I can now formulate a principle subordinate to the principle of indifference that defines probability measures.

Definition: The Full Principle of Indifference for Sets (FPOIS)

1. *A full epistemic state over event space* $(E, \mathscr{E})$: For all pairs of events in $\mathscr{E}$, either we have no reason to discriminate epistemically between them or we have reason to so discriminate (i.e. either their balances of reasons are equal or unequal).
2. *The input to the POI*: E is the set of atomic events, $\mathscr{E}$ is the σ-algebra of events, over which a full epistemic state is given.
3. *Representation of events by a name space*: Event space $(E, \mathscr{E})$ is represented by a measurable space (Ω, Σ), iff there is a naming $n: \mathscr{E} \to \Sigma$.
4. *Equal ignorance for* (Ω, Σ): For all $\sigma, \tau \in \Sigma$, σ and τ have equal epistemic status iff their preimages $n^{-1}(\sigma)$, $n^{-1}(\tau) \in \mathscr{E}$ have equal epistemic status.[22]
5. *Principle of Indifference for Sets* (POIS): For a set Ω, given a σ-algebra, Σ, on Ω and a non-zero finite measure, μ, on Σ, such that we have no reason to discriminate between members of Σ with equal measures under μ, the members of Σ with equal μ-measure have equal probability.

A full epistemic state over an event space $(E, \mathscr{E})$ partitions $\mathscr{E}$ into what I shall call *equal reason cells*, where the members of the same cell have equal reason but the members of distinct cells do not. We saw an example of this in the last chapter when discussing the book paradox. Then equal ignorance for the name space partitions Σ likewise. I shall call such

---

21 As will shortly be evident, we need a non-zero finite measure since we want to proportion the probabilities to this measure and if it is infinite the proportions will end up at zero or undefined.

22 σ is a subset of Ω so preimage $n^{-1}(\sigma) = \{e \in E: \exists \omega \in \sigma, \omega = n(e)\}$.

partitions *equal reason partitions*. At the end of §14.8 I shall show that there is a significantly stronger FPOIS available because we can weaken the input.

Since POIS determines the probabilities, the converse direction holds, that members of Σ having equal probability have equal μ-measures. The probabilities defined by FPOIS are conditional on the full epistemic state but are not conditional probabilities in a standard sense. That being said, views which take conditional probability to be philosophically basic will be compatible with FPOIS if the basic conditionality implies a full epistemic state over the events of interest. See §§2.10–2.12 for more on the relation to conditional probability when the latter is taken to be philosophically basic.

The way this gives us the mathematical representation of the POI is traceable via the definitions.

Principle of Indifference Representation Thesis 5

The FPOIS defines the mathematical representation of the principle of indifference: the possibilities are represented by the σ-algebra of the name space; the equal ignorance is represented by a measure on that σ-algebra that gives equal measure to the representations of events of which we are equally ignorant; the equality of probability of possibilities of equal ignorance is then stated by POIS in terms of the measure on the σ-algebra representing the possibilities.

Explanation: This is called a thesis rather than a theorem since it is about the relation of the principle of indifference, which is a proposition in the philosophy of probability, to how the elements of the principle are represented by the mathematical objects.

Recall the principle: probabilities of possibilities between which you lack any reason to discriminate epistemically are equal. The possibilities and the equal ignorance are inputs and the probabilities are the outputs.

The first definition may appear trivial, but it is intended to exclude the kind of meteor/unicycle problem mentioned earlier, either because a domain restriction is in place or because my preferred metaphysical/methodological solution applies. It works by excluding any vagueness about equality of epistemic status for the given events.

The full epistemic state ensures that possibilities of equal reason and unequal reason are determinate in the event space, which event space is then defined by the second definition as the input of the possibilities, which input is then represented by the naming function in the third definition, giving the mathematical representation of the possibilities by the name space. Equal ignorance is carried forward as an input from the first and second definitions to its mathematical representation by the fourth definition. POIS then follows the POI by restating the various elements of the POI in the terms given by the first four definitions.

Since the application of FPOIS that will occupy us is mostly to continua, and for sake of exhibiting the subsumption of my earlier Principle of Indifference for Continuum Sized Sets (Shackel 2007:159), I articulate a principle subordinate to FPOIS restricted to continua:

> *Principle of Indifference for Continua* (POIC): For a continuum sized set Ω, given a σ-algebra, Σ, on Ω and a non-zero finite measure, μ, on Σ, and given that we have no reason to discriminate between members of Σ with equal measures under μ, the members of Σ with equal μ-measure have equal probability

A complication for continua is Ulam's theorem (Bruckner *et al.* 2008:94): that the only finite measure with all singletons having zero measure and its σ-algebra being the *power set* of a continuum is the zero measure. We, however, need a non-zero finite measure. The theorem implies that for such a measure, there will be unmeasurable subsets of Ω, which therefore do not belong to the measure's σ-algebra. If E is continuum sized this means that the event σ-algebra must be a strict subset of the power set of E. Consequently, there is a danger that a bijection $f:E \rightarrow \Omega$ maps a compound event to an unmeasurable set, which would mean it wouldn't have a probability downstream. This problem doesn't arise for our definition of a naming but it will produce some restrictions on the functions we can use in Lemma 3. to define a naming, since we can't afford f to map any members of $\mathscr{E}$ to unmeasurable subsets. This problem is ignored in the literature, and having raised it, we will assume the suitable restriction is in place whenever we use Lemma 3.

Since the POIC is a special case of POIS, the preceding thesis and later proofs about or involving POIS will thereby hold for POIC. This perhaps rather obvious point being made, unless there is a particular reason to refer to the POIC, I usually will speak of the POIS.

I have laid out these subordinate principles in this rather prolonged way because most of the time we will not distinguish the events from the mathematical space representing them, especially when the events are items in that mathematical space, and hence the principles of indifference we spend our time applying is the POIS rather than the full principles which are distributed over these definitions. This is reflected in the way that POIS is stated, which also allows for $(\Omega, \Sigma) = (E, \mathscr{E})$.

The next theorem shows that when convenient, we do not need to keep track of whether a measure used in POIS is on the events themselves or their representation, because the naming and measure on the naming space together give a measure on the event space suitable for use in POIS. In short, when convenient we can neglect the detail of FPOIS. For this reason, from hereon we will use measures and probability measures on event spaces or on their name spaces as suits without further mention of this theorem. On other occasions, when it matters we will return to distinguishing all the elements in FPOIS.

Theorem 6

Given a naming $n:(E,\mathscr{E})\rightarrow(\Omega,\Sigma)$ and non-zero finite measure on Ω, μ, then $(E,\mathscr{E},\mu\circ n)$ is a non-zero finite measure space. We have no reason to discriminate between members of Σ with equal measures under μ iff we have no reason to discriminate between members of $\mathscr{E}$ with equal measures under $\mu\circ n$. We can use $(E,\mathscr{E},\mu\circ n)$ as the measure space in POIS.

Proof: $\mathscr{E}$ is a σ-algebra and $\mu\circ n:\mathscr{E}\rightarrow\mathbb{R}^+$ is a non-zero finite measure on E. The biconditional and the final proposition follow from the definitions in FPOIS. QED

I now prove the derivation of probability measures from POIS.

POIS Probability Measure Theorem 7

Given a naming $n:(E,\mathscr{E})\rightarrow(\Omega,\Sigma)$ and a non-zero finite measure on Ω, μ, then $P:\Sigma\rightarrow[0,1]$, $P(\sigma)=\mu(\sigma)/\mu(\Omega)$ is a probability measure.

Proof: If μ is a non-zero finite measure on Ω then for all $c\geq 0$, $c\mu:\Sigma\rightarrow[0,\infty)$, $(c\mu)(\sigma)=c\times\mu(\sigma)$ is a measure on Ω. Let $c=1/\mu(\Omega)$. Then $P=c\mu$ and so it is a measure. $P(\Omega)=1$ so it is a normal measure. QED

Since μ in FPOIS is a non-zero finite measure, Theorem 7 suffices to define probability measures on the name space and the event space.

Corollary 8

Probability measure space $(\Omega, \Sigma, P_\Omega)$ from POIS for the name space:

$$P_\Omega : \Sigma \rightarrow [0,1]$$

$$P_\Omega(\sigma) = \frac{\mu(\sigma)}{\mu(\Omega)}$$

Corollary 9

Probability measure space $(E,\mathscr{E},P_E)$ from FPOIS for the event space:

$$P_E : \mathscr{E} \rightarrow [0,1]$$

$$P_E(e) = P_\Omega(n(e)) = \frac{\mu(n(e))}{\mu(n(E))}$$

Events get the same probability whether via their names or directly just because P_E is defined by P_Ω. Note how the two routes for using POIS via (Ω, Σ, μ) or $(E,\mathscr{E},\mu\circ n)$ from Theorem 6 also amount to the same thing. FPOIS and the corollaries map probability measures on a naming space to probability measures on its event space. The next corollary maps probability measures on an event space to probability measures on a naming space.

Corollary 10

Given a probability measure $(E, \mathscr{E}, P_E)$ on an event space and a naming $n:(E,\mathscr{E}) \to (\Omega, \Sigma)$, there is a probability measure, P_Ω, on the naming space that gives the same probabilities to events.

Proof: The probability measure on a name space gives a probability to an event by giving a probability to the name of that event. $\sigma \in \Sigma$ iff there exists $e \in \mathscr{E}$, $\sigma = n(e)$. We define $P_\Omega(\sigma) = P_E\,(n^{-1}(\sigma)) = P_E\,(e)$ when $P_\Omega(n(e)) = P_E\,(e)$. QED

At the end of §3.5 it will be explained how the output of FPOIS with these theorems and corollaries are instances of the Theorem of Induced Measure 16. As I have formulated things, the bearers of probability are events or names of events.

> **Notation and terminology:** The probability measure, P, got from using a measure μ in POIS and then applying the aforementioned theorem and corollaries will be spoken of as the probability measure begotten by μ in POIS, or just begotten by μ or begotten by POIS; we will also speak of μ begetting the probability measure P. Where convenient, we will speak of being begotten by the measure space (Ω, Σ, μ). When we are interested in discussing different measures, μ, ν, say, of the same event space or name space, each begetting a probability measure, we will index the probability measure so begotten by the measures, that is P_μ and P_ν, rather than indexing by E or Ω. Unless we have a contextual reason to distinguish the events and their representation, or to distinguish the measures, we drop the subscripts.

> **Definition:** *Equal ignorance probability measure* is a probability measure that gives equal probabilities to events between which we have no epistemic reason to distinguish

For brevity we will usually say that an equal ignorance probability measure gives equal probabilities to events of which we are equally ignorant.

Theorem 11

The measures P_Ω and P_E defined by FPOIS are equal ignorance probability measures.

Proof: Follows immediately from FPOIS and Corollaries 8 and 9. QED

Paradoxical probabilities are when an event space has a pair of inconsistent probability measures, which means there will be at least one event for which the two probability measures give distinct probabilities. *Consistent probabilities* are non-paradoxical probabilities. I now show that paradoxical probabilities for events are equivalent to paradoxical probabilities from renamings.

Paradoxical Probabilities Equivalence Theorem 12

Let $\mu:\mathscr{E}\to\mathbb{R}^+$ and $v:\mathscr{E}\to\mathbb{R}^+$ be measures on event space $(E,\ \mathscr{E})$ used in the POIS to beget probability measures P_μ and P_v on the events.[23] P_μ and P_v are inconsistent probability measures for events iff there is a naming and renaming to measures spaces that give the same inconsistent probability measures.

Proof: Let $M=\left\{x\in\mathbb{R}^+:x=\mu(e),e\in\mathscr{E}\right\}$ and $N=\left\{x\in\mathbb{R}^+:x=v(e),e\in\mathscr{E}\right\}$. Functions

$$m:\mathscr{E}\to\mathscr{E}\times M, m(e)=\left(e,\mu(e)\right)\text{and } n:\mathscr{E}\to\mathscr{E}\times N, n(e)=\left(e,v(e)\right)$$

are both namings. Define measures on those namings by

$$j:\mathscr{E}\times M\to\mathbb{R}^+, j\left(e,\mu(e)\right)=\mu(e)\text{ and } k:\mathscr{E}\times N\to\mathbb{R}^+, k\left(e,v(e)\right)=v(e),$$

and define a renaming

$$r:\mathscr{E}\times M\to\mathscr{E}\times N, r\left(e,\mu(e)\right)=\left(e,v(e)\right)$$

Then m is a naming to σ-algebra $\mathscr{M}=\mathscr{E}\times M$ with measure j and r is a renaming of m to σ-algebra $\mathscr{M}'=\mathscr{E}\times N$ with measure k (since $r\circ m=n$). The result follows because the diagram in Figure 1 commutes.

For any $e\in\mathscr{E}$,

$$P_j\left(m(e)\right)=\frac{j\left(m(e)\right)}{j\left(m(E)\right)}=\frac{\mu(e)}{\mu(E)}=P_\mu(e)$$

$$P_k\left(r\circ m(e)\right)=P_k\left(n(e)\right)=\frac{k\left(n(e)\right)}{k\left(n(E)\right)}=\frac{v(e)}{v(E)}=P_v(e)$$

Consequently P_j and P_k on the names of events give the same probabilities respectively as P_μ and P_v on the events and

$$P_\mu(e)=P_v(e)\quad\text{iff}\quad P_j\left(m(e)\right)=P_k\left(r\circ m(e)\right)\text{ QED}$$

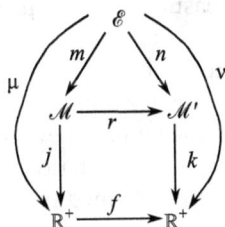

Figure 1 Commuting diagram of functions.

---

23 Theorems **6** and **7**. Which namings and measure spaces they were begotten by originally is irrelevant here.

Remark: The existence of the structure of functions in the commuting diagram, where m and n are namings, r is a renaming, and all of μ, j, k and ν are non-zero finite measures, is independent of the name spaces involved and this feeds through to the generality of the following corollary.

Non-linear Function Produces Paradoxical Probabilities
Corollary 13

Let a function $f\colon \mathbb{R}^+ \to \mathbb{R}^+$ be such that for all $e \in \mathcal{E}$, $f(\mu(e)) = \nu(e)$, that is f is a function mapping one event space measure to the other (as in the diagram). The probability measures begotten from μ and ν by POIS are inconsistent iff f is a non-linear function.

Proof: LTR: Suppose f is linear, that is $f(x) = cx + b$. Then for the empty event $\varnothing \in \mathcal{E}$, $0 = \mu(\varnothing) = \nu(\varnothing) = f(\mu(\varnothing)) = c\mu(\varnothing) + b = 0 + b$ so $b = 0$ and $f(x) = cx$. For all $e \in \mathcal{E}$:

$$P_\mu(e) = \frac{\mu(e)}{\mu(E)} = \frac{c\mu(e)}{c\mu(E)} = \frac{\nu(e)}{\nu(E)} = P_\nu(e)$$

RTL: If the probabilities are consistent then for all e

$$P_\mu(e) = \frac{\mu(e)}{\mu(E)} = \frac{f(\mu(e))}{f(\mu(E))} = \frac{\nu(e)}{\nu(E)} = P_\nu(e) \quad \text{so} \quad f(\mu(e)) = \frac{f(\mu(E))}{\mu(E)}\mu(e)$$

where the final fraction is a constant so f is linear. QED

2.7 The two different roles for measures in POIS

POIS articulates the general fact that the principle of indifference needs a measure to get a measure by stating the requirement for a measure, μ, that represents equal ignorance in a way that is logically prior to the posteriorly defined equality of probability.

Call the logically prior measure in (Ω, Σ, μ) that represents equal ignorance the *normative measure*. Many measures may appear to play that role: when we need to be careful about whether they do, we will say a non-zero finite measure μ *fulfills the normative measure role* iff members of Σ of equal epistemic status (i.e. the names of events of equal epistemic status) have equal μ-measure. The logically posterior measures are the probability measures $(\Omega, \Sigma, P_\Omega)$ and $(E, \mathcal{E}, P_E)$ defined from the normative measure by Theorem 7 and its corollaries.

The distinction between the logically prior normative measure role and the logically posterior probability measure role is crucial. This is the mathematical representation of the distinction between the normative input to the principle of indifference, equal ignorance, and the probabilistic output, equal probabilities for events of equal ignorance. Any principle of indifference for sets, whether explicitly or implicitly, must therefore distinguish the roles of being the prior normative measure and being the posterior probability measure. Consequently FPOIS and POIS and Theorem 7 are not restricted principles but are general principles for sets of which any other principles for sets must be instances.

There is a danger of confusing or conflating the distinction in roles. The danger arises because of the relation of the measures (Ω, Σ, μ) and $(\Omega, \Sigma, P_\Omega)$, in particular because the latter is defined in terms of the former and we usually pass to the latter without ever making explicit the former. This leads us ignore the difference between them. For example, in the dice case the counting measure on $\{1,2,3,4,5,6\}$ is rarely explicitly attended to, just used without thinking. We might use a normal measure as the normative measure and then the probability measure will be identical to it. For example, instead of the counting measure for the die we could have used a measure giving the value 1/6 for each singleton set and that measure would also fulfil the normative measure role.

In general, the significance of the normative measure is neglected because, as we shall see, what authors call a procedure for choosing an event at random is often a matter of selecting from a name space. In then using the name space's standard measure to define probabilities, or even worse, in using its normalized standard measure, they both conflate and confound the distinct roles. They thereby overlook the distinction even though the distinct roles have not been eliminated.

2.8 The Full Principle of Indifference for Sets is Unavoidable

I will finish the chapter with some sections on FPOIS and the wider context, including the question of alternative axiomatizations. First I want to prove FPOIS to be unavoidable.

Although I have argued in Chapter 1 that the principle of indifference as there formulated is the correct general formulation and exemplified in the chapter appendix other formulations that it subsumes, I cannot prove that there are no independent alternative proposals. However, I can now prove a theorem that means that probability measures from *any proposed principle of indifference whatsoever* are equivalent to those produced by the FPOIS subordinate principle and that all the features of that principle can be derived from any such proposed principle of indifference.

That is a powerful result. For whatever reasons the principle of indifference and the generality of the way I formulate equal epistemic status is rejected, whatsoever replacement principle or replacement definition of epistemic equality is proposed, the principle cannot be applied to produce a probability measure without being committed to the whole panoply of features of my subordinate principles. They will be committed to a full epistemic state over a σ-algebra of events, to the input to the POI, to a representation of events by a measurable space for which equal ignorance is defined, and finally to a measure on that which satisfied the Principle of Indifference for Sets, which when the relevant corollary is applied gives their probability measure. The generality of equal epistemic status cannot be avoided: whatever is meant by epistemic equality is what defines equal ignorance in that panoply of features.

For this reason, and although it is not the subject of this book, any such proposal faces evaluation by examination of the panoply it produces. For example, if the equal epistemic status implied got the epistemic situation wrong, that would cast doubt on the proposal. We will see an example of this in §13.6.

Consequently, the formal foundation of the rest of the book is unavoidable and that means that all the results of the book from hereon are unavoidable too, since the rest of the book is built on that foundation. No alternative proposals for the principle or for equal ignorance can get out of this.

To be clear of just how extreme this consequence is, consider a proposed principle that reverses the order of epistemic explanation by rejecting the priority of equal epistemic status, takes probability to be prior and defines equal epistemic status by equal probability. As I said in the previous chapter, one might wonder why anyone would make such a proposal, since if you have probabilities as an input you have already satisfied Salmon's Ascertainability criterion, so why are you bothering with a principle of indifference at all? On top of that, such a position has difficulty in explaining why the so defined equal epistemic status is genuinely normative, taking us into a highly controversial area in metanormativity. None of that matters from our point of view: for whatever reason anyone wishes to advance such a proposal and whatever answers can be given to those questions, it cannot save them from the results because of this theorem:

FPOIS is Unavoidable Theorem 14

Let a probability measure Q on events be begotten from an application of a principle of indifference and let equal epistemic status be defined by whatever definition is used to define the principle. Then there exists a full epistemic state over a σ-algebra of those events, an event space $(E, \mathcal{E})$ for the input to the POI, a representation of events by a measurable space (Ω, Σ) for which equal ignorance is defined and a measure μ on Ω for which members of Σ with equal μ-measure have equal epistemic status. (Ω, Σ, μ) satisfies POIS and Theorem 7 gives probability measure Q.

Proof: Q is a probability measure on events begotten by a principle of indifference and therefore gives equal probabilities to events of equal epistemic status. Q is therefore an equal ignorance probability measure. We have a σ-algebra of events, $\mathcal{E}$, on which Q is defined. For every pair in $\mathcal{E}$, either their measure under Q is the same or different, and since Q is an equal ignorance probability measure, they are the same iff they have the same epistemic status, hence we have a full epistemic state over $\mathcal{E}$. The base set, E, for that σ-algebra, is the set of events in their finest discrimination by Q and is therefore the set of atomic events for Q. So event space $(E, \mathcal{E})$ is the input to the POI. Take an Ω of the same cardinality as E, whence there exists a bijection $f: E \rightarrow \Omega$. Applying Lemma 3 gives us a naming $n: \mathcal{E} \rightarrow \Sigma$. Since Σ is defined by n correlating subsets of Ω with subsets of E, for all σ, τ $\in \Sigma$, σ and τ have equal epistemic status iff $n^{-1}(\sigma)$, $n^{-1}(\tau) \in \mathcal{E}$ have equal epistemic status. Hence equal ignorance for (Ω, Σ) is defined. We define μ on Ω by $\mu(\sigma) = Q(n^{-1}(\sigma))$. Suppose $\mu(\sigma) = \mu(\tau)$. Then $Q(n^{-1}(\sigma)) = Q(n^{-1}(\tau))$, hence $n^{-1}(\sigma)$ and $n^{-1}(\tau)$ have equal epistemic status, whence σ and τ have equal epistemic status. Hence (Ω, Σ, μ) satisfies POIS and consequently if $\mu(\sigma) = \mu(\tau)$ then σ and τ have equal probability. Applying Corollary 9

$$P_E(e) = P_\Omega(n(e)) = \frac{\mu(n(e))}{\mu(n(E))} = \frac{Q(n^{-1}(n(e)))}{Q(n^{-1}(n(E)))} = \frac{Q(e)}{Q(E)} = Q(e) \text{ QED.}$$

POIS is Unavoidable Corollary 15

Given a probability measure Q on a mathematical representations of events $(\Psi, \mathcal{M})$, there exists measurable space (Ω, Σ) for which equal ignorance is defined and a measure μ on Ω for which members of Σ with equal μ-measure have equal epistemic status. (Ω, Σ, μ) satisfies POIS and Corollary 8 gives probability measure Q.

Proof: This is obvious since the probability measure space to which Q belongs satisfies POIS with Q playing both the normative measure role and probability measure role. Alternatively let $(E, \mathcal{E})$ in the foregoing theorem be $(\Psi, \mathcal{M})$, then the naming function is the identity function. Either way, the result follows immediately. QED

2.9 November on the mathematical representation of the principle

Only one paper in the literature discusses in any detail the mathematical representation of the principle of indifference in measure theory. November analyses the principle of indifference thus:

IP [Indifference Principle] has the following three components:

1. Presumption—Events are comparable and thus can be considered equivalent (or equal) in some sense.
2. Assertion—Equivalent events have (or should have) equal probabilities.
3. Conditionalization—There is no information indicating otherwise. Thus, equivalent events have equal probabilities if (sometimes iff) there is no information indicating otherwise.

(2019a:6)

November worries that many different σ-algebras may be used for the same events and that even if we confine ourselves to a single one, many different probability measures can be given for it.[24] November holds that

> mathematical formalization of IP in Kolmogorov's theory must be a set of constraints on same-events spaces that is accompanied by a suitable order relation on the σ-algebra component.
>
> (2019a:21)

Same-event spaces are not what we mean by event spaces but are probability spaces:

> take the set... of all probability spaces which have the same σ-algebra component as a representative case of the set of all probability spaces which have equivalent σ-algebra components
>
> (2019a:17)

November has not noticed the significance of the Need-a-Measure-to-Get-a-Measure principle and has consequently failed to see the necessity of explicitly distinguishing the roles of the normative measure and the probability measure.

November concludes

> IP's mathematical formalization has not been widely covered in the literature.... This does not mean that IP does not have a mathematical formalization; On the contrary, since almost all writers about IP claim that they apply IP in different circumstances, it seems that each of them has in mind some implicit mathematical formalization of it. These formalizations are used implicitly in each of the writers' calculations when they apply IP. However, since commonly IP is not explicitly formalized, it is not clear whether it has one specific (implicit) mathematical formalization which can be considered the mathematical formalization of IP. More importantly, this means that currently the existence question [of the set of constraints on same-events spaces] is still unanswered.
>
> (2019a:26)

I agree with his remarks about widespread use of implicit formalizations and the complete absence of explicit ones. In particular, the failure to notice either the significance of the Need-a-Measure-to-Get-a-Measure principle and the consequent conflation of the roles that should be distinguished

---

24 See also November 2019b for more of his concerns.

is not confined to November: it is as widespread as the use of implicit formalizations, which is to say, it is universal in the literature.

In the preceding sections culminating in FPOIS I have given a set of definitions that define the mathematical representation of the principle of indifference and proved that FPOIS is unavoidable for any and all applications of the principle of indifference to derive probabilities.

November's worries do not trouble the FPOIS. There is no problem, at least so far as representation goes, in different measurable spaces representing the event spaces because the naming functions are bijections and because renamings are bijections. We could even define an equivalence relation and work everything out that way, although the theoretical benefit would not be worth the work of the additional complexity. Event comparability ('Presumption') is defined by the order on the real numbers applied to the normative measure μ on the measurable space (Ω, Σ) that represents the event space. FPOIS satisfies 'Conditionalization' and 'Assertion' because equivalence of events is defined by equal epistemic status and is represented by the normative measure giving equal measure to the representations of events of equal epistemic status, whereby POIS then defines equal probabilities for events whose representations have equal normative measure.

November thinks that what ought to be settled is more than what is settled by FPOIS because we need

> a set of constraints (C) which manages to constrain every set of same-events spaces in such a way that *only one space in each of these sets satisfies C.*
>
> (2019a:24 my emphasis)

It is probability spaces that he wants to be constrained, so the emphasized proposition means that the formalization guarantees a unique probability measure for the events.[25] On the basis of some of the things he says, he regards this condition as necessary for a proper formalization of the principle of indifference. In so doing he is addressing the paradoxes and offering a solution of a kind I discuss later, so rather than get ahead of ourselves, I will return to his solution in Chapter 12.

2.10 Basic probabilities and conditional probabilities

The way I have set up FPOIS may appear to commit me to what Climenhaga calls orthodoxy about the structure of epistemic probabilities.

---

25 Part of Chapter 13 is about explicit mathematical constraints that entail constraints on November's same-events spaces.

The premise of the structural project is that… the values of some probabilities are determined by the values of other probabilities…. Basic probabilities… are the elementary quantities out of which other probabilities are built; they are the 'atoms' of probability theory. Given values for basic probabilities, we can compute values for all non-basic probabilities.

(2020:3215)

Climenhaga uses an example for his discussion:

an urn… was selected by coin flip from two urns, U_1 and U_2. U_1 contains 1 black ball and 2 white… U_2 contains 2 black balls and 1 white…. In this problem there are two variables: the contents of the urn, and what colour ball we draw…. each variable has an associated partition, that is, set of mutually exclusive and jointly exhaustive possibilities: $\{U_1,U_2\}$, $\{B,W\}$…. Of particular interest are the following complex propositions: $U_1\&B$ $U_1\&W$ $U_2\&B$ $U_2\&W$. These propositions are state-descriptions…. maximally complete descriptions of the world of our problem.

(2020:3216)

Orthodoxy holds that

the basic probabilities are the unconditional probabilities of state-descriptions: $P(U_1\&B)$, $P(U_1\&W)$, $P(U_2\&B)$, and $P(U_2\&W)$.

(2020:3218)

The contending view offered by Climenhaga is explanationism

According to Explanationism, basic probabilities are the probabilities of atomic propositions conditional on propositions directly explanatorily prior to them.[26]

(2020:3219)

I'm not going to attempt to explain explanationism and refer the reader unfamiliar with the background he is deploying to the original paper. What I shall say is that Climenhaga makes a persuasive case with which I am in sympathy. Whilst FPOIS is compatible with orthodoxy, it is not incompatible with explanationism due to a subtlety that may not be immediately obvious, a subtlety that originates in what I called the generality of the principle of indifference.

---

26 See 2020:3234 for his final formal definition.

My atomic events are indeed mutually exclusive and jointly exhaustive and can therefore correspond to Climenhaga's state descriptions (propositions) or Carnap's (sentences). But it does not follow that the basic probabilities from FPOIS are unconditional probabilities *of state descriptions*, for the reason that we have already seen: standardly for continua, the measure of atomic events does not determine the measure of compound events. We shall see in Chapter 13 that a premeasure on a semialgebra over the atomic events suffices to determine a normative measure, when in this sense the basic probabilities would be the probabilities of *all* members of the semialgebra, not just the atomic events.

That being said, in the kind of finite events case Climenhaga gives, an integer premeasure on the atomic events will usually suffice to determine a normative measure on all the events, and hence in this sense the basic probabilities would be the probabilities of the atomic events. The question at this point is in what sense of determination do the basic probabilities determine the rest.

In a footnote to the first of the preceding quotations Climenhaga says

Plausibly, this determination relation is metaphysical grounding, but one need not assume this to pursue the structural question. I briefly discuss the possibility of other kinds of non-causal explanatory priority relations in §2.3.

(2020:3234)

The later amplification he gives is this:

Causal priority, as in the above urn examples, is one kind of explanatory priority, and the most common kind to which Bayesian networks have been applied. Schaffer (2016) also uses Bayesian networks to formalize metaphysical grounding. Whether causal and metaphysical priority are really the only two kinds of direct explanatory priority is disputable. Mathematical priority might be distinct from metaphysical grounding. Huemer… discusses temporal, part-whole, in-virtue-of, and supervenience priority. Henderson et al.… speak of more specific theories as being "constructed" out of more general theories, giving examples in which the probability of the specific theory conditional on the general theory is apparently treated as basic by scientists. I leave the question of whether these are really (distinct) kinds of explanatory priority, and whether there are other kinds, as an area for further research.

(2020:3222)

It only requires mathematical priority and the other senses of determination to come apart for FPOIS to be compatible with Climenhaga's view that his conditional probabilities are basic in these other senses. For although FPOIS makes basic the probabilities of atomic events in the finite case and the probabilities of at least a semialgebra of events for continua, it makes them mathematically prior only in the sense of being able to calculate the rest from them, including conditional probabilities. That is compatible with the conditional probabilities being basic in the other senses. So provided FPOIS does not give the wrong conditional probabilities it need not advance the unconditional probabilities as basic in those other senses.

2.11 Climenhaga's conditional probabilities objection to the principle

Climenhaga argues that the principle of indifference does, in fact, gets probabilities wrong.

> I have an urn in front of me that contains 1 black ball and 1 white ball.... I am going to sample from the urn twice, and... the outcome of the first draw will influence the outcome of the second... if I draw the black ball the first time, I will set it aside, and so be ensured to draw the white ball the second time. If I draw the white ball the first time, I will set it aside, but also add a green ball to the urn. Now we have two partitions: $\{B_1, W_1\}$, $\{B_2, W_2, G_2\}$. This gives us six state descriptions: $\{B_1 \& B_2, B_1 \& W_2, B_1 \& G_2, W_1 \& B_2, W_1 \& W_2, W_1 \& G_2\}$. Your background knowledge that $B_1 \leftrightarrow W_2$ and $W_1 \leftrightarrow B_2 \vee G_2$ allows you to eliminate the first, third, and fifth outcomes, leaving you with $\{B_1 \& W_2, W_1 \& B_2, W_1 \& G_2\}$. If you apply the Principle of Indifference to those state-descriptions not excluded by your knowledge, they each get 1/3 probability. This implies that, before either draw has been made, $P(B_1) = 1/3$ and $P(W_1) = 2/3$. So without giving you any new knowledge about how I make the first draw and without telling you about any actual (as opposed to merely possible) effects of that draw, *I have made it more initially likely for you that the first draw is white*. This is the intuitively wrong result.

(2020:3231)

Although he hasn't shown this to get the conditional probabilities wrong, it does. Evidently $P(B_2|W_1) = 1$ but if we calculate from the state descriptions we get

$$P(B_2 \mid W_1) = \frac{P(W_1 \& B_2)}{P(W_1)} = \frac{\dfrac{1}{3}}{\dfrac{2}{3}} = \frac{1}{2}$$

Climenhaga's argument fails if we apply FPOIS. Here we meet the afore-mentioned subtlety. It does *not* say give equal normative measure to each of (the names of) the atomic events, it says the normative measure gives equal measure to members of the σ-algebra *with equal epistemic status*. Using ≈ for equal epistemic status, and using μ for the normative measure, knowing the set up as we do gives us the premisses of equal epistemic status ignored by him.

1. $B_1 \approx W_1$ (premiss)
2. $B_1 \leftrightarrow W_2$ (premiss)
3. $B_1 \& W_2 \approx W_1$ (1, 2)
4. $W_1 \leftrightarrow B_2 \vee G_2$ (premiss)
5. $B_1 \& W_2 \approx (W_1 \& B_2) \vee (W_1 \& G_2)$ (3, 4)
6. $B_2 \approx G_2$ (premiss)
7. $W_1 \& B_2 \approx W_1 \& G_2$ (6)
8. $\mu(B_1 \& W_2) = \mu((W_1 \& B_2) \vee (W_1 \& G_2))$ (5, FPOIS)
9. $\mu(W_1 \& B_2) = \mu(W_1 \& G_2)$ (7, FPOIS)
10. $P(B_1 \& W_2) = P((W_1 \& B_2) \vee (W_1 \& G_2)) = 1/2$ (8, Theorem 7)
11. $P(W_1 \& B_2) = P(W_1 \& G_2) = 1/4$ (9, 10, Theorem 7, additivity of P)
12. $P(W_1) = P((W_1 \& B_2) \vee (W_1 \& G_2)) = 1/2$ (10)

Lines 10 and 11 show that FPOIS agrees with the probabilities Climen-haga gets for these state-descriptions from his basic probabilities and a simple calculation from our values shows we agree with his basic probabilities (2020:3231–2). In particular, Lines 10 and 12 show that we get the correct probability for W_1 and for the conditional $P(B_2|W_1) = P(B_1 \& W_2)/P(W_1) = \frac{1}{2}/\frac{1}{2} = 1$.

The capacity to handle this kind of problem is one of the reasons I made explicit the distinction between the event space and the naming measure space, articulated a full epistemic state over the event space and then de-fined how that carries over to the name space and then to the normative measure. The condition required to overcome the failure of the measures of continuum many singletons to determine measures for their σ-algebra created the subtlety used here. FPOIS need not blindly give a single equal probability to all the atomic events but allows us to partition them accord-ing to their equal epistemic status and then attribute equal probabilities to members of the same partition. It is able to do this just because it success-fully implements what I called the generality of my principle of indifference (the feature that allowed us to avoid the book paradox).

2.12 Other axiomatizations

I mentioned previously some of the ways in which a probability measure fails to fully represent probability. For a broader survey of the limitations see Lyon 2016 and for a more detailed mathematical survey see Fine, T. 1973:chapter 3. These limitations have driven the search for alternative axiomatizations. Hajek remarks that

> There are other formalizations that give up normalization; that give up countable additivity, and even additivity; that allow probabilities to take infinitesimal values (positive, but smaller than every positive real number); that allow probabilities to be imprecise—interval-valued, or more generally represented with sets of precise probability functions; and that treat probabilities comparatively rather than quantitatively.
>
> (Hájek 2019:6–7)

For example, the cause of the difference between certainty and probability of 1 is that for uncountable event spaces the probability measures fail *regularity* (any event has a non-zero measure). The failure of regularity has knock on effects, such as leaving probabilities conditioned on events of zero probability undefined. There are technical ways of addressing this. Kolmogorov's original definition of probability for infinite spaces in terms of expectations can avoid this problem.[27] Non-standard analysis with infinitesimals preserves regularity (Loeb 1979). Carnap 1971 addresses the issue, and Rényi 1955 and Popper 2002 offer axiomatizations that take conditional probability as basic.

Renyi's axiomatization straightforwardly defines Kolmogorovian probability just because its second axiom defines conditional probability on whatever condition to be a countably additive measure and so gives a Kolmogorovian unconditional probability when conditioned on the whole space. With some additional complexity we might therefore define POIS via an intermediary Renyi conditional probability. This is not the case for Popper's, as is shown by Spohn's representation, which embeds it in a measure-theoretic context and demonstrates that

> all and only σ-additive Popper measures can be... uniquely... represented by...dimensionally well-ordered families of σ-additive probability measures.
>
> (Spohn 1986:69)

---

27 Roughly, for a continuous random variable X, despite $P(X=x)=0$, the Radon-Nikodym theorem promises a function f_A satisfying for all Borel B: $P(B)>0, \int_B f_A(x)dP_X = P(A \wedge X^{-1}(B))$. Kolmogorov defines $P(A|X=x)=f_A(x)$. This simple case can be extended to define probability conditional on sub-algebras of the σ-algebra. See Nualart 2004:§1.2 and for the original Kolmogorov 1933/1956: chapter 5.

Hájek 2003 gives a prolonged examination of problems with which conditional probability confronts Kolmogorov's axiomatization and Lyon points out that to some extent the problems are 'not Kolmogorov's alone' (Lyon 2016:161). Here is not the place to go into them further.

Subjectivists such as Ramsey are not required to reject Kolmogorov. Fishburn points out that

> The theory of subjective probability attempts to make precise the connection between coherent dispositions toward uncertainty and quantitative probability as axiomatized by Kolmogorov.
>
> (Fishburn 1986:353)

Ramsey proves that his definition of rational degrees of belief as betting odds makes it a probability measure over his σ-algebra of beliefs. Alternate axiomatizations from subjectivists, such as those of de Finetti and the later Savage, accept finite additivity but deny countable additivity, and this amounts to adopting a segment of the theory of finitely additive measures.

Fine 1973 is a valuable and detailed examination of mathematical axiomatizations of the comparative and quantitative aspects of probabilities, and he often traces their relation to Kolmogorov's axiomatizations. His recent Fine, T. 2016 is a more approachable update.

I am not aware of any attempts to solve Bertrand's paradox that locate its origin in the axiomatization of probability rather than the principle of indifference. For that reason, I shall not attempt to prove that alternate axiomatizations make no difference to the conclusions of the book. The prior work of §2.8 supports my opinion that they would not. Consequently, the mathematical work done herein is almost entirely within standard measure theory, whose precision and explicitness provides sufficient materials for translation into alternative theories if it is possible to do so, or to determine where divergences occur if it is not possible. I will not be addressing alternative axiomatizations except insofar as a proposed solution rests on such an alternative, and my treatment will address the solution rather than worrying about the axiomatization. Where appropriate, I shall also comment on the relation of various philosophical theories of probability to the subject of our book without concerning myself with whether they make use of alternative axiomatizations.

3   Bertrand's Paradoxes

3.1 Bertrand's paradoxes

Joseph Louis François Bertrand (1822–1900) was a French mathematician who wrote an influential book on probability theory. In *Calcul Des Probabilités* he argued (among other things) that the principle of indifference is *not* applicable to cases with *infinitely* many possibilities because

> [To be told] to choose *at random*, between an infinite number of possible cases, is not a sufficient indication [of what to do].
>
> (1889:4, my translations throughout the book)

For this reason, Bertrand holds, deriving probabilities using the principle of indifference in such cases gives rise to contradiction. As proof, he offers a number of examples. The first I shall call his Square paradox:

> We ask, for example, the probability that a number, integer or fractional, rational or irrational,[28] chosen at random between 0 and 100, is greater than 50. The answer seems obvious: the number of favourable cases is half of that possible cases. The probability is ½.
>
> Instead of the number, however, we can chose its square. If the number is between 50 and 100, the square will be between 2500 and 10,000. The probability of a number chosen at random between 0 and 10,000 greater than 2500 seems obvious: the number of favourable cases is three quarters of the number of possible cases. The probability is ¾.
>
> (1889:4)

---

28 He says 'commensurable ou incommensurable', commensurable or incommensurable in English.

DOI: 10.4324/9781003456308-3

His Horizon paradox:

A plane is randomly chosen in space; what is the probability that it makes an angle with horizon of less than π/4 [radians].[29]

We can say: all angles are possible between 0 and π/2, [so] the probability that the choice falls on an angle less than π/4 is ½.

We can say also: let us draw a radius perpendicular to the plane in question from the center of a sphere. To choose the plane at random is to choose at random the point where this perpendicular pierces the sphere.

For the angle of the plane with the horizon to be less than π/4 it is necessary that the perpendicular intersect the sphere in the interior of a region [i.e. in the cap in Figure 2] whose area is

$$2\pi R^2 \left(1 - \cos\frac{\pi}{4}\right) = 4\pi R^2 \sin^2\frac{\pi}{8} 2$$

The ratio of the area of this region to that of the hemisphere is

$$2\sin^2\frac{\pi}{8} = 0.29$$

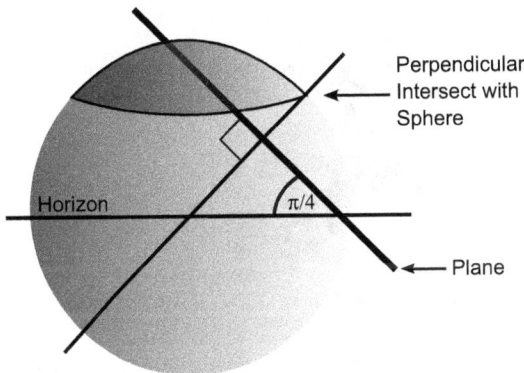

Figure 2 Horizon paradox.

29 I.e. it divides the angle from horizon to zenith at less than π/4 radians = 45°. We use radians rather than degrees throughout. 2π radians = 360°.

The probability is therefore 0.29.[30] This question, like the previous one, is badly posed and the two contradictory answers are proof of that.

(1889:6)

His Celestial Sphere paradox:

Two points are randomly fixed on the surface of a sphere[31]; what is the probability that their distance is less than 10′.[32]

The first point can be assumed to be known, the position it occupies, whatever it is, does not change the sought probability. The great circle which unites the two points can also be assumed to be known, the possible chances being the same in all directions. If this circle is divided into 2160 arcs of 10′ in such a way that the common point is a dividing point, the points located in the two arcs separated by the given point fulfil only the required condition, the probability is therefore 2/2160 = 1/1080.

We can also say the first point being known, so that the second is at a distance less than 10′, it must be located in a region (see Figure 3) whose area is[33]

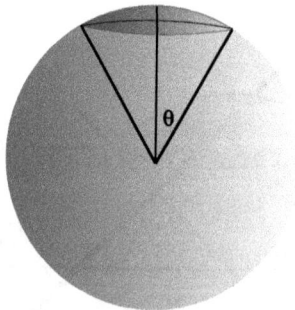

Figure 3 Celestial Sphere paradox.

30 This derivation is peculiarly long-winded, since one can simply take his first expression for the area of the cap and divide by the area of the hemisphere, $2\pi r^2$, giving $1 - \cos(\pi/4) = 1 - \sqrt{1/2} \approx 0.29$.
31 I.e. the locations of two stars on the celestial sphere.
32 10 minutes of arc. Each degree has 60′ of arc, giving 21,600′ in a circle.
33 Area of spherical surface defined by cone from centre with apex angle 2θ is $2\pi R^2(1 - \cos\theta) = 4\pi R^2\sin^2\theta/2$. Here $\theta = 10′ = 2\pi/2160$ radians.

$$4\pi R^2 \sin^2 5' = 4\pi R^2 \sin^2 \frac{\pi}{2160}$$

The ratio of the area of this region to that of the sphere is

$$0.0000042308 = \frac{1}{236362}$$

more than two hundred times smaller than 1/1080.

The probabilities relating to the distribution of stars, assuming them to be randomly placed on the Celestial Sphere, are impossible to assign unless the question is further specified.

(1889:6–7)

Finally, we have the paradox which is renownedly Bertrand's paradox: his Chord paradox

We trace *at random* a chord in a circle. What is the probability that it would be smaller than the side of the inscribed equilateral triangle?

We can say if one of the ends of the chord is known, this information does not change the probability; the symmetry of the circle does not allow any influence to be attached to it, favourable or unfavourable to the arrival of the event asked about.

One end of the chord being known, the direction must be set by chance. If we draw the two sides of the equilateral triangle having the given point for vertex, they form between them and with the tangent three angles of 60°. The chord, to be greater than the side of the equilateral triangle, must be in that one of the three angles which is included between the other two. The probability that the chance between three equal angles which can receive it directs it in that one seems, by definition, equal to ⅓ (see Figure 4).

We can also say if we know the direction of the chord, this information does not change the probability. The symmetry of the circle does not allow any influence to be attached to it, favourable or unfavourable to the arrival of the event asked about. The direction of the chord being given, it must, in order to be greater than the side of the equilateral triangle, intersect one or other of the radii which compose the perpendicular diameter in the half nearest the center. The probability of this being so seems, by definition, to equal to ½ (see Figure 5).

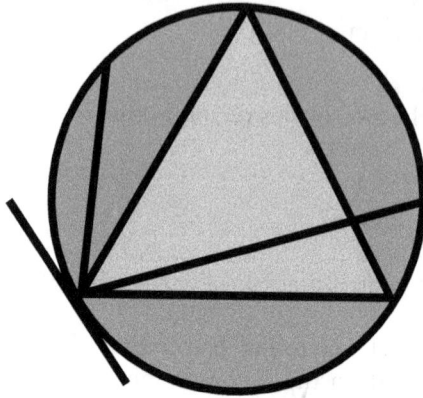

Figure 4 Chord paradox: Angle case.

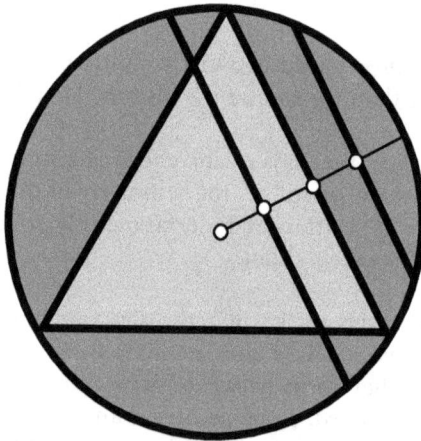

Figure 5 Chord paradox: Direction case.

We can also say: to choose a chord at random is to choose its midpoint at random. For the chord to be greater than the side of the equilateral triangle, it is necessary and sufficient that the midpoint be at a distance from the center smaller than half the radius, that is, inside a circle four times smaller in area. The number of points located in the interior of an area four times less is four times less.[34] The probability that the chord whose midpoint is chosen at random is greater than the side of the equilateral triangle seems, by definition, equal to 1/4 (See Figure 6).

Which of these three answers is the genuine one? None of the three is false, none is true, the question is badly posed.

(1889:4–5, figures are my interpolations)

Amusingly, having asked the question of the probability of the chord being shorter than the side of the inscribed triangle, he answers by giving the probability of it being longer. We will join Bertrand and the entire subsequent literature in being concerned with what I shall hereafter call Bertrand's question 'What is the probability that a random chord of a circle is longer than the side of the inscribed equilateral triangle?' I shall speak of the answer to Bertrand's question as the probability of a longer chord or for brevity the probability of longer.

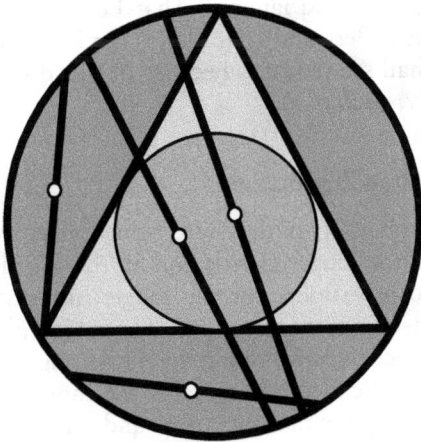

Figure 6 Chord paradox: Midpoint case.

34 The original is, indeed, 'Le nombre des points...'. The cardinalities are the same and I find it difficult to believe Bertrand was unaware of this fact. So this is an odd remark from Bertrand. As I understand his intention, he is expressing misleadingly his application of the POI, under which application probabilities for a point chosen at random in a region are proportionate to the area of the region. See previous chapter Cricket Square example. This gives a the uniform probability measure over the circle. Against this, see an hilarious fallacy, advanced in aid of Bertrand saying that infinity is not a number, which takes a proof that there are as many numbers between 1 and 2 as there are greater than 1 to be a reductio of the proposition that there are numbers greater than 2 (Jackson 1903).

The essence of Bertrand's procedure to answer his question is first to appeal to geometric intuitions of equal ignorance over *properties* of the chords. Second, that equal ignorance and the properties as possibilities are the input to the principle of indifference. The output is the normalized Lebesgue measures on those properties being the probability measures on those properties. Finally he defines the probabilities for sets of chords by the probabilities of their properties.

We will eventually discuss Bertrand's explicit deployment of the paradoxes against the principle of indifference when applied to infinitudes of possibilities rather than when applied to finitudes of possibilities. Until then, we treat the paradoxes as claimed refutations of the principle of indifference in general.

Our focus throughout the book is on the chord paradox but we shall see that some discussions of Bertrand's paradox look more like discussions of the square paradox and for that reason on various occasions we will also concern ourselves with the latter. I shall eventually have more to say about all four paradoxes. We will see that there is an important, but unremarked, division between them, and later still see that the significance of that division has been neglected and ignored. That Bertrand's Chord paradox is renownedly known as Bertrand's paradox is not knowingly because of this division, but we shall see that it is because of this division that it is quite rightly, renownedly, so known.[35]

3.2 Frailties of Bertrand's procedure

To determine the probability of the chord being longer we want to measure two sets of chords, the longer chords and all the chords—whose ratio of measures gives the probability, or the longer and not longer chords—whose ratio of measures gives the odds.[36] There are questions to be raised about Bertrand's procedures for doing the measuring.

1. Let $\mathbb{C}$ be the set of chords. To measure the chords Bertrand uses measures on the real line in the first two cases and a measure on the plane in the third. To do this he is implicitly using partial functions from $\mathbb{C}$ into $\mathbb{R}$ or $\mathbb{R}^2$, and then the Lebesgue measure on the image is used to calculate the probabilities. Furthermore, only in the midpoint case (the third) is the

35 Crofton's section on geometrical probabilities (Crofton 1885) raises the question of probabilities of chord with various features but doesn't pose a paradox. He is acknowledged by recent authors (Holbrook and Kim 2000; Chiu and Larson 2009) for introducing some of the methods used in calculating the probability of longer. Borel (1909) spent more time discussing the celestial sphere paradox than the chord paradox, but he also thought it worth discussing both rather than just one of them.

36 Odds determine probabilities and vice versa. The odds of A to ¬A are $x{:}y$ iff $P(A)=x/(x+y)$ and $P(\neg A)=y/(x+y)$.

image an image of $\mathbb{C}$ itself.[37] None of these functions are namings. In the angle and direction cases the image is of a proper subset of $\mathbb{C}$ (hence partial functions), which subsets are taken to be representative. But are they?

2. The principle of indifference has not been applied to $\mathbb{C}$ but (1) to the image of a function that is not even a bijection (angle and midpoint cases), (2) which is an image of merely a proper subset of $\mathbb{C}$ (angle and direction cases), (3) and which probability measure on the image has been got by applying the POIS to the Lebesgue measure of that image. It is obscure how this amounts to applying the principle of indifference to $\mathbb{C}$. Really, this is a kind of blunder, since $\mathbb{C}$ is a subset of the power set of $\mathbb{R}^2$ and so a measure on $\mathbb{C}$ has to be a measure on a subset of the power set of $\mathbb{R}^2$. No measure on $\mathbb{R}$ or $\mathbb{R}^2$ is a measure of such a subset.

3. Arkadani and Wulff explicitly reject the adequacy of Bertrand's treatment of the direction case:

> the chosen radius cannot be considered as a proper sample space to represent random chords.
>
> (Ardakani and Wulff 2014:30)

Earlier, Rowbottom complained more generally

> in each case, a chord was drawn (or a cut was made) at random *from a proper subset of the possible chords that might be drawn.*
>
> (Rowbottom 2013:112)

When I made the same point a while ago (Shackel 2007:156–7), I pointed out that 'Bertrand explicitly mentions the symmetry fact' to justify letting a particular case go for the general case. Klyve's discussion of Rowbottom concurs in this opinion stating that

> Once we fix an orientation, each chord can be considered as a representative of the class of all chords of equal length.
>
> (Klyve 2013:368)

Klyve, however, is too sanguine in concluding that

> We cannot escape, therefore, the paradox simply by claiming that Bertrand fails to consider all possibilities.
>
> (Klyve 2013:368)

---

37 See Shackel 2007, Rowbottom 2013 and Ardakani and Wulff 2014 for complaints based on this point. Brentano made the same complaint some time ago and thought it solved the paradox (Brentano 1915, 1917). My thanks to Adrian Maître for sharing in personal correspondence his recent rediscovery of this.

The worry here needs spelling out, when we find it has two elements; the first I explain in this item and the second in the next.

The direction case implicitly partitions ℂ and considers a measure on one equivalence class. The relation which partitions ℂ is being parallel.[38] For each diameter of the circle, there is one such partition of parallel chords, all members of which are perpendicular to that diameter of the circle. The angle case doesn't partition ℂ. The subsets are defined for each point x on the circumference of the circle as the chords having that point as an endpoint. Since each chord has two ends there are two such subsets that each chord belongs to. So each chord belongs to two of the subsets measured in terms of the angle with the tangent at one end. The midpoint partitions ℂ unevenly, since each chord is in its own partition except the diameters, which are all in the same partition. Consequently, the procedure over and undercounts the chords because none of the functions is a naming.

4. Bertrand's procedures amount to measuring ratios of abstract cross sections of ℂ in order to determine ratios in the whole measure space—rather like measuring the ratio of the volume of pink and white sugar in seaside rock by measuring the pink and white areas on a slice (see Figure 7). This assumes that the way of producing the abstract cross section produces cross sections that have uniform sizes. Klyve appears to worry about this, although not in these terms, when saying

> Once we fix an orientation, each chord can be considered as a representative of the class of all chords of equal length. All chords (including the diameter) are now included in the argument. The only thing that changes is that the method of selecting one (class of) chord from this set may be

Figure 7 Brighton Rock.

---

38 A chord is parallel to itself; if a chord is parallel to another then that other is parallel to it; if a chord is parallel to another and that one to a third, then the first is parallel to the third. Hence the relation of being parallel is an equivalence relation and therefore it partitions ℂ into sets of parallel chords.

biased.... Some of Bertrand's methods are biased toward choosing more short chords, and some to choosing more long chords.

(Klyve 2013:368–9)

That the cross sections are uniform in the angle and direction cases has initial plausibility, but in the midpoint case it is not initially plausible when all the diameters are mapped to the same point. In all cases there is a danger of a proliferating paradox or begging the question.

To determine whether the cross sections have uniform sizes requires having a measure on $\mathbb{C}$ to measure the cross sections. If two measures on $\mathbb{C}$ can disagree about the size of the cross sections they can disagree about the ratios and therefore the probabilities, multiplying the number of probabilities of longer for each method of taking a cross section. If there is a measure that is definitive on the cross sections, it is definitive on $\mathbb{C}$. But then, that would be the one to use for the normative measure in the POIS, thereby determining the probability measure, and because it was definitive it would rule out the alternatives and hence solve the paradox. But whether there is such a definitive measure is the whole challenge of Bertrand's paradox, so to assume there is one begs the question. Call this the *question of uniform cross section*. Whether the symmetries in the angle and direction cases can answer this question is unclear, since not all symmetries preserve measures.

5. Bertrand justifies his procedure by plausible appeals to geometrical intuitions, such as his mention of the symmetry facts for why he can address the angle case in general on the basis of the particular case of chords with the same endpoint in common, and likewise for the direction case. Klyve also raises

the question of when we can reason from a specific example to reach a general conclusion.... [Bertrand] stresses the point that the symmetries of the circle allow him to do so.... the probability that a chord perpendicular to a fixed radius will... [be longer]... is the same, regardless of which radius is chosen. Sometimes, however, mathematicians tread on dangerous ground when making such assumptions. For example, when Leonhard Euler first proved his 'polyhedral formula'... he claimed that the theorem is true for any three-dimensional body bounded by planes. He then offered a proof, considering a 'general example'. Today students learn that in fact the formula holds for only some polyhedra—for example, it will not hold if the polyhedron has a hole.

Philosophers know of this history of Euler's theorem through its role in Lakatos's *Proofs and Refutations* (1976).

Furthermore, and more importantly for our purposes, we know geometrical intuitions can lead us astray when it comes to measure. For centuries mathematicians got into difficulties attempting to use geometrical intuitions and implicit bijections for measuring areas, for example, by 'adding' up the 'lines' from which they were 'composed'. Consider Cavalieri's method of proving the equality of the area of the triangles.

> If two plane figures have equal altitudes and if sections made by lines parallel to the bases and at equal distance from them are always in the same ratio, then the plane figures are also in this ratio.
>
> (Andersen 1985:316)

To illustrate with a simple case, two right triangles of equal heights and bases have correlated sections, shown by dotted, dashed and solid line, as in Figure 8 on the left.

That they have the same area is evident by each being half the rectangle they make up, so Cavalieri's method gets the right answer. Consider, however, if one dropped the condition of equal distance from the base: and why shouldn't one, since if areas are really determinable by adding up the lines why should their distance from the base make a difference? Then consider a rectangle with a convex curve running from opposite corners as in Figure 8 on the right. In the latter case the areas are different, yet by the method of comparing the lines from which the area is constituted, they come out the same. Cavalieri succeeded because he found ways round the obvious problems that plagued previous attempts at getting areas from lines, but in doing so he was really leaving behind the geometrical intuitions that were being appealed to in those attempts. On the other hand, we should not forget the success of Newton's geometrical intuition that the height of a curve is the rate at which the area underneath it is increasing, which thought contains the essence of the fundamental theorem of calculus. Again, however, it required the development of analysis in the 19th century before mathematicians stopped committing errors on the basis of this intuition.

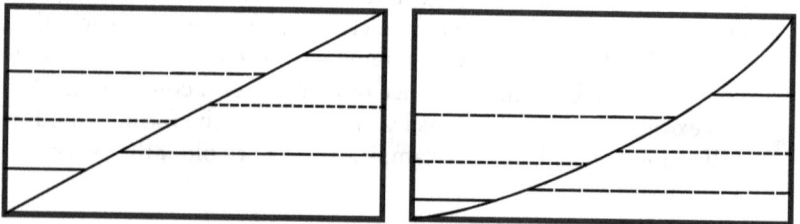

Figure 8 Cavalieri's method.

Furthermore, we know that even a bijection between sets is insufficient for equality of length or area or volume. All line segments have the same cardinality, and hence between any two line segments there exists a bijection, including between line segments of differing lengths. More dramatically, we have the Banach-Tarski theorem (Banach and Tarski 1924), a consequence of which is that a sphere can be decomposed into a finite partition and then that partition can be recomposed by translations and rotations into two distinct spheres, each the volume of the first sphere! Both being continuum sized entities entails that there is a bijection from the single sphere to the pair of spheres, yet it has half the volume. The appeal to geometric intuition cannot be wholly relied upon.

6. It is possible to prove that Bertrand's procedure, the use of equal ignorance over properties of the chords and the related measures in R or R^2 of those properties to define probabilities, is not internally consistent. Instead of concerning ourselves with angles with tangents, directions with points of perpendicular intersection or midpoint locations, we could use the areas occupied by chords. Evidently equal ignorance over the chords implies equal ignorance over chords occupying equal areas. Each chord is in the Lebesgue σ-algebra on R^2 with an area of 0. Sets of chords are not in that σ-algebra but their unions generally are,[39] and the unions of the chords, longer chords and not longer chords have areas given by the Lebesgue measure. Following Bertrand's procedure, then, we can use the areas of these unions to determine the probabilities. Let R be the radius of the circle. The union of the chords = the union of the longer chords = the closed disk $D = \{(x,y) : x^2 + y^2 \leq R^2\}$ and so both have the same area of πR^2, whilst the union of not longer chords has area $3\pi R^2/4$. This gives us two further probabilities of longer: taken as a proportion of the chords, it is 1, and taken in ratio to the not-longer chords, it is 4/7. So Bertrand's way of determining the probability is not necessarily consistent with itself! This internal inconsistency of the area case is a proof that the procedure of measuring subsets of chord by measuring their unions did not produce a probability measure on the chords because it fails the additivity condition on measures.[40] But in doing that we were following

---

39 I say only generally since the set of Lebesgue measurable sets of R^2 is a strict subset of $P(R^2)$ so some unions might not be in the σ-algebra. This point also applies to Bertrand's cases without the 'might'. I say might for the unions of chords because I both conjecture and doubt that every union of chords is Lebesgue measurable. Potential proof sketch: every chord is the image of a continuous function from $[0,1] \to R^2$. Show every union is the image of a continuous function $[0,1] \times [0,1] \to R^2$ (this is the step I doubt). $[0,1] \times [0,1]$ is compact so its image is compact and therefore measurable.

40 Sum of measures of longer and not longer should equal measure of their union, the closed disk, but $\pi r^2 + 3\pi r^2/4 \neq \pi r^2$.

Bertrand's procedure. So we have shown that his procedure is internally inconsistent and its output need not be a probability measure at all, in which case the production of inconsistent probabilities need not be a challenge to the principle of indifference. One can reject the paradox for this reason.

So scratching the surface of Bertrand's procedure reveals a list of frailties of procedure:

1. Partial functions from $\mathbb{C}$ into $\mathbb{R}$ or $\mathbb{R}^2$ using the latter's Lebesgue measures as if they were measures on the chords.
2. The principle of indifference has not been applied to the chords.
3. Over- and undercounting chords.
4. The question of uniform cross sections.
5. Bertrand justifies his procedure by appeals to geometrical intuitions but we know such intuitions can mislead us, even very badly as shown by the Banach-Tarski theorem, when attempting to measure sets.
6. Bertrand's procedure is internally inconsistent because its output is not guaranteed to be a probability measure.

There is a final frailty, of validity, that I list here for the sake of completeness, but since it is treated in the next chapter I omit its explanation here.

7. Bertrand's paradox may refute Bertrand's question rather than the principle of indifference.

I readily concede that Bertrand's cases are suasive. The question is whether they are justifiably so despite the frailties. For insofar as Bertrand's paradox is claimed to refute the principle of indifference, the refutation is incomplete unless the frailties can be removed. We need a rigorous account of how exactly the Principle of Indifference for Sets can be deployed to produce the contradictory probabilities needed to establish the paradox.

3.3 A theorem in aid of Bertrand's cases

To apply the POIS to define the probability measure we need a measure on $\mathbb{C}$ to be the normative measure, that is the measure μ for which we have no reason to discriminate between members of the σ-algebra having equal μ-measures. In the angle case, Bertrand has used the Lebesgue measure on the interval $(0, \pi)$ for the normative measure and thereby defined the uniform probability measure on $(0, \pi)$. In the direction case, he has done the same for the interval $(0, R)$. In the midpoint case, Bertrand has used the

Lebesgue measure on the open disk $D=\{(x,y): x^2+y^2<R^2\}$ as the normative measure and thereby defined the uniform probability measure on the disk.

In each case, the POIS has been applied, but not applied to $\mathbb{C}$, and so what we are given are not measures on $\mathbb{C}$ at all. This is obvious for the angle and direction cases. In the midpoint case, we must not let $\mathbb{C}$'s close association with $\mathbb{R}^2$ blind us to the fact that $\mathbb{C}$ is not a subset of $\mathbb{R}^2$ but of $P(\mathbb{R}^2)$, so a σ-algebra for a measure on $\mathbb{C}$ is a subset of $P(P(\mathbb{R}^2))$ not of $P(\mathbb{R}^2)$. By contrast, the σ-algebra for the Lebesgue measure on the disk is a subset of $P(\mathbb{R}^2)$. Hence the Lebesgue measure on the disk is not a measure on $\mathbb{C}$.

We can, however, reconstruct Bertrand's cases with the aid of a theorem by which we can use a measure on one set to induce a measure on another.

Theorem of Induced Measure (TIM) 16

Given a set, Y, with a σ-algebra, $\mathscr{S}$, and a measure, ν, any function $f:X \rightarrow Y$ induces a measure on the set X, giving measure space (X, Σ, μ) where $\Sigma = \{\sigma \subset X : \sigma = f^{-1}(S), S \in \mathscr{S}\}$ and $\mu:\Sigma \rightarrow \mathbb{R}^+$, $\mu(\sigma)=\nu(f(\sigma))$. The function from Σ to $\mathscr{S}$ defined by f is a bijection. The image of f has non-zero finite ν-measure iff μ is a non-zero finite measure.

Proof: A map between sets yields a morphism of set systems (Jost 2015:Lemma 4.3.1). Essentially what is going on is that preimages of subsets of codomains define functors (Jost 2015:104) and such functors preserve set system type. Here, the set system on the codomain of the map is a σ-algebra so the functor determines a set system on the domain that is also a σ-algebra; then we can define a measure on one by a measure on the other via the functor's correlation of the σ-algebras. Hence in the statement of the theorem we have defined Σ to be the set of pre-images of members of $\mathscr{S}$, and defined the measure under μ of an element in Σ to be the measure under ν of its image in $\mathscr{S}$. Then Σ is a σ-algebra by the theorem from Jost.

μ is non-negative with $\mu(\varnothing)=0$ so to show it to be a measure we need only check countable additivity: Let $\sigma=\{\sigma_1, \sigma_2, \sigma_3,...\sigma_n,...\}$ be a sequence in Σ. $y \in f(\bigcup \sigma)$ iff $\exists n \, \exists x : x \in \sigma_n$ and $y=f(x)$ iff $y \in \bigcup_{n \in \mathbb{N}} f(\sigma_n)$ hence $f(\bigcup \sigma)=\bigcup_{n \in \mathbb{N}} f(\sigma_n)$, giving

$$\sum_{n \in \mathbb{N}} \mu(\sigma_n) = \sum_{n \in \mathbb{N}} \nu(f(\sigma_n)) = \nu\left(\bigcup_{n \in \mathbb{N}} f(\sigma_n)\right) = \nu\left(f\left(\bigcup_{n \in \mathbb{N}} \sigma_n\right)\right) = \mu\left(\bigcup_{n \in \mathbb{N}} \sigma_n\right).$$

Suppose $\sigma, \tau \in \Sigma$ and $f(\sigma)=f(\tau)$. Then there are $S, T \in \mathscr{S}$: $\sigma = f^{-1}(S)$ and $\tau=f^{-1}(T)$. Hence $S=f(f^{-1}(S))=f(\sigma)=f(\tau)=f(f^{-1}(T))=T$. So $\sigma=f^{-1}(S)=f^{-1}(T)=\tau$. So f restricted to Σ is an injection. By the definition of Σ, it is a surjection, so it is a bijection. Since $\mu(X)=\nu(f(X))$, the image of f has non-zero finite ν-measure iff μ is a non-zero finite measure. QED

Remark: $f:X \rightarrow Y$ defines a function between the σ-algebras $\mathscr{S}$ and Σ and it is convenient to abuse our notation and call that function 'f' as well. Then μ is the composition of ν and f, where $f: \mathscr{S} \rightarrow \Sigma$.

Terminology: When a TIM begotten measure is used as the normative measure in POIS we will also call the subsequent POIS begotten probability measure the TIM begotten probability measure.

Corollary 17

Given measure space $(Y,\mathscr{P},v)$, an injection $f{:}\mathrm{P}(X){\to}\mathrm{P}(Y)$ induces a measure space (X, Σ, μ). f restricted to Σ is a bijection.

Proof: $\Sigma = \{\sigma \subset X : \sigma = f^{-1}(S), S \in \mathscr{P}\}$ is a σ-algebra and $\mu{:}X{\to}\mathbb{R}^+$, $\mu(\sigma) = v(f(\sigma))$ is a measure by essentially the same proof. QED

Corollary 18

Let $(Y,\mathscr{P},v)$ be a measure space. Let f: $\mathrm{P}(X){\to}\mathrm{P}(Y)$ be such that $f(A)=\varnothing$ iff $A=\varnothing$, $f(X)=Y$, $f(A - B)=f(A)-f(B)$, and if W is a sequence in $\mathrm{P}(X)$ then $f(\bigcup W)=\bigcup_{w \in W}f(w)$. Let $f^{\text{invert}} : \mathscr{P} \to \mathrm{P}(X)$, $f^{\text{invert}}(S)=\bigcup\{A{\subset}X : f(A)=S\}$. Let $\Sigma = \{\sigma \subset X : \sigma = g(S), S \in \mathscr{P}\}$ and $\mu(\sigma)=v(f(\sigma))$. Then (X, Σ, μ) is a measure space induced by f.

Proof: By its definition, μ is a measure provided Σ is a σ-algebra. To avoid clutter, let $g=f^{\text{invert}}$. By the definition of g and the distribution of f over set unions, for all $g(S){\in}\Sigma, f[g(S)]=S$. Note also that $g(Y)=X$.
Applying these, let $S{\in}\Sigma$. $Y - S \in \mathscr{P}$ so $g(Y - S) \in \Sigma$.

$$f\big[g(Y)-g(S)\big]= f\big[g(Y)\big]-f\big[g(S)\big]=Y-S$$

and $g(Y-S) = \bigcup\{A \subset X : f(A)=Y-S\}$ so $g(Y)-g(S) \subset g(Y-S)$.

$$f\big(g(Y-S)-[g(Y)-g(S)]\big)= f\big[g(Y-S)\big]-f\big[g(Y)-g(S)\big]$$
$$=(Y-S)-(Y-S)=\varnothing$$

Hence $g(Y-S)-[g(Y)-g(S)]=\varnothing$ and so $g(Y-S)=g(Y)-g(S)=X-g(S)$. So Σ is closed under complements.
Let W be a sequence in Σ. Then there exists a sequence $S=\{S_1, S_2, S_3,..., S_n,...\}$ in $\mathscr{P}$ such that $W=\{g(S_1), g(S_2), g(S_3),..., g(S_n),...\}$.

$$f(\bigcup W) = \bigcup_{n \in \mathbb{N}} f\big(g(S_n)\big) = \bigcup_{n \in \mathbb{N}} S_n = \bigcup S \in \mathscr{P}.$$
$$g(\bigcup S) = \bigcup\{A \subset X : f(A)=\bigcup S\} \text{ so } \bigcup W \subset g(\bigcup S).$$
$$f\big[g(\bigcup S)-\bigcup W\big]= f\big[g(\bigcup S)\big]-f(\bigcup W) = \bigcup S-\bigcup S = \varnothing.$$

Hence $\bigcup W=g(\bigcup S)$ and $\bigcup S \in \mathscr{P}$ so $\bigcup W{\in}\Sigma$ and Σ is closed under countable unions. Finally, $g(\varnothing)=\varnothing$ so Σ is a σ-algebra. QED

Remark: We have this rather odd corollary for the sake of the analysis of the area case later. Because f need not be an injection there need not be an inverse and so in its place I had to define f^{invert} so that $f{\circ}f^{\text{invert}}$ is the identity on $\mathscr{P}$.

Corollary 19

Suppose $f:X \to Y$ induces by TIM a measure (X, Σ, μ) from $(Y, \mathscr{S}, v)$. Then (X, Σ, μ) and $(Y, \mathscr{S}, v)$ are measure isomorphic.

Proof: Let U be the set of singletons in Σ, $h:X \to U$, $h(x) = \{x\}$, let V be the set of singletons in $\mathscr{S}$, let $k:V \to f(X)$, $k(\{y\}) = y$, let $g:U \to V$ be the restriction to singleton sets of the bijection defined by f between Σ and $\mathscr{S}$ and let $w = k \circ g \circ h$. g, h and k are bijections so $w:X \to f(X)$ is a bijection. For all $\sigma \in \Sigma$

$$w(\sigma) = \{y \in Y : x \in \sigma, y = k \circ g \circ h(x)\} = \{y \in Y : x \in \sigma, y = k \circ g(\{x\})\}$$

$$= \{y \in Y : x \in \sigma, y = k \circ f(\{x\})\} = \{y \in Y : x \in \sigma, y = k(\{f(x)\})\}$$

$$= \{y \in Y : x \in \sigma, y = f(x)\} = f(\sigma)$$

Consequently w is the same bijection between Σ and $\mathscr{S}$ as is defined by f and hence for $\sigma \in \Sigma$ and $S \in \mathscr{S}$, $w(\sigma) \in \mathscr{S}$ and $w^{-1}(S) \in \Sigma$. From TIM's definition of μ,

$$\mu(\sigma) = v(f(\sigma)) = v(w(\sigma)).$$

For all $S \in \mathscr{S}$ there exists a $\sigma \in \Sigma$: $\sigma = w^{-1}(S)$, and so

$$\mu(w^{-1}(S)) = \mu(\sigma) = v(f(\sigma)) = v(w(\sigma)) = v(w(w^{-1}(S))) = v(S)$$

So w is a measure isomorphism. QED

Lemma 20

The inverse of a naming can be used for the function in TIM.

Proof: A naming is a bijection and hence its inverse exists. QED

Lemma 21

Without loss of generality, the function in TIM can always be a bijection: Suppose $f:X \to Y$ induces by TIM a measure space (X, Σ, μ) from (Y, A, m). Then there exists a measure space (Z, B, k) and a bijection $g:X \to Z$ that induces (X, Σ, μ) from (Z, B, k).

Proof: There exists a set Z with the same cardinality as X and hence there exists a bijection $g:X \to Z$. Since g is a bijection we can define a function $h:Z \to Y$, $h(z) = f(g^{-1}(z))$. Let (Z, B, k) be the TIM begotten measure space induced on Z by h from (Y, A, m), so $C \in B$ iff $h(C) \in A$ and $k(C) = m(h(C))$. Now consider the measure (X, Γ, v) induced on X by g from (Z, B, k). $S \in \Gamma$ iff $g(S) \in B$ iff $h(g(S)) \in A$ iff $f(g^{-1}(g(S))) \in A$ iff $f(S) \in A$ iff $S \in \Sigma$. $v(S) = k(g(S)) = m(h(g(S))) = m(f(g^{-1}(g(S)))) = m(f(S)) = \mu(S)$. So $(X, \Gamma, v) = (X, \Sigma, \mu)$. Hence the bijection g induces (X, Σ, μ) from (Z, B, k). QED

Remark: Later theorems using TIM do not require f to be a bijection unless they say so, but the reader may find it easier to follow many proofs assuming a bijection, which this lemma shows to be a true assumption.

Necessity of TIM Lemma 22

Let (X, Δ, δ) be a measure space. Then there exists a measure space $(Y, \mathscr{S}, v)$ and a function, $f: X \to Y$ such that the measure space (X, Σ, μ) got from them using the Theorem of Induced Measure is (X, Δ, δ).

Proof: Let Y have the cardinality of X. Then there exists a bijection $f: X \to Y$. Define $\mathscr{S} = \{S \subset Y: f^{-1}(S) \in \Delta\}$ and $v : \mathscr{S} \to \mathbb{R}^+$, $v(S) = \delta(f^{-1}(S))$. $\mathscr{S}$ is a σ-algebra and v is a measure.[41] Then applying the Theorem of Induced Measure gives

$$\Sigma = \{\sigma \subset X : \sigma = f^{-1}(S),\ S \in \mathscr{S}\} = \{\sigma \subset X : \sigma = f^{-1}(S),\ f^{-1}(S) \in \Delta\} = \Delta \text{ and}$$

$$\mu(\sigma) = v(f(\sigma)) = \delta(f^{-1}(f(\sigma))) = \delta(\sigma).$$

Hence $(X, \Sigma, \mu) = (X, \Delta, \delta)$. QED

Remark: This lemma is trivial since the identity function could always be used if we start with a measure space. It gains its significance from how we use it later as a test of *whether* we have a measure space to start with. I leave further explanation until then.

3.4 The Theorem of Induced Measure is Unavoidable

Any non-zero finite measure on a set suffices as a normative measure for the POIS and also for Theorem 7 to apply, so once we have such a measure induced by the TIM, we have sufficient for a probability measure on the set.

We now show that there is a certain sense in which the Theorem of Induced Measure is unavoidable for probability measures and that the probability of longer can take any value in [0,1] by an application of TIM to give a normative measure for POIS. Initially this last may appear trivial, since obviously we can simply assign any value to the probability of longer without using TIM. However, as we see in the coming sections and chapters how and why TIM being unavoidable permeates attempts to determine the probability of longer, its apparent triviality will evaporate.

TIM is Unavoidable Theorem 23

A function, $\phi: \Delta \to [0, \infty]$, $\Delta \subset \mathbb{P}(X)$ is a probability measure on Δ iff there exists a measure space $(Y, \mathscr{S}, v)$ and a function $f: X \to Y$ such that $v(f(X)) < \infty$ and $\phi = \mu/\mu(X)$ where μ is the measure $\mu: \Delta \to \mathbb{R}^+$, $\mu(\delta) = v(f(\delta))$ induced on X by TIM from $(Y, \mathscr{S}, v)$ and f.

Proof: LTR: (X, Δ, ϕ) is a measure space so by Lemma 22, there exist $(Y, \mathscr{S}, v)$ and a function $f: X \to Y$ such that TIM using them gives (X, Δ, ϕ) and

---

41 The proofs of this are essentially the same as those given in the proof of TIM.

so φ is the measure induced on X by TIM. φ is a probability measure so by the application of TIM, $1 = \phi(X) = v(f(X)) < \infty$ and $\phi = \phi/\phi(X)$.

RTL: From TIM we have $\mu(X) = v(f(X)) < \infty$ so μ is a non-zero finite measure and hence by Theorem 7 $\phi = \mu/\mu(X)$ is a probability measure. QED

This theorem will be used in a variety of ways. It has an additional great importance in the face of any disputes over formulating the principle of indifference for sets. It means that whatever principles of indifference are proposed, and however much any such principle is supposed to supplant my principle, any probability measure defined by such a rival principle can be begotten from the Theorem of Induced Measure and my Principle of Indifference for Sets.

Deluge (of probability measures) Theorem 24

Given event space $(E, \mathscr{E})$, let e be an event in $\mathscr{E}$. For any x in [0,1], there exists a measure space $(Y, \mathscr{S}, v)$ and a function $f : E \to Y$ which, if used in TIM to induce a non-zero finite measure on events $\mathscr{E}$, which is then used as the normative measure in POIS, results in the probability of any event in $\mathscr{E}$ being x.

Proof: Let $e \in \mathscr{E}$. For any $x \in [0,1]$, if e has cardinality of the continuum, there exists a surjection $f : E \to [0,1]$ such that $f(e) = [0,x)$. Using f and the Lebesgue measure restricted to [0,1] in the Theorem of Induced Measure gives the measure on $\mathscr{E}$, μ, such that $0 < \mu(E) < \infty$ and $\mu(e) = x\mu(E)$. Corollary 8 gives $P_E(e) = x$. If e does not have the cardinality of the continuum we use an injection $f : \mathbb{P}(E) \to \mathbb{P}([0,1])$ for which $f(e) = [0,x)$ and use the Corollary 17 of TIM. QED

Remark: This theorem is unsurprising, in the sense that we already knew we could assign probabilities to events arbitrarily, although using TIM is not itself arbitrary. It gains its significance from the preceding theorem, especially when we must use TIM because we have nowhere internal to the events to start from for a normative measure, examples of which we will be seeing. It also serves a purpose when we later speak of meta-indifference. Applied to the chords it gives $P(L) = x$ for any x in [0,1], but only this much says nothing about whether the measure μ induced by TIM goes beyond merely *playing* the role of the normative measure and actually *fulfills* that role by giving equal measure to events of which we are equally ignorant. We turn to that requirement in the next section.

In the meantime, to illustrate the threat of the Deluge theorem in the ease of finding more probabilities of longer, here is the place to note some additional appeals to a geometric induction of a measure on the chords. Garwood and Holbrook (2016) consider the chords defined by picking two points inside the circle. In common with many others, they do it all in terms of probability density functions. For us, they thereby define a measure, μ, on $[0, \pi/2]$ by

$$\mu(\sigma) = \frac{16}{3\pi} \int_\sigma \cos^4 d\lambda, \quad \sigma \in \Sigma$$

where Σ is the σ-algebra of Lebesgue measurable sets of $[0, \pi/2]$.[42] The function in TIM is $f{:}C{\to}[0, \pi/2]$, $f(c)=\pi/2-\alpha$, where α is the chord's angle with the tangent. This results in the probability of longer being $\frac{2}{3} - \frac{3\sqrt{3}}{4\pi}$.

Funkenbush for the chords defined by a point on the circumference and a point inside the circle gives probability of longer $\frac{1}{3}+\frac{\sqrt{3}}{2\pi} \approx 0.609$ (1962:144). Chiu and Larson come up with a few further methods, such as defining the probability by the areas into which a chord divides the circle. Two give the same probability as Funkenbush, one of which is an unacknowledged repetition of him and one of which is the same method as Garwood and Holbrook but they calculate a different result (0.7468). They finish with a robot choosing a chord by going round the circle and stopping with a Poisson process, the chord being defined by the starting and stopping points. This can give any answer but converges to 1/3 as the stopping rate tends to zero (2009:9–18). Vidovic using different geometric procedures comes up with the same discrete range of probabilities and adds a new one: 0.4454 (2021:438). In a more recent paper (2022) he has come up with one further value, 0.4694. Soranzo and Volčič (1998) offer an account under which the probability of longer can take any value in the interval [⅓, ½] based on essentially the same geometry as used by Marinoff (1994) for a random process defined by a point outside the circle. Finally, many of these authors have also found variant geometrical approaches producing the same results as Bertrand. In particular, Ardakani and Wulff asked about the probability of lengths of chords generally and produce Bertrand's results as special cases (Ardakani and Wulff 2014:28).

3.5 The equal ignorance function and the sufficiency principle

If we are going to use the TIM to provide a normative measure to be the input to the POIS applied to the chords, we need to see that applying the theorem to a set can represent equal ignorance about that set. This is straightforwardly done by defining:

---

42 See Bruckner *et al.* 2008: theorem 5.9 (iii): Given a measure space, here $(\mathbb{R}, \mathscr{S}, \lambda)$, the integral of a non-negative measurable function defines a measure. The function $\cos^4$ is an even power and hence non-negative.

An *equal ignorance function, f, on a set* X is a function from X to a measure space $(Y, \mathscr{S}, v)$, $f(X)$ having non-zero finite measure, and subsets of X of which we are equally ignorant are mapped to subsets of that measure space with equal v-measure.

The *equal ignorance measure* of an equal ignorance function $f:X \rightarrow Y$ is the measure v in $(Y, \mathscr{S}, v)$.

This gives us a principle of the sufficiency of the combination of the Theorem of Induced Measure and the POIS to provide a probability measure that gives equal probabilities to possibilities of which we are equally ignorant:

Sufficiency Principle 25

Let f be an equal ignorance function on X. Let the normative measure in the POIS, μ, be the measure on X induced by the Theorem of Induced Measure using f. Then the probability measure, P, thereby defined by the POIS is an equal ignorance probability measure.

Proof: $\mu(X)=v(f(X))<\infty$. Suppose we are equally ignorant between σ, $\tau \in \Sigma$. Then $\mu(\sigma)=v(f(\sigma))=v(f(\tau))=\mu(\tau)$. The middle equality holds because by the definition of f, v gives equal measures to images of subsets of X of which we are equally ignorant. Consequently events of which we are equally ignorant get equal μ-measure and so we can use μ as the normative measure in the POIS. Applying the Corollary 8, the definition of the probability measure on X from POIS gives $P(\sigma)=\mu(\sigma)/\mu(X)=\mu(\tau)/\mu(X)=P(\tau)$. So P gives equal probabilities to events of which we are equally ignorant. QED

Property Sufficiency Corollary 26

Let the function in the Sufficiency principle be defined by a property, Φ, of the members of a set, of which property we are equally ignorant. Then the probability measure thus defined is an equal ignorance probability measure that gives equal probabilities to subsets of which we are equally ignorant over property Φ.

Finally, with the theorems of this section in view, we can see that FPOIS requires a naming to be an equal ignorance function (but an equal ignorance function is not required to be a naming). Consequently we could reformulate the cumulative effect of FPOIS and Corollary 9 in terms of the naming of item 3 in FPOIS and TIM. The naming used as the function in TIM begets a normative measure for use in POIS. That normative measure from TIM would now be a measure on the events rather than the naming space and the probability measure defined by Corollary 8 would already be a probability measure of the events rather than of their names. This also explains why Theorem 6 and others that allow us when convenient to ignore whether we are speaking of events or of their names were provable, although proving them via this route would have been more long-winded.

3.6 Bertrand's three cases: the rigorous construction

We return to the chords. The points on the circumference are not chords. In what follows it would involve additional explanations about other literature that counts them as chords and also notational complexities to avoid treating them as if they were. Such explanations tend to clutter up the discussion without adding illumination. For this reason, from hereon we include circumferential points in C, the set of chords, as degenerate cases of chords. I now apply the Theorem of Induced Measure in the angle, direction and midpoint cases, in a manner intended to put Bertrand's original calculations on a sound footing as applications of the POIS.

Let $\mathscr{L}_1 \subset P(\mathbb{R})$ and $\mathscr{L}_2 \subset P(\mathbb{R}^2)$ be the σ-algebras of Lebesgue measurable sets and λ_1 and λ_2 be the corresponding Lebesgue measures. Let $L \subset C$ be the set of longer chords.

The tangent angle of a chord is the angle the chord makes with the tangents to its endpoints. For the angle case we use the function $\alpha : C \to [0, \pi/2]$, $\alpha(c) =$ the tangent angle of c. We have σ-algebra $\mathscr{S} = \{[0, \pi/2] \cap A, A \in \mathscr{L}_1\}$ and measure $v = \lambda_1$ restricted to $\mathscr{S}$. We define $\Sigma_\alpha = \{\sigma \subset C : \sigma = \alpha^{-1}(S), S \in \mathscr{S}\}$ and $\mu_\alpha : \Sigma_\alpha \to \mathbb{R}^+$, $\mu_\alpha(\sigma) = v(\alpha(\sigma))$. By TIM, $(C, \Sigma_\alpha, \mu_\alpha)$ is a measure space. We now use μ_α as the normative measure in the POIS giving the probability space $(C, \Sigma_\alpha, P_\alpha)$ from Theorem 7. The probability of longer is

$$P_\alpha(L) = \frac{\mu_\alpha(L)}{\mu_\alpha(C)} = \frac{v(\alpha(L))}{v(\alpha(C))} = \frac{v\left(\left[\frac{\pi}{3}, \frac{\pi}{2}\right]\right)}{v\left(\left[0, \frac{\pi}{2}\right]\right)} = \frac{\frac{\pi}{6}}{\frac{\pi}{2}} = \frac{1}{3}.$$

For the direction case we use the function $\rho : C \to [0, R]$, $\rho(c) =$ the radial polar coordinate of the point of perpendicular intersection of the chord with the diameter. We have σ-algebra $\mathscr{S} = \{[0, R] \cap A, A \in \mathscr{L}_1\}$ and measure $v = \lambda_1$ restricted to $\mathscr{S}$. We define $\Sigma_\rho = \{\sigma \subset C : \sigma = \rho^{-1}(S), S \in \mathscr{S}\}$ and $\mu_\rho : \Sigma_\rho \to \mathbb{R}^+$, $\mu_\rho(\sigma) = v(\rho(\sigma))$. By TIM, $(C, \Sigma_\rho, \mu_\rho)$ is a measure space. Using μ_ρ as the normative measure gives probability space (C, Σ_ρ, P_ρ). The probability of longer is

$$P_\rho(L) = \frac{\mu_\rho(L)}{\mu_\rho(C)} = \frac{v(\rho(L))}{v(\rho(C))} = \frac{v\left(\left[0, \frac{R}{2}\right]\right)}{v([0, R])} = \frac{R}{2R} = \frac{1}{2}.$$

For the midpoint case we use the function $\kappa : C \to D$, $\kappa(c) =$ the midpoint of the chord. We have σ-algebra $\mathscr{S} = \{D \cap A, \ A \in \mathscr{L}_2\}$ and measure $v = \lambda_2$ restricted to $\mathscr{S}$. We define $\Sigma_\kappa = \{\sigma \subset C : \ \sigma = \kappa^{-1}(S), \ S \in \mathscr{S}\}$ and $\mu_\kappa : \Sigma_\kappa \to R^+$, $\mu_\kappa(\sigma) = v(\kappa(\sigma))$. By TIM, $(C, \Sigma_\kappa, \mu_\kappa)$ is a measure space. Using μ_κ as the normative measure gives probability space $(C, \Sigma_\kappa, P_\kappa)$. The probability of longer is

$$
P_\kappa(L) = \frac{\mu_\kappa(L)}{\mu_\kappa(C)} = \frac{v(\kappa(L))}{v(\kappa(C))} = \frac{v\left(\left\{(x,y) : x^2 + y^2 < \left(\frac{R}{2}\right)^2\right\}\right)}{v\left(\left\{(x,y) : x^2 + y^2 \le R^2\right\}\right)} = \frac{\pi R^2 / 4}{\pi R^2} = \frac{1}{4}.
$$

The measure μ (standing for each of μ_α, μ_ρ and μ_κ) is being used for the normative measure. We need to check that this represents equal ignorance about the chords. Since these cases are so important let me spell out exactly how it is working. For each chord or set of chords, the value of the function represents a property of its argument: the angle with the tangent, the distance from the centre to the points of perpendicular intersection, the location of midpoints. Call these the angle, distance and location properties.

μ is defined by v, where v, being the restriction of a Lebesgue measure, is a uniform measure on the image of C. So if we are equally ignorant about the properties had by the chords, properties represented by the values taken in that image, that uniform measure on the image represents that equal ignorance. Since these are quantitative properties, equal ignorance is ignorance over individual quantities and ranges of quantities and v gives zero to individual properties and ranges over which we are equally ignorant have equal values under v. For example, we are equally ignorant over locations on one side of the disk and on the other and v(one side)$= \pi r^2/2 = v$(other side). And since μ is defined by the uniform measure v on those images, then subsets of chords of which we are equally ignorant of the properties represented by those images will have equal measures under μ. For example, we are equally ignorant of chords having the location of their midpoints in one side or the other of the disk. Since the function κ maps those sets of chords to one side or the other side of the disk and μ is the composition of v and κ, $\mu = v \circ \kappa$, their measure under μ is also $\pi r^2/2$ and therefore equal.

Hence the image of sets of chords of whose angle, distance or location properties we are equally ignorant will receive equal measures under their respective v-measure and hence α, ρ and κ are equal ignorance functions. Whence, sets of chords of whose angle, distance or location properties we are equally ignorant will have equal measures under their respective TIM induced μ-measure. Therefore possibilities between which we are equally ignorant, where those possibilities are subsets of chords with certain angle, distance or location properties, will be represented by equal measures under their respective μ-measure. Since α, ρ and κ are equal ignorance functions, by the Property Sufficiency Corollary 26 each of P_α, P_ρ and P_κ is an equal ignorance probability measure that gives equal probabilities to subsets of which we are equally ignorant over their respective properties.

To what extent are Bertrand's original cases now justifiable as applications of the POIS? Examining the list of frailties, numbers 1, 2, and 5 have been addressed. In each case the function is a surjection from the entirety of $\mathbb{C}$, so no more partial functions on $\mathbb{C}$. The image of the function is an image of the entire set of chords rather than a subset of it. The principle of indifference has been applied to the set of chords and in using the Property Sufficiency Corollary 26 we have given a justification of the results without appealing to geometrical intuitions.

The functions used are not one-one and so frailties 3 and 4, Bertrand's procedure over and undercounts the chords and the question of uniform cross section, remain. We have addressed one, but only one, source of undercounting by no longer using partial functions. The original uniformity of cross section question remains, although now buried more deeply, and there are additional cross sections whose uniformity is assumed. For example, in the angle case the original cross sections were defined by chords having the same endpoint and in now measuring the subsets of angles we still assume that those cross sections are uniform. In addition, and perhaps undergirding the latter assumption, we assume uniformity for cross sections formed by making the same angle with the tangent. We will return to these frailties shortly.

Frailty 6 is that Bertrand's procedure is internally inconsistent and its output is not guaranteed to be a probability measure. We have at least shown that Bertrand's measures, when rigorously constructed via the Theorem of Induced Measure, are probability measures. With the tools so far developed, we are in a position to identify the flaw in his procedure that produces internal inconsistency. Before we do so, I remove a potential distraction.

3.7 The relation to random variables

It might be thought that what we have just done amounts to defining random variables on the chords. That is not true. A random variable is a function from a probability measure space, whose codomain is a measurable space and which function is itself a measurable function. Had we a probability measure space $(\mathbb{C}, \Sigma, P)$ on the chords to start with, each of $\alpha:\mathbb{C}\to(0,\pi/2)$ $\rho:\mathbb{C}\to(0,R)$, $\kappa:\mathbb{C}\to D$ from the last section would be a random variable provided that each was a P-measurable function. There is consequently a temptation to treat Bertrand's procedure backwards with this thought in mind, that probabilities got from the Lebesgue measure on the random variable codomain give us his probabilities. This makes a certain amount of sense apart from the fact that random variables are defined on prior probability measures.

If they were random variables we would be able to calculate the probability of, for example, random variable $\alpha \in (\pi/3, \pi/2]$ *from* the probability of the chord being longer under probability measure P. But Bertrand's procedure is in the other direction. Bertrand is calculating the probability of longer *from* the probability of 'random variable' $\alpha \in (\pi/3, \pi/2]$ under the uniform probability measure on $(0, \pi/2]$.

What the last amounts to, if we are going to think of Bertrand's procedure in the angle case as getting a probability from an angle random variable, is that α is defined not on (C, Σ, P) but from the normalized Lebesgue measure space $([0,1], \mathcal{L}, 2\lambda/\pi)$ as $\alpha:[0,1] \rightarrow (0, \pi/2]$. We will not be following this particular rabbit down his hole to look at his buried can of worms because whatever can be made of it is irrelevant to the various propositions I have proved and it is upon their truth that my analysis proceeds.

3.8 The internal inconsistency in Bertrand's procedure

The area case is an application of Bertrand's procedure based on equal ignorance over properties that shows the procedure to be internally inconsistent. Yet TIM is Unavoidable **23** so why should not it be given a rigorous reconstruction using TIM?

The area case uses a function not from C to D but $f:\mathbb{P}(C) \rightarrow \mathbb{P}(D)$, mapping sets of chords to their union in $\mathbb{R}^2$. It didn't then use a measure$(\mathbb{P}(D), \mathcal{S}, \nu)$, $\mathcal{S} \subset \mathbb{P}(\mathbb{P}(D))$—as required by TIM—to define a measure on C. That is not accident, since the TIM begotten measure so produced is not a measure of area, but the point of the area case is to use equal ignorance over areas. Instead it uses the measure $(D, \mathcal{S}, \nu)$, where $\mathcal{S} = \mathbb{P}(D) \cap \mathcal{L}_2$ and $\nu = \lambda_2|_D$. Applying the machinery of TIM, the definitions of Σ from $\mathcal{S}$ and μ on C from ν are

> For all $K \in \mathbb{P}(C), K \in \Sigma$ iff $f(K) \in \mathcal{S}$
>
> $\mu(K) = \nu(f(K))$.

On this definition both L and $C \in \Sigma$ because $f(L) = f(C) = D \in \mathcal{S}$ and then $\mu(L) = \nu(f(L)) = \nu(D) = \nu(f(C)) = \mu(C)$. Suppose μ is a measure. Then by additivity

$$\mu(C) = \mu((C - L) \cup L) = \mu(C - L) + \mu(L) = \nu(f(C - L)) + \mu(C)$$
$$= 3\pi R^2 / 4 + \mu(C) > \mu(C).$$

So μ is not a measure because it fails additivity and this is why, if used to play the role of the normative measure in POIS, it produces inconsistent probabilities.

Since TIM is Unavoidable but the area case uses the function f:$P(\mathbb{C}) \to P(\mathbb{D})$, so we must use the corollaries to apply TIM correctly. We cannot use Corollary **17** to induce a measure because f is not an injection. Applying Corollary **18**, we define Σ by

$$K \in \Sigma \text{ iff } K = f^{\text{invert}}(J) \text{ for some } J \in \mathscr{S}.$$

This is equivalent to the area case definition only if f is an injection. Consequently, since f is not an injection, $f(K) \in \mathscr{S}$ does not imply $K \in \Sigma$. It is this that removes the inconsistency. If we follow Corollary **18**: the set of longer chords, L, is *not* in Σ since there is no $J \in \mathscr{S}$ such that $L = f^{\text{invert}}(J)$. The obvious candidate, $J = f(L)$, fails because $f^{\text{invert}}((f(L))) = f^{\text{invert}}(\mathbb{D}) = \mathbb{C} \neq L$.

The consequence of all this is that although following Corollary **18** for the area case and using its output for the normative measure in the POIS will produce an equal ignorance probability measure on the chords, it is not one that gives a probability to the longer chords and so cannot be used to answer Bertrand's question.

We can now state the flaw in Bertrand's procedure. His procedure consists in appealing to geometric intuition to indicate loosely a function to play the role of an equal ignorance function from the set of chords to measure space a $(Y, \mathscr{S}, \nu)$. He then defines a set of subsets of the chords by using the function in its forward direction, that is $K \in \Sigma$ iff $f(K) \in \mathscr{S}$. He then takes measures on the images of that function to define a function, ϕ, on the Σ so defined. This procedure is flawed in defining the set of subsets of the chords, Σ, by the forward direction of the function rather than the backwards direction using preimages as in TIM. When Σ is defined by the forward direction of the indicated function ϕ need not be additive and consequently when ϕ is used to play the role of the normative measure in POIS, the function on Σ then defined by the POIS need not be a probability measure.

3.9 A criterion for equal ignorance probability measures

We can now demonstrate a criterion for an equal ignorance probability measure that separates the sheep from the goats in the flock bred of Bertrand's procedure. We already have the Sufficiency principle. I now demonstrate the

Necessity Principle 27

Let X be a set on which on which we have a function, ϕ, mapping subsets of X into $\mathbb{R}^+$. Suppose ϕ is an equal ignorance probability measure. Then there exists a measure space $(Y, \mathscr{S}, v)$ and an equal ignorance function $f\!:\!X \to Y$ such that, letting the normative measure in the POIS, μ, be the measure on X induced by the Theorem of Induced Measure using f, the probability measure got from the POIS thereby is φ.

Proof: This follows immediately from Necessity of TIM Lemma **22**. QED

Remark: The contrapositive is often the useful version of this principle: If there does not exist an equal ignorance function $f\!:\!X \to Y$ such that the measure got via the Theorem of Induced Measure and the POIS is φ, then φ is not a probability measure.

As I mentioned earlier, the lemma which proves this principle is trivial in itself. It gains its significance for us as a test and diagnosis in cases such as:

(1) when we have a triple that is claimed to be a measure space, we can test whether it is by whether there is another measure space and a function on which the use of TIM gives us the triple.
(2) when we have a function that purports to be playing the role of an equal ignorance function from a set of interest to a measure space on a different set being used to produce 'probabilities', we can test whether it does so by whether it can be formulated as an instance of applying TIM;
(3) when we have a function that purports to be playing the role of an equal ignorance function from a set of interest to a measure space on a different set being used to produce 'probabilities', we can diagnose what is going right or wrong by contrasting how the function and the carrying over of the measure is conducted with how it works by applying TIM.

The last was what we did for the area case.

We can now separate Bertrand's sheep from his goats: Bertrand's procedure works if and only if the final function, φ, satisfies the Necessity principle, that is could have been produced via the TIM using the function loosely indicated in the procedure. If instead it amounts to a loose way of using a different measure space to produce 'probabilities' that *cannot* be an instance of applying the Theorem of Induced Measure, then it hasn't produced a probability measure even if it appears to have done so.

So despite the triviality of the theorem, the Necessity principle based on it has significance for us. In a later chapter we look at the methods of measure theory for defining a measure on a set. If we have a set X on which those methods are, for some reason, arbitrary for our purposes, then the Necessity principle says there is no other way of producing the normative measure needed for the POIS to produce a probability measure on X than to find an equal ignorance function and the Sufficiency principle says if we can find an equal ignorance function then we can have a probability measure that gives equal probability to possibilities of equal ignorance.

Putting the necessity and sufficiency principles together we can characterise:

Equal Ignorance Functions Are Unavoidable Theorem 28

A function, ϕ, mapping subsets of a set into $\mathbb{R}^+$, is an equal ignorance probability measure iff there exists a measure space $(Y, \mathscr{S}, v)$ and an equal ignorance function $f{:}X{\rightarrow}Y$ such that the measure on X then begotten by TIM and POIS is ϕ

Proof: This follows immediately from the Necessity and Sufficiency principles 27 and 25. QED

What this means is that any equal ignorance probability measure is begotten of TIM and POIS and any equal ignorance function begets an equal ignorance probability measure. So this theorem shows yet again that the Theorem of Induced Measure Is Unavoidable. Anyone proposing that the principle of indifference issues in their equal ignorance probability measure candidate is committed to the use of this theorem as a criterion for whether they have in fact done so.

I have now proved three unavoidability theorems, FPOIS is unavoidable, which includes POIS being unavoidable, TIM is unavoidable and equal ignorance functions are unavoidable.

3.10 The remaining two frailties

The remaining two frailties of Bertrand's procedure are:

3. Over and undercounting chords.
4. The question of uniform cross sections.

What produces these frailties is that Bertrand's procedure doesn't use namings, and in our rigorous reconstruction none of α, ρ and κ are bijections and so do not define namings by Lemma 3 either. We now attempt to fix these frailties for Bertrand's cases.

The abstract cross sections of the chords are defined by geometric features from Bertrand's appeal to symmetry, which is an appeal to rotational symmetry. As mentioned when explaining the question of uniform cross section, if we had a definitive measure on the chords to measure Bertrand's cross sections by, there wouldn't be a paradox. In the absence of a definitive measure, all we can do is see whether cross sections are uniform internally to a relevant name space. We shall therefore be measuring the size of a cross section of the chords by the length of the cross section in the corresponding name space, the points of which latter cross section are the names of the chords in the chord cross section.

Cross sections of the chords defined in terms that correspond to simple Cartesian cross sections of the circle to which the chords belong, that is cross sections defined by $x =$ constant or $y =$ constant, are not relevant to the question of uniform cross sections. They are not relevant because such cross sections of the chords do not relate to the geometric symmetries. Instead, as we will see, it is polar coordinates for the circle that define the abstract cross sections of the chords that are named by cross sections of name spaces.

3.11 Angle case

Starting with the angle case: the tangent angle of a chord is half the angle it subtends at the centre, and the latter is a simple function—we see it below—of the difference between the angular coordinates of the polar coordinates of the chord endpoints. So a natural approach to include all the chords rather than only all those sharing a fixed endpoint is to represent the chords with the square $[0, 2\pi] \times [0, 2\pi]$, whose points (θ, ϕ) each represent a chord with endpoints (R, θ) and (R, ϕ). This gives us the measure space for the chords of Figure 9:

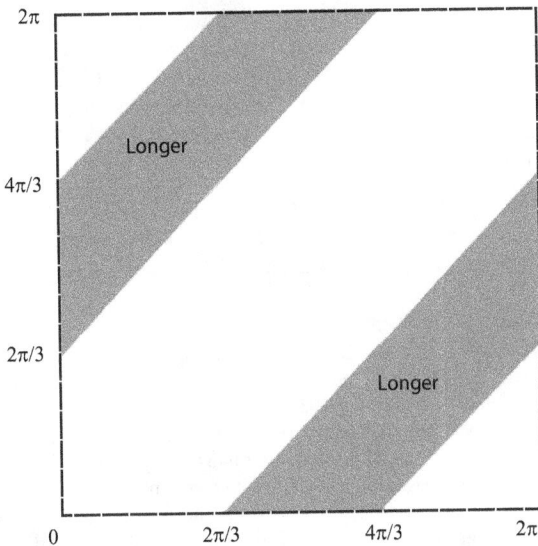

Figure 9 $[0, 2\pi] \times [0, 2\pi]$ measure space. Longer chords represented by shaded area.

Since $0 = 2\pi$ radians, the top and bottom edges and the left and right edges are in fact the same edges, whence this measure space is a torus.[43] Longer chords are represented by the points in this space, (θ, ϕ), for which $2\pi/3 < |\theta - \phi| < 4\pi/3$.

Marinoff notes that 'an isomorphism obtains between chords and circumferential pairs of points' (1994:6), which is true so far as it goes. His presentation of the solution is a diagram essentially the same as Figure 9. His method of randomly choosing a chord is analogised to throwing darts at the square. So he is using the function $f: [0, 2\pi] \times [0, 2\pi] \to \mathbb{C}$, $f(\theta, \phi) =$ the chord whose endpoints have polar coordinates (R, θ) and (R, ϕ). But f is not an injection since $f(\theta, \phi) = f(\phi, \theta)$ but $(\theta, \phi) \neq (\phi, \theta)$. So f is not an isomorphism from the torus to the chords, the torus double counts the chords and so the torus is not a name space. This frailty is not uncommon in the literature (Aerts and Sassoli de Bianchi 2014:13, Ardakani and Wulff 2014:29).

What we need to do is take the quotient space, $[0, 2\pi] \times [0, 2\pi]/\sim$, by the equivalence relation $(\theta, \phi) \sim (\theta', \phi')$ iff $(\theta, \phi) = (\theta', \phi')$ or $(\theta, \phi) = (\phi', \theta')$. The effect of this quotient is to fold the space $[0, 2\pi] \times [0, 2\pi]$ onto itself along the diagonal $\theta = \phi$, giving us the measure space of Figure 10.[44]

We now define $\alpha: \mathbb{C} \to [0, 2\pi] \times [0, 2\pi]/\sim$, $\alpha(c) = [(\theta, \phi)]$[45] for a chord with endpoints (R, θ) and (R, ϕ). The quotient removes the double counting and α is a bijection.

Lemma 29

α is a bijection.

Proof For each $[(\theta, \phi)] \in [0, 2\pi] \times [0, 2\pi]/\sim$ there is a chord with endpoints (R, θ) and (R, ϕ) so α is a surjection. Suppose $\alpha(c) = \alpha(c') = [(\theta, \phi)]$. The chord c has endpoints (R, θ) and (R, ϕ) and so has chord c' so $c = c'$, so α is an injection. QED

We continue with TIM as before: We have σ-algebra $\mathscr{S} = \{[0, 2\pi] \times [0, 2\pi]/\sim \cap A, A \in \mathscr{L}_2\}$ and measure $\nu = \lambda_2$ restricted to $\mathscr{S}$.[46] We define $\Sigma_\alpha = \{\sigma \subset \mathbb{C} : \sigma = \alpha^{-1}(S), S \in \mathscr{S}\}$. By Lemma 3, since α is a bijection it

---

43 The torus is homeomorphic to the quotient space of the square that identifies the horizontal and vertical edges by the equivalence relation $(x, y) \sim (x', y')$ iff $(x - x')/2\pi$ and $(y - y')/2\pi$ are integers. See Crossley 2010:81 and Willard 2004:62–3.

44 This triangle lives on a Möbius band with the diagonal forming its edge and the horizontal edge is joined to the vertical edge by a half twist. See picture in §6.15.

45 $[(\theta, \phi)]$ is the equivalence class of (θ, ϕ) under the just defined equivalence relation $\sim$.

46 Strictly speaking, we should use the Lebesgue measurable sets and the Lebesgue measure for $\mathbb{R}^2/\sim$, which latter is the plane folded along the line $x = y$, but the latter is homeomorphic to the portion of $\mathbb{R}^2$ to the left of that line and likewise $[0, 2\pi] \times [0, 2\pi]/\sim$ is homeomorphic to the triangle formed of $y = 2\pi$, $x = y$ and the y-axis, so we omit these complications and use $\mathscr{L}_2$ and λ_2.

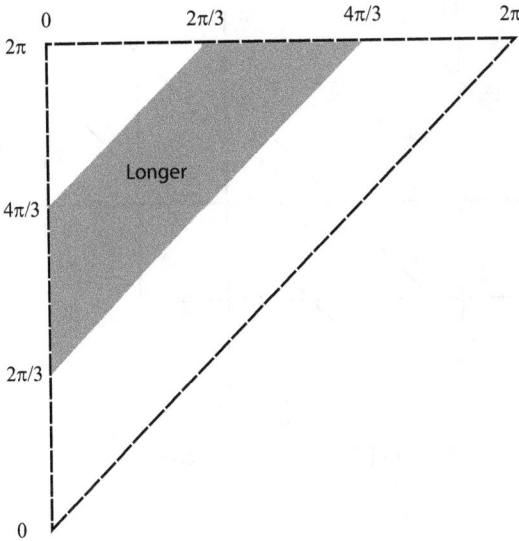

Figure 10 $[0, 2\pi] \times [0, 2\pi]/\sim$ quotient measure space. Longer chords represented by shaded area.

defines a naming of the chords, $\alpha{:}\Sigma_\alpha \to \mathscr{S}$. α being a naming means we have now dealt with the third frailty for the angle case.

Continuing, $\mu_\alpha{:}\Sigma_\alpha \to \mathbb{R}^+$, $\mu_\alpha(\sigma) = \nu(\alpha(\sigma))$. By TIM, $(\mathbb{C}, \Sigma_\alpha, \mu_\alpha)$ is a measure space. We now use μ_α as the normative measure in the POIS giving the probability space $(\mathbb{C}, \Sigma_\alpha, P_\alpha)$. The probability of longer is

$$P_\alpha(L) = \frac{\mu_\alpha(L)}{\mu_\alpha(C)} = \frac{\nu(\alpha(L))}{\nu(\alpha(C))}$$

$$= \frac{\nu\left(\left\{[(\theta,\varphi)]: \dfrac{2\pi}{3} < |\theta - \varphi| < \dfrac{4\pi}{3}\right\}\right)}{\nu\left([0,2\pi] \times [0,2\pi]/\sim\right)}$$

$$= \frac{2\pi^2/3}{2\pi^2} = \frac{1}{3}$$

What about the fourth frailty, the question of uniform cross sections? Since TIM is unavoidable, the question of uniform cross sections for the chords becomes here the question of uniform cross sections of this measure space. In Figure 10 it looks as if we have failed to fix this, since straightforward cross sections have different lengths. But we need to remember that this is a quotient space, which makes the cross sections rather special. Consider for example the cross sections of the original space and observe what happens to them under the quotient operation (Figure 11).

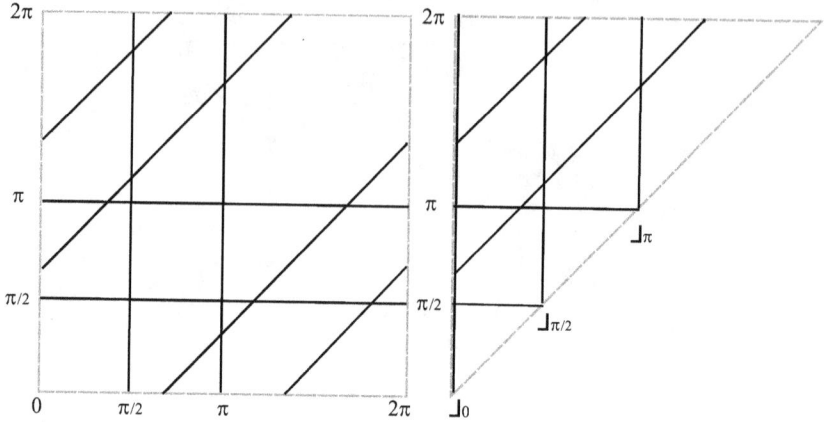

Figure 11 $[0, 2\pi] \times [0, 2\pi]$ and $[0, 2\pi] \times [0, 2\pi]/\sim$ cross sections.

What is happening is that where the original cross section meets the diagonal, the diagonal acts as a mirror. So, for example, the cross section of $x = \pi/2$ meets the cross section of $y = \pi/2$ at the mirror.

Now here is the important point. Every point on that ⌐ shape, $⌐_{\pi/2}$, is the name of a distinct chord, one of whose endpoints is at $\pi/2$. The horizontal part of the ⌐ has the names of the chords having the other endpoint in $[0, \pi/2]$ and the vertical part of the ⌐ is the names of the chords having the other endpoints in $[\pi/2, 2\pi]$. The corner of the ⌐ is the name of a point on the circumference. Consequently, each ⌐ defines a cross section corresponding to Bertrand's original treatment, where the corner of the ⌐ names a circumferential point to be Bertrand's 'one of the ends of the chord is known' (1889:4). Each chord (apart from the degenerate circumferential ones) appears in two cross sections, exemplified here by where $⌐_{\pi/2}$ and $⌐_{\pi}$ intersect at $(\pi/2, \pi)$. So the cross sections of the abstract measure space of names that we want for the angle case, the cross sections that correspond to the cross sections of the chords assumed in Bertrand's original treatment of the angle case, are *exactly* these ⌐s.

So it is just these ⌐s that are the cross sections of the chords about whom the question of uniformity arises. A simple calculation shows that the length of each ⌐ is the same as all the others, namely 2π and so we have now proved the (internal) uniformity of the abstract cross sections Bertrand used for the angle case. (A further interesting fact is that the cross sections of this name space that name the sets of parallel chords are also internally uniform. This is proved in Chapter 6, where its further significance is developed.)

And we can go further. Each ⌐ shaped cross section is itself a quotient space of $[0, 2\pi] \times [0, 2\pi]/\sim$. The ⌐ defined by φ is the quotient space of $[0, 2\pi] \times [0, 2\pi]/\sim$ defined by the function[47]

$$f_\varphi: \ [0, 2\pi] \times [0, 2\pi]/\sim \longrightarrow \, \rfloor_\varphi, \ f_\varphi(x,y) = \begin{cases} (x,\varphi) \text{ if } x \in [0,\varphi] \\ (\varphi,y) \text{ if } x \in (\varphi,2\pi) \end{cases}$$

$\rfloor_0$, in particular, is the vertical interval of the y-axis $\{(x,y): x=0, \ y \in [0,2\pi]\}$, that is the left hand edge of our measure space. Chords with names in $\{(x,y): x=0, \ y \in [0,\pi]\} \subset \rfloor_0$ subtend the angle y at the centre and chords with names in $\{(x,y): x=0, \ y \in (\pi,2\pi]\}$ subtend $2\pi-y$. Now we take a further quotient

$$g: \ \rfloor_0 \rightarrow [0,\pi], \ g(x,y) = \begin{cases} y, \text{if } y \in [0,\pi] \\ 2\pi - y \text{ if } y \in (\pi,2\pi] \end{cases}$$

Evidently I could have done this for any of the $\rfloor_\varphi$, but this is the easiest to follow. Finally recall that the tangent angle of a chord is half the angle subtended at the centre, and so $\frac{1}{2} \, g$ maps the names in $\rfloor_0$ to the tangent angles of the named chords.

Consider the composition $\frac{1}{2} \, g \circ f_0 \circ \alpha : C \rightarrow [0, \pi/2]$. This maps each chord to its tangent angle and so this is the equal ignorance function for the tangent angle property of chords we used in the earlier rigorous construction of §3.6, only now defined in a way that has made use of proven uniform cross sections.[48]

I have in this way reconstructed Bertrand's original treatment in terms of an abstract cross section of the chords, $\rfloor_0$, a cross section that is uniform with all the other cross sections $\rfloor_\varphi$. α is a naming. f_0 takes us to the names of a cross section of the chords that corresponds with the cross section of chords that Bertrand takes if 'the end of the chord [that] is known'(1889:4) is the point with polar coordinate $(R, 0)$. And it does it in a way that justifies Bertrand's suasive but undefined appeal to rotational symmetry, since not only are the abstract cross sections uniform, but each chord in the cross section goes for all the others with the same tangent angle in the way defined by f_0. $\frac{1}{2} \, g$ takes us to that angle with the tangent. $\frac{1}{2} \, g \circ f_0 \circ \alpha$ then plays the role of the equal ignorance function in TIM (in the way shown in §3.6), giving us λ_1 restricted to $[0, \pi/2]$ as the normative measure for use in the

---

47 Here we use Willard's definition of a quotient space in terms of a function (Willard 2004:61). This can also be used to prove that the $\rfloor_\varphi$ are homeomorphic.

48 For precision, let the angle case function in §3.6 be $\alpha':C \rightarrow [0, \pi/2]$, $\alpha'(c)=$the tangent angle of c. I have just shown that $\alpha' = \frac{1}{2} \, g \circ f_0 \circ \alpha$.

POIS, and finally delivering once more the probability of longer as 1/3. On this basis I have, for the angle case, removed the over- and undercounting problems, conclusively answered the question of uniform cross section and replaced Bertrand's original treatment with an unimpeachable treatment.

3.12 Direction case

As we noted before, the direction case partitions the chords by their direction and so we can give a unique name to each chord in terms of an index for the direction and an index for each chord in that partition. The chords having the same direction all perpendicularly intersect the same diameter at their midpoint, which point is unique on the diameter, so an index of their midpoints discriminates them (see Figure 12).

So we can determine unique names thus: Let (r, θ) be the polar coordinate of chord midpoints. For $r > 0$, if $\theta \in [0, \pi)$ then θ is the angle the of diameter the chord intersects perpendicularly and if $\theta \in [\pi, 2\pi)$ then $\theta - \pi$ is the angle of that diameter. The diameters all have $r = 0$ and we discriminate them by the angle of the diameter they intersect perpendicularly. So using the polar coordinates of the midpoints gives us the bijection: $\rho : \mathbb{C} \to [-R, R] \times [0, \pi)$,

$$\rho(c) = \begin{cases} (r, \theta) & \text{if } r > 0 \text{ and } \theta \in [0, \pi) \\ (0, \theta) & \text{if } r = 0 \text{ and } \theta \text{ is the angle of the diameter} \\ & \quad \text{intersected perpendicularly} \\ (-r, \theta - \pi) & \text{if } r > 0 \text{ and } \theta \in [\pi, 2\pi) \end{cases}$$

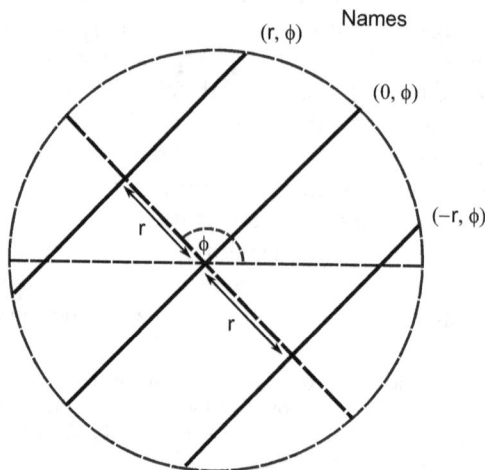

Figure 12 Naming by direction: diameter in dashed line, chords in solid line.

Let us now apply TIM. We continue exactly as before: We have σ-algebra $\mathscr{S} = \{[-R, R] \times [0, \pi) \cap A, A \in \mathscr{B}\}$ and measure $\nu = \lambda_2$ restricted to $\mathscr{S}$. We define $\Sigma_\rho = \{\sigma \subset \mathbb{C} : \sigma = \rho^{-1}(S), S \in \mathscr{S}\}$. By Lemma 3, since ρ is a bijection it defines a naming of the chords, $\rho : \Sigma_\rho \to \mathscr{S}$. ρ being a naming means we have dealt with the third frailty for the direction case.

Continuing, $\mu_\rho : \Sigma_\rho \to \mathbb{R}^+$, $\mu_\rho(\sigma) = \nu(\rho(\sigma))$. By TIM, $(\mathbb{C}, \Sigma_\rho, \mu_\rho)$ is a measure space. We now use μ_ρ as the normative measure in the POIS giving the probability space $(\mathbb{C}, \Sigma_\rho, P_\rho)$. The probability of longer is

$$P_\rho(L) = \frac{\mu_\rho(L)}{\mu_\rho(C)} = \frac{\nu(\rho(L))}{\nu(\rho(C))} = \frac{\nu\left(\left(-\frac{R}{2}, \frac{R}{2}\right) \times [0, \pi)\right)}{\nu([-R, R] \times [0, \pi))} = \frac{\pi R}{2\pi R} = \frac{1}{2}$$

What about the fourth frailty, the question of uniform cross sections? Since ρ is a naming of the chords, when ρ is used in TIM as was just done, it is the measure space of these names that induces the measure on the chords. Let us look at the name space (Figure 13).

In so far as TIM is unavoidable, the question of uniform cross sections for the chords becomes here the question of uniform cross sections of this name space. The abstract cross section that concerns us is the one that corresponds to Bertrand's original cross sections of parallel chords. Each such cross section has their midpoint intersecting the same diameter, which diameters are defined by the angle ϕ they make with the x-axis. Consequently,

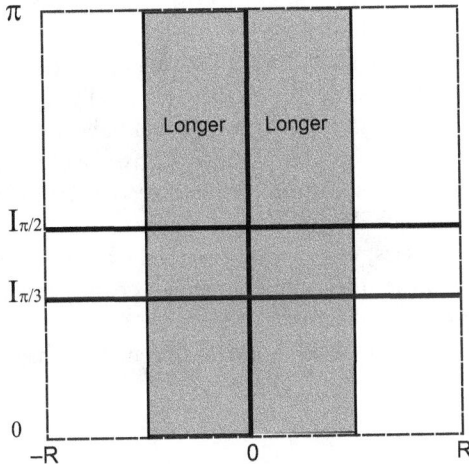

Figure 13 $[-R, R] \times [0, \pi)$, parallel chord cross sections in horizontal, diameters mapped to the vertical cross section, x=0, longer chords represented by shaded area.

each cross section of $[-R, R] \times [0,\pi)$ defined by $\phi \in [0,\pi)$ gives us the names of the chords in one of Bertrand's original cross sections of parallel chords. Evidently they have the same length of $2R$ and so we have now proved the (internal) uniformity of the abstract cross sections for the directions case.

We can go further, once more, only more simply, since each line cross section corresponding to $\phi \in [0,\pi)$ is a quotient space of $[-R, R] \times [0,\pi)$ defined by the function

$$f_\phi : [-R,R] \times [0,\pi) \to I_\phi, \ f_\phi(r,\theta) = (r,\phi)$$

$f_\phi \circ \rho$ maps each chord to the name of a chord whose midpoint is the same distance from the centre and is on the diameter with angle ϕ to the x-axis. Let $g: I_\phi \to [0, R]$, $g(r, \phi) = |r|$. Then $g \circ f_\phi \circ \rho : \mathbb{C} \to [0, R]$ maps each chord to the radial polar coordinate of its midpoint. So $g \circ f_\phi \circ \rho$ is the equal ignorance function for the direction property of the chords we used in the earlier rigorous construction of §3.6, only now defined in a way that has made use of proven uniform cross sections.[49]

Once again, I have in this way reconstructed Bertrand's original treatment in terms of an abstract cross section of the chords, I_ϕ, a cross section that is uniform with all the other cross sections. ρ is a naming. For any $\phi \in [0,\pi)$, $f_\phi \circ \rho$ takes us to a cross section of the name space that corresponds with the cross section of parallel chords that Bertrand takes 'if we know the direction of the chord' (1889:4). And it does it in a way that justifies Bertrand's suasive but undefined appeal to rotational symmetry, since not only are the abstract cross sections uniform, but each chord in the cross section goes for all the others whose midpoint is the same distance from the centre. $g \circ f_\phi \circ \rho$ then plays the role of the equal ignorance function in TIM (in the way shown in §3.6), giving us λ_1 restricted to $[0, R]$ as the normative measure for use in the POIS, and finally delivering once more the probability of longer as 1/2. On this basis I have, for the direction case, removed the over- and undercounting problems, conclusively answered the question of uniform cross section and replaced Bertrand's original treatment with an unimpeachable treatment.

Due to these successes, at this point it might be thought that, once sufficient care is taken, the question of uniform cross sections is always successfully answerable. In the next section we see that we cannot fix the midpoint case and this leads us to two different reasons to reject the midpoint case altogether.

---

49 For precision, let the direction case function in §3.6 be $\rho':\mathbb{C} \to [0, R)$, $\rho'(c) =$ the radial polar coordinate of the point of perpendicular intersection of the chord with the diameter. I have just shown that $\rho' = g \circ f_\phi \circ \rho$.

3.13 Excluding the midpoint case

In the midpoint case the chords are represented by the function mapping chords to their midpoints. Bertrand's appeal to rotational symmetry manifests here in that function mapping midpoints of parallel chords to the same diameter and chords of the same length to a concentric circle. Let us look at the measure space (Figure 14):

Chords of the same length are rotationally symmetric and these abstract cross sections of the chords are represented in the measure space by their midpoints being mapped to the radial cross sections, which are concentric circles. All the longer chords of the same length are mapped to radial cross sections within the turquoise solid circle, which represents chords whose length is that of the inscribed equilateral triangle, and the shorter chords of the same length are mapped to radial cross sections outside the turquoise circle. So for the midpoint case it is these cross sections that matter.

For the sake of the comparison I am about to make, the radial cross sections are shown in halves. The comparison is given in the following measure spaces (Figure 15). The colours of corresponding cross sections match across all three spaces.

The triangle midpoint measure space on the left of Figure 15 is homeomorphic to the disk midpoint measure space of Figure 14, from its quotient by the equivalence relation that identifies the LHS of the upper triangle with the LHS of the lower triangle, and identifies the RHS of the upper triangle with the RHS of the lower triangle (shown by orange arrows).

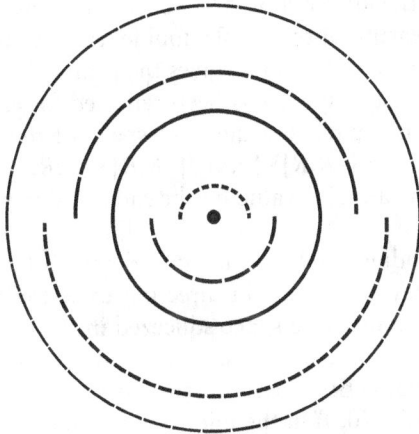

Figure 14 Disk midpoint measure space. Dashed semi-circles: radial cross sections. Centre dot the cross section containing the diameters. Turquoise solid circle: cross section representing chords on inscribed equilateral triangles.

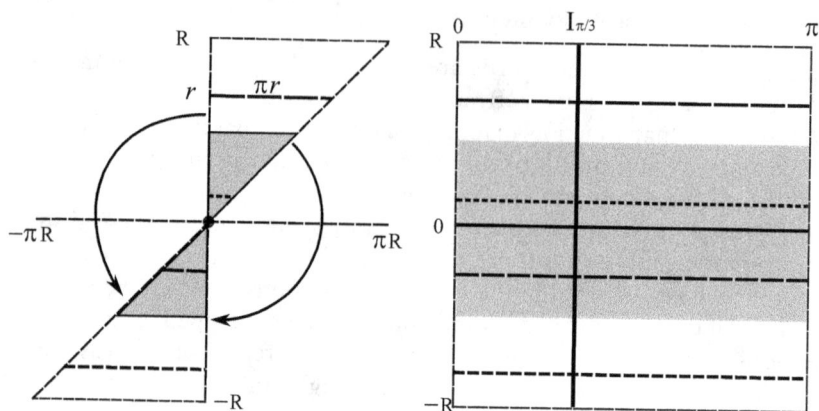

Figure 15 Double triangle midpoint measure space and (reflected) direction measure space. (Original direction case cross sections now vertical.)

Each half radial cross section of radius r corresponds with a horizontal cross section of the same length in the triangle midpoint measure space (both have the length πr), with those in the upper half of the circle in the upper triangle and in the lower half the lower triangle. Double triangles in the triangle measure space have the same measure as corresponding disks in the disk measure space. Using the Lebesgue measure restricted to this double triangle gives the same probability for the midpoint case as before when using the function κ in TIM for this case above. The turquoise double triangle representing the longer chords has Lebesgue measure $\pi R^2/4$, a quarter of the Lebesgue measure of the whole double triangle and so when used as the normative measure in POIS this gives the probability of longer ¼.

The direction measure space has been reflected for ease of comparison. The triangle measure space is a quotient space of the direction measure space by the function $f:[-R,R] \times [0,\pi) \to [-R,R] \times [-\pi R, \pi R)$, $f(r, \phi) = (r, \phi r)$, which multiplies the angular value by the radial value producing the double triangle.

Essentially, attending to the polar coordinates of the midpoints and comparing how they were used in direction case, the midpoint measure space is the direction measure space squeezed in the middle. In particular, all the diameters, which were each uniquely named by points on the purple cross section of the direction measure space at $r=0$, are now all mapped to the single purple point $(0,0)$ in the midpoint measure space.

The consequence of this is that in the midpoint space, sets of chords of lengths longer than the inscribed equilateral triangle have a smaller cross section than sets of chords of shorter lengths. The set of chords of length $2\sqrt{(R^2-r^2)}$ have abstract cross section of length $2\pi r$, that is, a linear function of the distance, r, of their midpoint from the centre. So these cross sections are not uniform.

We could have proved what now follows more awkwardly for the disk. Instead I have deliberately mapped the disk to a space which preserves the measure of radial cross sections and disks, which allowed us to make the comparison with the direction measure space and for which Cavalieri's method of determining the areas of triangles from the line lengths applies. It allows us to see immediately that non-uniform horizontal cross sections matter because it skews the measure of the chords that is determined by the lengths of the triangle measure space horizontal cross sections.

The horizontal cross sections represent items between which we are equally ignorant, namely the lengths of chords, between which chords of the same length we are indeed equally ignorant. If they are not uniform, then the non-uniform cross section skews the very measure space that we want to use as the normative measure, because it determines different normative measures for sets of chords of which we are equally ignorant. Yet the normative measure must give equal measure to equal ignorance. Consequently, a measure whose relevant cross sections are not uniform cannot fulfil the normative measure role.

In the direction measure space the horizontal cross sections represent chords of the same length and we can see that each cross section is the same length. Consequently the longer chord and the shorter chord cross sections are measured uniformly.

In the triangle measure space, the measure of the longer chords is determined by the lengths of the horizontal cross sections in the turquoise area, all of which are shorter than the length of the cross sections of the shorter chords. The measure space got from the midpoints of chords produces cross sections that are not uniform, whose lack of uniformity feeds through directly—in the way made evident by Cavalieri's method—into the measure that plays the role of the normative measure in the POIS. So we have proved that that the midpoint case cannot answer the question of uniform cross sections successfully and consequently the measure space it produces cannot fulfil the normative measure role.

From this discussion, we can see that because all the diameters share the same midpoint, we are not going to be able to find a naming based on the midpoints alone. To fix the diameter problem we can only distinguish them by an index that distinguishes them by direction, whether it be their angle with an axis or their point of intersection with the upper circle or whatever. And that means that a naming based on midpoints but supplemented suitably to distinguish the diameters is going to be a composition of the direction naming with a bijection from the direction name space. Consequently the midpoint case cannot fix the diameter problem to produce a naming without turning into a variant of the direction case.

The upshot of all this is that the midpoint case itself cannot be vindicated with a naming to avoid the undercounting chords. It is true that the generality of TIM means a naming is not essential to determining probabilities. But that hardly helps, since we have seen that the mapping used by TIM in the midpoint case manifests exactly the hindrance that non-uniform cross sections threaten. So the representation of the chords by their midpoints falls to both the third and fourth frailties and this is a reason to reject it.

There is yet a further reason for rejecting the midpoint case, a reason based in measure theoretic considerations. A null set in a measure space is a set which can be covered by a sequence of other sets whose total measure is arbitrarily small. Consequently, null sets have measure zero. Nullity indicates a kind of sparseness within the measure space as a whole. For example, the rational numbers are null in the real line.[50] However, nullity is not correlated with cardinality. Cantor's ternary set is null yet uncountable.[51] The sparseness here is more that the rationals and Cantor's ternary set are scattered like dust over the continuous real line-segment. In general, subsets of a measure space which are contiguous in a relevant sense are not sparse in these ways and do not have measure zero.

The third frailty remains because our applications of the Theorem of Induced Measure to the set of chords on Bertrand's behalf still doesn't count each chord once and only once. There is, however, a measure theoretically defensible relaxation of this restriction. It could be argued that it would not matter if chords that are sparse in C, in the sense of sparseness shared by the rationals or Cantor's ternary set, were not represented at all, or if represented, they got mapped to images which had measure zero. The frailty would be objectionable only if a set of chords in C which was not sparse got an induced measure of zero.

The set of diameters, D, of a circle is a subset of C, and is a continuum sized set. There is a clear sense in which D is not sparse in C. What makes the Cantor set sparse despite being continuum sized is that it is nowhere dense, which means that any open interval has an open subinterval whose intersection with the Cantor set is empty. We might be able to define a topology on the chords under which D is nowhere dense, but it would have to be very peculiar geometrically, since considered geometrically D is contiguous in a relevant sense. By contrast, suppose we determine a Cantor set of chords perpendicular to a single diameter got by a chord being in that set iff its point of perpendicular intersection belongs to a Cantor set of points on that diameter. That set is not contiguous in the same relevant sense. So in a topology that is geometrically natural for the chords, D is not

50 See Weir 1973:18.
51 See Weir 1973:20.

nowhere dense and so not sparse in a way that justifies or makes sense of it having an induced measure of zero. Any induced measure on ℂ under which it does is therefore unacceptable.

Now the midpoint case when fully spelt out maps members of ℂ onto their midpoints in the open disk *D*, and then assigns probability measures for sets of chords on the basis of the area occupied by their midpoints. In so doing, it maps the entire set of diameters onto the origin, a single point, which is null in the disk. Consequently, the midpoint case amounts to assigning measure zero to the set of diameters. That is objectionable for the reasons just given, and so the midpoint case is ruled out.

3.14 Conclusion

Bertrand's angle and direction cases have now been reformulated with neither reliance on Bertrand's own flawed procedures nor vulnerability to any of his frailties. His midpoint case, on the other hand, has been proved to be irreparably flawed. We are now in a position to reject various diagnoses of the paradox.

Norton diagnoses the source of the paradoxes thus:

> All the paradoxes have the same structure. We are given some outcomes over which we are indifferent and thus to which we assign equal probability. The outcomes are redescribed... [by]... a disjunctive coarsening, in which two outcomes are replaced by their disjunction; or... [by]... a disjunctive refinement, in which one outcome is replaced by two of its disjunctive parts.... A... class of examples employs a continuous outcome space, indexed by a continuous parameter, and the coarsening and refinement arises through a manipulation of this parameter. These examples are often associated with so called 'geometrical probabilities' (Borel 1950...), since these cases commonly arise in geometry; the locus classicus is Bertrand.
>
> (Norton 2008:52)

We can now see that, even if this coarsening and refining diagnosis fitted Bertrand's original procedure (and it is not obvious that it does), it is certainly incorrect for where we have now arrived. Neither of our final versions of the angle and direction cases is a coarsening or refining of the other because they are each using a naming from the chords to a name space.

Norton also mentions the other standard diagnosis: 'The device used to generate the paradoxes of indifference, the rescaling of variables' (Norton 2008:55). We will see that he is not alone in this diagnosis. Such a rescaling is either between the name spaces which produces a rescaling between the probability measures, or between the probability measures directly. Corollary **13** showed that, although one might *generate* a paradox in this way, the mere appearance of a non-linear rescaling in a case of paradoxical probabilities is *inevitable* and therefore rescalings as such can't be diagnostic. Whenever there are paradoxical probabilities there is a non-linear scaling of the normative measures, and by the commuting diagram in Theorem **12** (Figure 1) this is equivalent to one in the renaming. I shall explain this further and give the correct diagnosis of the source of the paradoxes in Chapter 13.

4 The Threat to the Principle and Four Kinds of Solution

4.1 Introduction

In the last chapter I proved that Bertrand's angle and direction cases can be well founded with neither reliance on Bertrand's own flawed procedures nor vulnerability to any of his frailties, and have thereby proved that the frailties of Bertrand's own procedure do not relieve the threat posed by his paradox to the principle of indifference. The reason it poses a threat is that application of the principle of indifference is supposed to suffice for satisfying Salmon's Ascertainability criterion and thereby solve questions of numerical probabilities. Such questions have, of their nature, unique solutions, because a solution is a *single* probability measure, which entails that each event has a single probability. For an event to have more than one numerical probability means we have the inconsistency of paradoxical probabilities. Yet in attempting to answer Bertrand's question, the two different ways of applying the principle here results in two distinct probabilities for the same event.

Bertrand himself never actually articulates an argument from his paradox to the falsity of the principle of indifference. To simply assume that the paradox refutes the principle commits the seventh frailty, the frailty of validity. Bertrand concludes 'the question is badly posed' (1889:5) and says nothing further. The danger here is that the question being ill-posed may undermine the refutation of the principle, for an ill-posed question is one that might be ill-formed in such a way as to make it not a proper question and therefore, whilst giving the appearance of being answerable, is for some reason unanswerable. In that case, however, the answers he has offered are the mere appearances of answers and so cannot refute the principle. In short, the refutation is not of the principle but of his question. The refutation of the question stands not against the principle of indifference but only against asking for probabilities for infinite possible events. Towards the end of the chapter we shall see that Bertrand himself advances it in exactly this way, which explains why he doesn't articulate an argument to the falsity of the principle of indifference.

DOI: 10.4324/9781003456308-4

Whether the target of the paradox can be *kept* restricted in this manner is unclear. After all, the principle, if it is true, is in principle unrestricted. So why exactly should it not apply to infinite possibilities? It won't do to say, as the sole response, that the paradox proves why, since that is question begging when what is at issue precisely is whether the inconsistency refutes the question rather than the principle. The restriction to finitism for probabilities is also much controverted. This is why in general the paradox has been taken to be a refutation of the principle rather than the question. Since the inconsistency originates in one or the other, the refutation does require that there be nothing wrong with the question itself: it must be answerable *in principle*, even if not in fact, if its ill-posing is to refute the principle of indifference rather than prove the question to be *unanswerable* in principle. For there is no fault in the principle of indifference if it is unable to give a consistent answer to a question that is unanswerable in principle.

I turn now to formulating the threat as an explicit argument and the accompanying analysis leads us to defining the kinds of solution that are possible. I shall identify exactly how the paradox threatens the principle rather than the question. An analysis of the distinction between determinate and indeterminate questions allows me to give an explicit argument from the paradox to the falsity of the principle, whose premisses articulate the crucial dependencies of the refutation.

4.2 Determinate and indeterminate problems

Bertrand's question poses a problem. It is not easy to specify criteria of identity for problems. For example, a problem is not identified by a specification of what counts as a solution because many distinct problems can share the same specification.[52] Nor is it identified by its answer, since many distinct problems can share the same answer. Nevertheless, I have no doubt but that we successfully pose and solve problems all the time, and so we have a practical grip on them even if we face difficulties in making that grip theoretically explicit. So I assume that problems have identity.

We distinguish persons from their names but in the case of problems we often use the same word, 'problem', to speak of the problem itself and the linguistic means of expressing it.[53] I am therefore going to use a distinction between determinacy and indeterminacy to distinguish for problems the three possibilities similar to those for reference that arise when using a name like 'Fred'. One person, several people or nobody may bear the name. In the first case the reference is determinate (it is Fred), in the second

52 See example shortly.
53 Likewise for laws, etc.

it is indeterminate but repairable (Fred *the baker*), and in the third it is empty because there is no one called Fred.

By a *determinate problem* I mean a problem whose identity has been fixed by the way it has been posed. For example, a question which has a single meaning (which singularity might depend not just on the words, but also the context and its background constraints and assumptions) will generally suffice for the problem posed to be determinate. Necessarily, if a problem is determinate then what would count as a solution is determinate. However, a determinate problem need not have a solution. In §6.2 we will see Nathan (1984) confusing exactly what I am distinguishing here, namely, confusing having a determinate probability problem with having a problem which has a determinate solution.

By an *indeterminate problem* I mean a problem whose identity has *not* been fixed by the way it has been posed. Two possibilities arise: In the first case, this could be because the way it is posed is vague or ambiguous or underspecified, or because what is to count as a solution has not been determined. We may be able to resolve such an indeterminate problem entirely into determinate problems. In such a case the indeterminacy may originate in nothing more than a useful or well understood way of referring generically to a class of determinate problems, or from a slip, omission or clumsiness in expression. In the second case, however, the indeterminacy may be entirely irresolvable, and in such a case there is in fact *no* problem whose identity has been picked out. This could arise because the way it is posed is covertly meaningless or rests on some false and irreparable presupposition.

In speaking of the determinacy of a problem I might have been speaking of an epistemological matter, a matter of knowing what the problem is or being able to solve it. Certainly, success in fixing the identity of a problem has implications for our epistemic relation to it, but it is the success in reference and not the epistemic relation that I am speaking about.

4.3 The Rigorous argument from Bertrand's paradox to the falsity of the principle of indifference

As mentioned earlier, Bertrand never articulates an argument from his paradox to the falsity of the principle of indifference. With these issues over the determinacy of Bertrand's question now clarified, we can give what I shall hereafter call *the Rigorous argument*.

In this argument I shall use 'single' in the ordinary sense of unique—again, in order to facilitate the later discussion of permissivism.

1. Suppose the principle of indifference is true. (Assume)
2. For each event space there is a single epistemically rational probability measure for that event space. (Premiss)

3. Bertrand's question about the chords poses a determinate probability problem. (Premiss)
4. Determinate probability problems have single solutions. (2)
5. Bertrand's question has a single answer. (3, 4)
6. In choosing a chord at random, we have no epistemic reason to discriminate among the chords. (Premiss)
7. The principle of indifference gives an equal ignorance probability measure for the chords (1, 6, Theorem 11)
8. The principle of indifference applies to Bertrand's question. (7)
9. Therefore the principle of indifference gives a single answer to Bertrand's question. (1, 2, 3, 5, 6, 8)
10. There are 2 mappings from the chords to 2 measure spaces which are each defined by a property of the chords of which we are equally ignorant. (Premiss, Chapter 3)
11. Each such mapping is an equal ignorance function. (10, definition of equal ignorance functions, Chapter 3)
12. Each such mapping gives an equal ignorance probability measure for the chords (11, Property Sufficiency Corollary 26).
13. Each such equal ignorance probability measure gives an answer to Bertrand's question distinct from the other. (12, Chapter 3)
14. Therefore the principle of indifference is false. (1, 9, 13 R.A.A.)

4.4 Kinds of ill-posed question

Bertrand's own conclusion, that the question is ill-posed, follows from lines 5 and 13 of the Rigorous argument. In general, what mathematicians mean by an ill-posed question or problem is one which requires but lacks a single solution.[54] There is, however, an ambiguity in the notion.

In what is, for our purposes, the primary sense, the fault of ill-posing is the fault of posing a *determinate* problem that requires but lacks a single solution. A classical example would be the problem of solving a simultaneous equation when the equations are not linearly independent. Such a problem is determinate and the solution required is a single tuple of numbers satisfying each of the equations.[55] But linear dependence implies that there are either no or infinitely many tuples that satisfy the equations, and

---

54 Mathematicians often speak of ill-posed problems as being ill-posed because they have no or many solutions—which might be objected to if having many solutions is also a way of having no solution. However, it can be a matter of mathematical significance whether failure of uniqueness is due to dearth or surfeit, and hence their vocabulary reflects their attention to these distinct modes of failure.

55 Continuing from fn. 52, so a unique satisfying tuple is the specification of what counts as a solution that is shared by many distinct simultaneous equation problems.

so this problem is ill-posed in the primary sense. This kind of ill-posing is not repairable. Nevertheless, there is nothing wrong with a question *as such* if it is ill-posed in the primary sense. It is a perfectly good question since it poses a determinate problem and is therefore *answerable in principle*, even if it is unanswerable in fact.

In the secondary sense, the fault of ill-posing is the fault of posing an *indeterminate* problem that requires but lacks a single solution. A question that is ill-posed in the secondary sense can fairly be said to be *unanswerable in principle* just because it fails to pose a determinate problem. A question can be legitimately rejected on this ground. We then distinguish between questions whose indeterminate problem can be entirely resolved into distinct determinate problems, none of which are ill-posed in the primary sense,[56] and those which can't. In the latter case we have an entirely unanswerable question.

Now we can articulate (numbers correlate with strategies below)

> Bertrand's paradox refutes the principle of indifference if and only if Bertrand's question is ill-posed in the primary sense.

RTL: If Bertrand's question is ill-posed in the primary sense then it poses a determinate probability problem which lacks a single solution. Yet applying the principle of indifference is supposed to be sufficient for us to solve a determinate probability problem for which we are equally ignorant over the relevant events, and such problems have single solutions. (1) Consequently, the paradox refutes the principle.

LTR: If Bertrand's question is not ill-posed in the primary sense it is either not ill-posed at all, in which case it doesn't refute the principle (2), or it is ill-posed only in the secondary sense. If it is ill-posed only in the secondary sense, then either it cannot be resolved into a determinate problem at all (4), or it can be entirely resolved into determinate problems none of which are ill-posed in the primary sense (3). If the former then no challenge to the principle of indifference has been offered; if the latter, the principle suffices for a unique solution to each problem; in either case, the paradox does not undermine the principle.

---

56 There are some complications that can arise because of the possibility of a regress if distinguishing problems within an indeterminate problem produces further indeterminate problems. Investigating whether an indeterminate problem is ill-posed only in the secondary sense may require investigating a 'tree' of problems, and there are the complications of infinite trees and of 'mixed' trees whose terminating nodes include both determinate and indeterminate problems. The former can be dealt with by the requirement that the length of any infinite branch be a limit ordinal and the latter by pruning all branches which are indeterminate at every node since each node poses an irreparably indeterminate problem which is therefore not a problem at all. I am going to ignore these complications because I don't think they apply to Bertrand's paradox.

4.5 Kinds of solution of Bertrand's paradox

We can now see that there are precisely four ways of solving Bertrand's paradox.

Directed against Bertrand's question as answerable in principle
1. The **Irrelevance strategy** grants that Bertrand's question is ill-posed in the primary sense but denies the implication to the principle's refutation on the ground that there is a subtle or covert error that vitiates counting the distinct probabilities as a falsifying inconsistency. *Diagnosis*: Bertrand's question produces his paradox but the appearance of falsifying the principle of indifference is founded on an error. The paradox is irrelevant to the truth of the principle and so can be rejected. *Response to the Rigorous argument*: Holds the straightforward interpretation of the premisses or conclusions to be misleading or equivocal due to the error. Targets: premisses 2 or 10 are false or are contraries or inferences from them to lines 5 or 13 respectively are invalid.
2. The **Well-posing strategy** asserts that Bertrand's question is well-posed, on the ground that it poses a determinate problem for which the principle of indifference is sufficient to determine a unique solution. *Diagnosis*: Bertrand's question merely appears to allow of distinct answers, due to mistaken interpretation, failure to attend to implicit constraints or failure to use sufficiently sophisticated mathematical tools to answer it. Absent these errors, there is no paradox. *Response to the rigorous argument*: Premiss 10 is false.

Directed against Bertrand's question as unanswerable in principle
3. The **Distinction strategy** grants that Bertrand's question is ill-posed, but only in the secondary sense, on the ground that the indeterminate problem it poses it can be entirely resolved into distinct determinate problems with unique answers. *Diagnosis*: Bertrand's question is unanswerable in principle *as it stands*, being a clumsy way of raising whilst confounding distinct well-posed questions with unique answers, so there is no paradox. *Response to the Rigorous argument*: Premiss 3 is false.
4. The **Entirely Unanswerable strategy** grants that Bertrand's question is ill-posed, but only in the secondary sense, on the ground that it poses an irresolvably indeterminate problem. *Diagnosis*: Bertrand's question cannot challenge the principle of indifference because it is irreparably unanswerable in principle. Consequently it poses no problem at all and so is entirely unanswerable. *Response to the Rigorous argument*: Premiss 3 is false.

It is perhaps unsurprising that most of the frailties offer a line by which to pursue the Irrelevance strategy. Partial functions, measures that are not measures of the chords, the principle of indifference applied to something other than the chords, the assumption of uniform cross section and the question of whether geometrical intuition can be relied on for that assumption, the internal inconsistency of Bertrand's procedure, are all faults which, if not fixed, show the paradox to be irrelevant. Fortunately I have now fixed all of them.

The borderlines between these strategies is sharp but it must be admitted that in the literature those advancing solutions sometimes use the ground of one but the rhetoric of another. We will see examples of Distinction strategies presented as Entirely Unanswerable strategies, which then offer distinct answerable replacement 'Bertrand's questions' (Nathan 1984 on Jaynes in Chapter 6 and Aerts and Sassoli de Bianchi 2014 in Chapter 9). This amounts to a merely terminological disagreement: any rejection of Bertrand's question as indeterminate which then offers answerable replacements is an instance of the Distinction strategy.

For that reason I know of only one instance in the literature of the Entirely Unanswerable strategy: Bertrand himself. I place Bertrand here because of his introduction to the paradoxes:

> infinity is not a number; one must not, without explanation, introduce it into reasoning. The illusory precision of words could give rise to contradictions. To choose at random, among an infinite number of possible cases, is not a sufficient indication.
>
> (1889:4)

This is, admittedly, an odd remark since it is hard to believe he was unaware of Cantor's work (Cantor 1955) which was being published from 1870. Nevertheless, I think we should understand it along these lines. Suppose I ask 'What is the parity of Eeyore?'. Now, excluding puns and poetry, donkeys are necessarily neither even nor odd so my question contains a false and irreparable presupposition. (By contrast, if I ask 'Where is Fred?', my false presupposition of 'Fred' having unique reference is repairable as earlier instanced.) So although I may appear to have posed a determinate problem (I know what an answer would be) since it is impossible for donkeys to have parity the question is entirely unanswerable. Similarly, Bertrand is denying and holding to be irreparable a presupposition of the question, that there are probabilities for infinite state spaces, and so for this reason saying that the question is entirely unanswerable.

We are thus able to make sense of Bertrand's own treatment as an instance of the Entirely Unanswerable strategy being used to defend the principle of indifference for finite state spaces alone. Finitists about event state spaces, who are usually deniers of countable additivity, might pursue this strategy but I do not know of one doing so. For anyone not a finitist, the ground of this strategy is difficult to defend, since Distinction strategists have offered obvious answerable questions that advert to specific ways of randomly choosing a chord.

4.6 Prospects

I turn now to the attempts to defend the principle of indifference by solving Bertrand's chord paradox, all of which fall under the first three strategies. I cover solutions that for a long time have been prime contenders and also the most recent solutions based on radical mathematical proposals, solutions that have been neither analysed nor rebutted previously. These attempts are supplemented with four of my own.

I start with chapters on prime contenders and fellow travellers in each of the first three strategies. Marinoff offers an instance of the Distinction strategy and is joined by many others. Jaynes' instance of the Well-posing strategy is based on the symmetry requirement and is joined by Wang and Jackson's instance based on requiring homogenous lines and by Rizza's instance based on Sergeyev's arithmetic of infinity. The chapter on the Irrelevance strategy addresses Bangu's claim of randomness transmission failure and Gyenis and Rédei's radical mathematical reconstruction of the principle in terms of the Haar measure. The next chapter is on the Maximum Entropy principle, which offers an instance of the Well-posing strategy posed by what I call meta-indifference. The following chapter is an explicit attempt at measure theoretic meta-indifference from Aerts and Sassoli de Bianchi and this is followed by a chapter on meta-indifference in general. I then turn to two chapters on rather more broadly based attempts at the Irrelevance strategy: permissivism and uniqueness a criterion of identity for POI. The final attempt I examine is an instance of Well-posing constructed from the completely general treatment of symmetry by group theory.

5 The Distinction Strategy

5.1 Introduction

In this chapter we look first at Marinoff's attempted solution, a solution that was historically and for a long time the prime contender for solving the paradox using the Distinction strategy. I show that Marinoff falls into what I call van Fraassen's trap. In this case, to fall into the trap is to claim that the ignorance of which random process is used to pick a chord justifies distinguishing distinct ways of doing so, each of which produces consistent probabilities. But the principle is supposed to deal with ignorance and re-placing Bertrand's question with others specifying random processes is not saving the principle but covertly abandoning it. This point is not specific to Marinoff's solution but applies to the Distinction strategy as such, and I identify the same error in other literature pursuing this strategy. I finish by showing that what I call meta-indifference, primarily a resource for the Well-posing strategy, cannot avoid the trap.

5.2 Marinoff's Distinction strategy and van Fraassen's trap

Marinoff's 1994 paper has been widely accepted as a successful resolution of Bertrand's paradox. He says

> Bertrand's original problem is vaguely posed… clearly stated variations lead to different but… self-consistent solutions…. [Thus] The principle of indifference appears consistently applicable to infinite sets provided that problems can be formulated unambiguously.
>
> (Marinoff 1994:1)

This is evidently an example of the Distinction strategy. The claim is that Bertrand's question is an unanswerable question as it stands because it poses a problem whose identity is indeterminate due to vagueness, but it

DOI: 10.4324/9781003456308-5

can be resolved into a number of distinct well-posed questions with unique answers.

Bertrand's question is not vague, since neither having a probability nor being longer than the side of an inscribed triangle are vague properties,[57] and we know what counts as a solution (a unique number in [0,1] being assigned as the probability of the chord being longer). Examining the detail of what Marinoff says, and despite his use of the word 'vague', it appears that Marinoff thinks Bertrand's question confounds distinct questions because it fails to specify a random process for selecting the chords:

> Our first task consists in clearly distinguishing three cases, which Bertrand's vague question... (perhaps deliberately) conflates.... When generating random chords, one clearly faces methodological alternatives.... Thus Bertrand's three answers can be construed initially... as replies to three different questions: What is the probability [of a chord being longer] given that the random chord is generated [by a procedure]
>
> Q1   ...on the circumference of the circle?
>
> Q2   ...outside the circle?
>
> Q3   ...inside the circle?
>
> (1994:4)

By the end of the paper, Marinoff has distinguished an additional four such questions, giving seven in all, and allows that there may be 'an infinite number' (1994:17). About these he concludes

> To dispute which of these questions—if any—"best" represents Bertrand's generic question... is to relinquish geometry for aesthetics. Although the randomizer in Q1 may be deemed elegant, and that in Q5 contrived, this distinction is arguably one of degree, not of kind. Each version of Q is so precisely because it foists a specific requirement, or set of requirements, upon the problem.... the escape from paradox lies in distinguishing between and among such requirements.
>
> (1994:22)

So Marinoff's position is that Bertrand's question confounds distinct problems. What is Marinoff's argument? He doesn't give one, but quotes Keynes and van Fraassen approvingly:

---

57 If having a probability is, in fact, vague, then Marinoff's replacement questions are also vague and yet well-posed, so its vagueness is irrelevant to the paradox.

Keynes concludes. "So long as we are careful to enunciate the alternatives in a form to which the Principle of Indifference can be applied unambiguously, we shall be prevented from confusing together distinct problems, and shall be able to reach conclusions in geometrical probability which are unambiguously valid".[58]

Response: This study has endeavoured to follow Keynes' positivistic prescription. Careful enunciations of alternatives, unambiguous applications of the principle of indifference, and clear demarcation between distinct problems together lead to conclusions in geometric probability that are self-consistent and therefore unparadoxical.

(1994:23)

Most writers commenting on Bertrand have described the problems set by his paradoxical examples as not well posed. In such a case, the problem as initially stated is really not one problem but many. To solve it we must be told *what* is random; which means, *which* events are equiprobable; which means, *which* parameter should be assumed to be uniformly distributed.

(van Fraassen 1989:305 as quoted in Marinoff 1994:4–5)

Marinoff states that he is 'implementing van Fraassen's recommended method' (1994:4–5), which is odd, since immediately following the passage quoted by Marinoff van Fraassen makes an objection:

But that response asserts that in the absence of further information we have no way to determine the initial probabilities. In other words, this response rejects the Principle of Indifference altogether. After all, if we were told as part of the problem which parameter should receive a uniform distribution, no such Principle would be needed. It was exactly the function of the Principle to turn an incompletely described physical problem into a definite problem in the probability calculus.

(van Fraassen 1989:305)

Marinoff offers no response to this objection from van Fraassen. One point available to him is that giving a parameter a uniform distribution is itself an application of the principle of indifference, so his approach is not a rejection of that principle altogether. But what will he say about van Fraassen's final sentence?

---

58 Keynes 1921/1973:68.

I shall call van Fraassen's point

van Fraassen's trap: Rejecting Bertrand's question and replacing it with others with more information is not saving the principle from refutation but covertly abandoning it.[59]

It is a trap because the ignorance requirement (§1.2) does not allow us to pick and choose convenient ranges of ignorance for the principle to deal with. Falling into the trap is an evasion rather than a solution of the paradox.

5.3 Bertrand's question: answerable or unanswerable?

For the sake of argument, grant that Marinoff's Q1, Q2 and Q3 (and his others) are well-posed distinct problems to which the principle of indifference can be applied successfully. If his object was the restricted one of making plausible the application of the principle in some infinite cases then he may have succeeded. Certainly, that is a rebuttal of Bertrand's rejection of probabilities for infinite cases.

> Significantly, the many versions of Bertrand's problem are solvable, and each solution relies upon the very procedure—namely, the consistent application of the principle of indifference to infinite sets—that Bertrand proscribed. Bertrand's former paradox of random chords is resolved by the expedient of providing what he, from the outset, withheld, namely, a 'sufficient specification' of such sets.
>
> (Marinoff 1994:22)

Here he makes explicit his view that what makes Bertrand's question unanswerable as it stands is that it is underspecified. But does distinguishing these several problems really get to grips with Bertrand's broader challenge? Bertrand might concede his finitism and still hold that his paradox embarrasses the principle of indifference by confronting us with distinct but contradictory ways of applying the principle to a determinate problem posed by his question.

Hence the bone of contention is whether Bertrand's question poses a determinate problem which lacks a unique answer, or poses an indeterminate problem which through underspecification confounds distinct determinate problems. In short, is his question answerable or unanswerable as it stands?

---

59 Keynes made essentially this criticism of Borel, see Keynes 1921/1973:52.

For Marinoff's resolution to succeed he must persuade us that Bertrand's question is unanswerable as it stands. Marinoff wants to be able to reply that if by choosing randomly you mean process X, then the probability is x, but if by choosing randomly you mean process Y, then the probability is y..., and if you don't specify what you mean by choosing randomly, then you haven't posed a determinate problem. For

> There exists a multiplicity, if not an infinite number, of procedures for generating random chords of a circle. The answers that one finds to Bertrand's generic question... vary according to the way in which the question is interpreted, and depend explicitly upon which geometric entity or entities are assumed to be uniformly distributed.
>
> (Marinoff 1994:17)

Recall our earlier example of the possibilities of reference for 'Fred' when we first sorted out determinate and indeterminate problems. Now if Bertrand's question is a generic singular question like 'what is the weight of Fred', then the question can be rejected as underspecified if no particular Fred is contextually salient and the asker refuses to identify which Fred 'Fred' stands for. If he goes on to say that his question is a general question, that he is interested in the weight of Freds in general, it can be rejected as unanswerable because it poses an indeterminate problem. Weights are properties of material individuals, but there are no such individuals as Freds in general. (I shall consider later what role the notion of the weight of the average Fred might play).

Marinoff's solution requires that Bertrand's question be similarly and *only* a generic singular question. But what are the grounds for insisting that Bertrand is confined to speaking of random choice in the singular when asking about chords chosen at random? Of course, Bertrand *could* ask about the chance of getting a longer chord when a particular way of choosing randomly is salient, but what he is interested in knowing is what is the chance of getting a longer chord given random choice *in general*. Marinoff would like to reject the general question, but the analogy doesn't carry through because, whilst there is no such thing as a Fred in general there is such a thing as random choice in general.

Furthermore, that a question has several answers doesn't of itself mean that distinct problems are being confounded. When neither the financial institution nor the riverside is contextually salient, the ambiguity of the question 'how can I get to the bank' leaves it indeterminate which problem is being posed. But if we know that the riverside is the goal, then there being different ways to get to the riverside doesn't mean that the question is confounding several distinct problems. It is just a single problem with several solutions.

Bertrand's question is not analogous to the former example but to the latter, because asking for the probability of getting a longer chord is not ambiguous. Some authors claim that 'choosing at random' is ambiguous (see below) but that phrase is not ambiguous in the way of the former example. What they mean is what Marinoff means, that it is underspecified if we are not told a specific way of choosing at random. What, then, is it about there being different ways of choosing at random which justifies taking his question to be confounding distinct problems which doesn't make the question of the way to the riverbank similarly confused? Without a good reply to that challenge, I do not see how Marinoff has resolved the paradox.

We have now identified the premiss underlying the Distinction strategy without which it must fail:

> *Underlying premiss*: Bertrand's question is unanswerable because it is a generic singular question which cannot be a general question.

The Distinction strategy is defined by holding Bertrand's question to be unanswerable as it stands but resolvable into distinct well-posed questions. His question is not vague or ambiguous and we know what counts as a solution, so the strategy requires that it be underspecified, which in turn requires that the Underlying premiss be true. Marinoff merely claims that the question is underspecified but gives no further argument. I have found many similar claims (see §5.6) but no arguments. The argument I have given shows the Underlying premiss to be false. If it is false, then Bertrand's question is not unanswerable but answerable, because it is a general question that poses a determinate problem.

5.4 The Distinction strategy fails

Provided Bertrand's question is answerable, Bertrand's point seems very well taken. The principle of indifference is supposed to deal with what is unknown by giving equal probabilities to the equally unknown. By choosing a set which lacks a uniquely salient measure he exposes a difficulty which we can now make explicit with the material from the earlier chapters: you need a measure to get a measure and the TIM is unavoidable.

Applying the principle of indifference to the chords requires a measure to fulfil the role of the normative measure by which equiprobability is assigned by the POIS. Any equal ignorance probability measure on the chords can be got via the TIM and the rigorous reconstruction of Bertrand's cases does so. Examination of the procedures Marinoff follows in his paper shows that he too is using TIM, albeit unknowingly, with functions he takes to be equal ignorance functions.

I mentioned Marinoff's angle case function in §3.11. His midpoint case uses my $\kappa:\mathbb{C}{\to}D$ for the midpoint case in §3.6. His direction case is rather gerrymandered. His function is only a partial function of the chords. Given a point, p, outside the circle on a given extended diameter at a distance h from the centre of the circle radius R, let $B=\{c\in\mathbb{C}: c$ extended contains $p\}$. Then his function is

$$Mar: B \to \left[0, \sin^{-1}(R/h)\right], \; Mar(c) = \sin^{-1}(r/h)$$

where r is the radial coordinate of the midpoint of c. Since the values of this function depend only on r and h, we can extend this to a full function on the chords. For each h we have the function

$$f_h: \mathbb{C} \to \left[0, \sin^{-1}(R/h)\right], \; f_h(c) = \sin^{-1}(r/h)$$

TIM and POIS then beget a probability measure on $\mathbb{C}$.[60]

To get probabilities for a set of chords parallel to the diameter containing p he has to take the limit as $h{\to}\infty$ of the probability measures *begotten from Mar*, since it is only the domain B which 'tends' to a set of parallel chords as h tends to infinity. (So there remain some difficulties to overcome to rescue Marinoff from the frailty of undercounting due to his partial function).

If we then disinter a function from underneath Marinoff's limit probability measure by using the LTR direction of our Equal Ignorance Functions are Unavoidable Theorem **28** we get a restriction of our function $\rho:\mathbb{C}{\to}[0, R]$ for the direction case in §3.6, namely the restriction whose domain is the set of chords parallel to the given diameter. Although Marinoff doesn't show this, this is the basis for his claim to identify Bertrand's direction case with 'the random chord [being] generated by a procedure outside the circle' (1994:4). It should be noted that, albeit unknowingly, Marinoff is essentially following Friedman 1975, who originates the treatment based on a rotating line around a point outside the circle and then takes a limit for the distance. (We return to Friedman in §6.3)

---

60 In his Table 7.1 he claims to be taking a uniform distribution over $[1, d]\times[-1, 1]$ (see Marinoff 1994:19–20 for how) but this is misleading. For each distance, d, of the point from the centre he uses a uniform distribution over the angles $[0, \sin^{-1}(R/d)]$ to define a probability measure (1994:8–9). His process of taking the limit as $h{\to}\infty$ is not a way of taking a uniform distribution over $[1, d]$. Yes, for small angles $\tan\theta \approx \theta$ and this he uses with a change of variables to get to $[-1, 1]$ out of the angles, but an approximately uniform distribution is still only approximate.

Although neither $[0, 2\pi] \times [0, 2\pi]$ nor D is a name space, when Marinoff speaks of choosing chords at random by throwing darts at a square (1994:7) or a dartboard (1994:12), he is doing the kind of thing I described when explaining why the significance of the normative measure is generally neglected. What he is calling a procedure for choosing a chord at random is in fact selecting from a quasi-name space and then using its standard measure to play the role of a normative measure which he then uses to define probabilities. Having drawn attention to this example, I leave it to the reader to spot examples in later authors.

Because you need a measure to get a measure, probability measures on the chords are relative to the measure playing the normative measure role. Since TIM is unavoidable, this is equivalent to being relative to the functions TIM deploys, whose only criterion of admissibility is that they be equal ignorance functions on the chords. If there is a well motivated restriction on that relativity, such as may be given by a question which gives more information, all is well and good. But if there isn't such a restriction, there isn't a principled way to get out of the difficulty Bertrand's paradox poses. Not knowing which way of choosing a chord at random is used shouldn't be a problem, since a state of ignorance is what the principle of indifference is supposed to allow us to deal with. But if ignorance does not give reason to discriminate between the distinct ways of applying the principle given by distinct admissible equal ignorance functions, and if those ways result in contradictory probabilities for the same event, the principle has failed.

The points I have just made do not apply only to Marinoff, but apply to any attempt at the Distinction strategy. Any such attempt requires the Underlying premiss that Bertrand's question is a generic singular question which cannot be a general question. I have found no grounds for rejecting it as a general question. It is just as meaningful to ask the question in the light of random choice in general as in the light of a particular method of random choice. Furthermore, as van Fraassen pointed out, if we are told which method of random choice to use we may not need the principle of indifference at all, so this strategy is in danger of merely evading the challenge Bertrand's paradox poses to the principle.

Secondly, even if the rejection of the general question could be maintained, generalizing statistically over a generic singular question is itself a procedure warranted by the principle of indifference. For example, although there is no such thing as a Fred in general, the principle of indifference warrants taking the statistical notion of the weight of the average Fred as representative. If we are ignorant of which method of random choice has been used, that is just more ignorance, and so equiprobability should be assigned to *those* possibilities. I'm going to call this *meta-indifference*. Later chapters are about applying principles that could be argued to amount to applying a kind of meta-indifference. I shall discuss meta-indifference itself at greater length in Chapter 10, prior to which we will also have examined a couple of attempts at meta-indifference in the literature. Here I shall consider only what bearing meta-indifference might have on the Distinction strategy.

5.5 Meta-indifference

Either consistent numerical probabilities can be derived by the application of meta-indifference to Bertrand's paradox or they cannot. If they can we find ourselves with a unique answer to the statistically generalised generic question in the same sense that the weight of the average Fred is a unique answer to the statistically generalised question of the weight of Fred. However, in that case, we have not vindicated the Distinction strategy, but the Well-posing strategy, for we have shown Bertrand's paradox to be well-posed in the sense that the principle of indifference is sufficient to turn a generic underspecified question into a determinate statistically general problem with a unique solution.

If meta-indifference doesn't entail consistent numerical probabilities, either it entails inconsistent numerical probabilities or it fails to entail any probabilities. If the former, Bertrand's paradox has recurred at the meta-level. If the latter, the supporters of the Distinction strategy may feel vindicated. They may argue as follows: So long as there seemed to be a viable notion of the probability of a longer chord in general, even just the etiolated sense got from the statistical generalization of the generic question, that possibility could be held up as a reproach to our strategy. However, just as there is no such thing as a Fred in general, the failure of meta-indifference to entail any probabilities *proves* that there is no such thing as a probability of the longer chord in general, not even in the etiolated sense. To demand that the principle of indifference be sufficient to calculate a probability *that does not exist* is no reproach. Consequently, all that there can be are the distinct particular problems into which we analyse Bertrand's generic question. Since the principle of indifference is sufficient to solve those problems, it is untroubled by Bertrand's paradox.

Even if we granted (which I do not) that there is no general Bertrand's question except for the statistical generalization of the generic question, I am unpersuaded that the failure of meta-indifference to entail any probability of a longer chord in the statistically general sense helps the Distinction strategist. It would only help him if it was a way of showing that the statistically general question was irresolvably indeterminate. But the statistically general question poses a determinate probability problem which the principle of indifference is supposed to suffice to solve. Saying that since meta-indifference fails to entail any probability of a longer chord the probability doesn't exist amounts to granting that the principle of indifference fails, and also to conceding Bertrand's wider point.

I would concede, however, that having pushed the argument this far, the matter is finely balanced, and my opponent has further resources to deploy. For example, he might argue that being a *statistical* generalization entails that a criterion of identity for the relevant probability is that meta-indifference suffices to calculate it, and so, since it doesn't suffice, the statistically general question is irresolvably indeterminate. For this reason I shall be showing in Chapter 10 that meta-indifference *cannot* fail to entail a probability of a longer chord. That argument constitutes the final part of my general refutation of the distinction strategy.

5.6 Ways of falling into van Fraassen's trap

I now exhibit by quotation other attempts at or allusions to the Distinction strategy that fail for the reasons we have seen.

> The problem has no solution until the meaning of "draw a chord at random" is made precisely a description of the procedure to be followed.
>
> (Funkenbusch 1962)

> Bertrand's paradox is of course not a logical paradox. The different results arise from assigning three different meanings to the phrase 'at random'
>
> (Tissier 1984:19)

> Intriguing problems whose solutions seem to run counter to intuition… under close scrutiny, they often appear to originate from an incomplete specification of the sampling procedures.[61]
>
> (Basano and Ottonello 1996:34)

> it is not hard to believe that his computations implicitly interpret "random chord" in three different ways. The paradox, though disconcerting, appears to be easily resolved.
>
> (Holbrook and Kim 2000:16)

> [A]ll so-called paradoxes to PI are simply disagreements and ambiguity in relation to sample space identification, saying nothing against PI as a logical principle.
>
> (Burock 2005:2)

---

61 These authors are not claiming a solution but discuss the ambiguities. One of their cases is essentially Bertrand's Celestial Sphere paradox, although they don't realize this.

Bertrand's paradox is perceived to be paradoxical because the conventional statement of the problem leaves undefined what "random" should be taken to mean in the context and, therefore, does not provide enough unambiguous information to let one come up with an unqualified answer that is obviously correct.

(Nickerson 2005:68)

Bertrand's paradox revisited: more lessons about that ambiguous word, *Random*.

(Chiu and Larson 2009:1)

[Randomness] is a notoriously difficult concept. In fact, it may not even be a single concept at all, but a cluster of concepts [...]

(Bangu 2010, 33)

Porto *et al.* might be thought to be an instance of the Well-posing strategy, since they offer 'an actual experiment, whose outcome is obviously unique' (2011:819) similar to Buffon's needle. However, they say

Bertrand's paradox... can be considered... a cautionary memento [sic]... of the possible ambiguous meaning of the term 'at random'... physics can help to remove the ambiguity by identifying an actual experiment, whose outcome is obviously unique.

(Porto *et al.* 2011:819)

So although though they offer a single answer, they do so on the basis of holding his question to be ill-posed, but only in the secondary sense.

Bertrand's paradox is not really a paradox since different solutions are based upon different methods of modelling the problem. This paper investigates the probability space of these modelling approaches to disentangle the inconsistency and settle on an appropriate approach.

(Ardakani and Wulff 2014:24)

[Bertrand's question] is vaguely posed... it is not the term "at random" which is the source of ambiguity in the BQ, but the nature of the entity which is randomized.... specifying the nature of the entity which is subjected to the random process, one obtains a well-posed problem, which can then be solved.

(Aerts and Sassoli de Bianchi 2014:14)

Solving this requires a precise definition of what "random chord" means.

(Zabell 2016b:334)

5.7 Conclusion

Marinoff's paper was historically and for a long time a prime contender for solving the paradox. He offers a solution which exemplifies the Distinction strategy. Our analysis of Marinoff's attempt at the Distinction strategy led us to a general refutation of this strategy.

The diagnosis of the Distinction strategy, that Bertrand's question is a clumsy way of raising whilst confounding distinct well-posed questions with unique answers, is false because the Underlying premiss is false. Even were it true, the principle of indifference warrants being indifferent over the distinctions to be made between the instances of a generic question and this takes us to a Well-posing solution by meta-indifference. Only if the premiss is true *and* meta-indifference fails to entail a probability of a longer chord does the strategy have any prospects. But even then, that failure may be as much a reproach to the principle of indifference as succour to the strategy.

I shall show in Chapter 10 that at least one method of being meta-indifferent, a method which defines a probability measure on the chords that is a measure-theoretically defined mean of all the probability measures on the chords, entails a probability of a longer chord. Consequently, the Distinction strategy fails because its diagnosis is false and it cannot even be rescued by a proof that I am mistaken about the falsity of the Underlying premiss.

6   The Well-posing Strategy

6.1 Introduction

The Well-posing strategy holds that Bertrand's question merely appears to allow of distinct answers. In fact, it poses a determinate problem for which the principle of indifference is sufficient to determine a unique solution. The Deluge Theorem **24** in Chapter 3 threatens that the probability of longer takes any value in [0,1]. The only levee to contain the flood is the requirement that the function in TIM be an equal ignorance function. This shows that piecemeal attempts at the Well-posing strategy are very unlikely to succeed. There are potentially *continuum* many options and therefore

> *Well-posing Requires a Principle* sufficient for restricting the continuum many potential probability measures.

But there is a

> *Well-posing Danger*: that the proposed principle, even if general in itself, imposes in its use a covert restriction on Bertrand's question, thereby *substituting* a restriction of the problem for the general problem rather than *comprehending* the general problem.

In this chapter we look at three attempted solutions that propose a principle and fall for the danger. Historically and for a long time a prime contender for resolving the paradox by the Well-posing strategy was the solution proposed by Jaynes, in which he applied his earlier defined principle of transformation groups (1968:§VII). In so doing, Jaynes is appealing to a principle of symmetry. His solution is a sophisticated application of this thought, and yet due to his idea of being indifferent over problems rather than events, it falls to the Well-posing Danger.

DOI: 10.4324/9781003456308-6

Our second is from Wang and Jackson, who rely on a principle about the homogenous distribution of chords. Lacunae in their proofs prove the necessity of deploying the resource of Chapters 2 and 3. Reliance on a distinction between chords chosen at random and random chords results in their answering a substitute question in place of Bertrand's question.

Our third is from Rizza, who applies a principle of representational adequacy, that the mathematical apparatus applied to a problem must be adequate to represent its intuitive content. Measure theory, he argues, fails at this, but Sergeyev's arithmetic of infinity applied to the chords gives an adequate mathematical determination of the intuitive content of Bertrand's question. Rizza's solution is the first radical mathematical treatment we address. It too falls for the Well-posing Danger.

For obvious reasons, I cannot prove that there is no principle by which the well-posing strategy can succeed. By the end of the chapter we will have seen three ways in which the Well-posing Danger defeats a use of a principle and the generality of that threat will be evident. After the next chapter on the Irrelevance strategy, we will see three chapters on Well-posing strategies using meta-indifferent principles. Our final chapter on the Well-posing strategy will be the chapter that examines whether symmetry of any kind can solve the paradox.

6.2 Jaynes' solution

Jaynes agrees that Bertrand's question is general, and argues that its very generality shows how to solve it:

> [1] the principle of indifference may... be applied legitimately at the more abstract level of indifference between problems; because that is a matter that is definitely determined by the statement of a problem.... [2] Every circumstance left unspecified in the statement of a problem defines an invariance property which the solution must have if there is to be any definite solution at all. The transformation group, which expresses these invariances mathematically, imposes definite restrictions on the form of the solution, and in many cases fully determines it.
>
> (1973:488)

I shall call this *Jaynes' Invariance Principle*. It starts with an application of what we would now call van Fraassen's

> *Symmetry Requirement*: problems which are essentially the same must receive essentially the same solution.
>
> (van Fraassen 1989:236)

It spells out the application in terms of a restatement of his principle of transformation groups from his earlier work (1968:§VII). Applied to Bertrand's question,

> If... the problem is to have any definite solution at all, it must be 'indifferent' to .... small changes in the size or position of the circle. This seemingly trivial statement... fully determines the solution.
>
> (1973:480)

Jaynes' point is that Bertrand is asking about circles in general, not about particular circles. Accidents of position and scale of the circle concerned give problems that are essentially the same and therefore the probability measure on the chords should be invariant over those accidents. The relevant invariances are therefore the symmetries/transformation groups of rotation, dilation (scaling) and translation.[62]

Jaynes offers an additional motivation for the invariances by referring to 'tossing straws onto a circle' (1973:478) and says that this empirical situation should give the same results for distinct observers, that is to say, for distinct frames of reference, for whom the circle may appear rotated, scaled or translated relative to each other. This is why he speaks in terms of small changes. However, examination of his mathematics show that his probability measure is quite generally rotationally, scale and translationally invariant. Furthermore, it turns out that

> the requirement of translational invariance is so stringent that it already determines the result uniquely.
>
> (1973:485)

The mathematical problem as Jaynes treats it is this:

> The position of the chord is determined by giving the polar coordinates (r, θ) of its center. We seek to answer a more detailed question than Bertrand's: What probability density $f(r, \theta)dA$... should we assign over the interior area of the circle?
>
> (1973:481)

In doing this Jaynes is not using a naming and so is not applying FPOIS directly but falls instead into some of Bertrand's frailties. He is not alone in so doing, of course. He uses our function $\kappa: C \rightarrow D$ (from §3.6) and the

---

62 Jaynes is foreshadowed here by Poincaré (Poincaré 1912:94–96) whose solution is based on requiring invariance under rotation and translation.

measure space $(D, \mathscr{S}, \mu)$ where $\mathscr{S}=\{D\cap B, B\in\mathscr{L}_2\}$ and $\mu(s)=\int_s f(r, \theta)dA$, $s\in\mathscr{S}$.[63] f will be determined by his invariance principle and the principle of indifference and scaled so that μ is a normal measure on D and is therefore also a probability measure on D, which is the probability measure Jaynes uses to give the probability of longer. This confounds the normative and probability measure roles and the probability measure Jaynes concerns himself with is therefore not a measure on the chords $\mathbb{C}$ but on the disk D. Likewise, his application of the principle of indifference is not to the chords but to the disk D. He then (only implicitly!) uses κ in TIM to carry back that probability measure to the chords.

We can reformulate this rigorously as we did for Bertrand by using κ and Jaynes' $(D, \mathscr{S}, \mu)$ in TIM to beget a normative measure and POIS to beget a probability measure on the chords, which measures would be the same just because μ is a normal measure. We won't go into all the detail but will use elements of that reformulation as appropriate.

The charitable construction of Jaynes relies on the Theorem of Induced Measure. But he is not using a uniform measure on that disk to be the measure μ. Rather, he is applying it like this. Take a small area, Γ, in one circle and the subset of chords, S, picked out by having their centres in that set. Then consider an offset circle and the set of chords, S', in that circle that are collinear with a chord in the first set. Their centres define a small area in the second circle, Γ'. The collinearity of chords defines a bijection between these two areas, Γ and Γ'. He then takes it that his invariance principle means that these two areas have equal epistemic status and hence the principle of indifference requires

> assign[ing] equal probabilities to the regions Γ and Γ', respectively, since (a) they are probabilities of the same event, and (b) the probability that a straw which intersects one circle will also intersect the other, thus setting up this correspondence, is also the same in the two problems.
>
> (1973:484)

This, then, is the justification that the map $\kappa:\mathbb{C}\to D$ is an equal ignorance function to the measure space $(D, \mathscr{S}, \mu)$: because the density function $f(r, \theta)$ he derives, and therefore the measure μ, is translationally invariant in the way just specified, κ maps subsets of chords of which we are equally ignorant to subsets of the measure space with equal μ-measure. Since κ is an

---

63 Here we follow Jaynes' notation with R for the radius and r the radial polar coordinate. Jaynes' 'dA' means integrating with respect to the two dimensional Lebesgue measure λ_2. By Fubini's theorem $\int f d\lambda_2 = \iint f(x, y)dxdy = \iint f(r, \theta)rdrd\theta$ using the Jacobian for the change of variables from Cartesian coordinates to polar coordinates—see Lax and Terrell 2017:246.

equal ignorance function, when used in TIM it begets a measure that fulfills the normative measure role and hence POIS begets an equal ignorance probability measure.

There is a unique density function which possesses this translational invariance and gives a normal measure (1973:485):

$$f(r, \theta) = 1 / (2\pi Rr), 0 \le r \le R, 0 \le \theta \le 2\pi$$

Since a chord is longer iff its centre is in the set $s \in \mathscr{S}$, s = the disk radius R/2 concentric with D, and recalling that μ is a normal measure so $\mu(D) = 1$, we get the probability that the chord is longer

$$P(L) = \frac{\mu(k(L))}{\mu(k(\mathbb{C}))} = \frac{\mu(s)}{\mu(D)} = \int_s f(r, \theta) dA = \int_0^{2\pi} \int_0^{R/2} \frac{1}{2\pi Rr} r dr d\theta = \int_0^{2\pi} \frac{1}{4\pi} d\theta = \frac{1}{2}.$$

Jaynes' solution is evidently an example of the Well-posing strategy. Jaynes denies that Bertrand's paradox is ill-posed at all, and asserts that it poses a determinate problem to which his invariance principle applies. The consequent invariance constraints mean that the principle of indifference is sufficient to determine a unique solution.

6.3 Friedman's critique of Jaynes

Friedman considers a line rotated about a point outside the circle, when the length of chord is a function of the angle between that line and the extended diameter through that point, over which angle he takes a uniform distribution as an instance of rotational invariance. He shows this gives the same result as Jaynes if one takes the limit as the distance of the point tends to infinity. Friedman is committing the frailty of undercounting the chords and is therefore depending on our only partial rescue of this treatment by TIM, as described earlier when discussing Marinoff's version of Friedman's treatment.

Friedman's purpose in his treatment is to criticise Jaynes because Friedman's 'general solution obtained is neither translationally nor scale invariant' (Friedman 1975:89). For this reason he concludes that

> Jaynes' proof vindicates this distribution only relative to the acceptability of these invariance principles. Yet the acceptability of these principles does not follow from the fact that they are invariance principles, but rather has to be judged on independent grounds.
>
> (Friedman 1975:90)

Jaynes offered independent grounds, namely that what is unknown determines invariances, so Friedman's critique is obscure. It would appear to be this: I derived my solution without using translational or scale invariance so those constraints cannot be essential to solving the paradox. And yet they are essential to Jaynes' solution, hence his solution cannot be based on what is essential to solving the paradox. Since his solution is based on the principle that whatever is unknown determines invariances, his use of that principle is not essential to solving the paradox.

When Friedman shows that his solution tends to Jaynes' in the limit,[64] he is not saying that Jaynes' solution is right. Rather, he is saying that Jaynes' and his solutions agree closely, within 6% when the distance of the external point is only twice the radius. Consequently, when Jaynes' claim empirical verification of his invariant solution by the experiment he conducted with straws, that is just as well explained by Friedman's variant solution

> The success of Jaynes' straw-tossing experiment may have been due to the motion of the straws being primarily translational... and to the similarity between [Jaynes' and Friedman's solutions] for even fairly small values of [the ratio of distance to radius].
>
> (Friedman 1975:91)

Friedman himself is using rotational invariance so he can't altogether reject that principle. Jaynes' use appears to be well motivated when we consider that, for example, observers with different view points and different measuring sticks shouldn't be coming out with different answers, which at least gives us both rotational and scale invariance. These two invariances still permitted two solutions, one agreeing with Bertrand's direction case and the other with Bertrand's midpoint case (Jaynes 1973:483). The translational invariance was needed to get the unique solution, which also turned out to be sufficient on its own. Despite that, Jaynes remarks

> the solution... would in any event have to be tested for scale invariance, and if it failed to pass that test, we would conclude that the problem as stated has no solution.
>
> (1973:485)

---

64 For those who want the details, the non-invariant general solution is Friedman's equation 4 and it tends in the limit to his equation 1, which is the same as Jaynes' equation 13 which Jaynes derives by a further change of variable from his $f(r,\theta)$ above to get a probability density function over the length of chords—see Jaynes 1973:485.

The defence of this criterion is the point about measuring sticks. So it is not clear quite what is wrong with Jaynes' use of scale invariance.

So Friedman's critique doesn't really engage with Jaynes' reasons for appealing to invariance but merely points out that Friedman's scale variant solution is close enough for empirical agreement. And of course, that is often the case. An inverse 1.99999999 law will usually agree empirically with Newton's inverse square law. A reason for accepting the latter is that area increases as the square of distance so something weakening by being spread out uniformly over an area will decrease as a square of the distance.

Similarly, Jaynes is appealing to a broader empirical cum theoretical consideration than simply agreement in observations when he appeals to scale and translational invariance. Moreover, the fact that Friedman 'is forced by the appeal to scale invariance and translational invariance' to take 'the limiting case' (Friedman 1975:91) of his solution and it then turns out that, so forced, his limit is Jaynes' solution, could equally be taken to be a support of Jayne's solution! For Friedman has thereby shown that, provided the invariances are a proper constraint, Jaynes' solution is not path dependent but can instead be reached by a method of approximation as well.

6.4 Nathan's critique of Jaynes

Nathan's critique is in part directed at Jaynes' solution but also directed at Jaynes' principle of transformation groups and the broader question of what he calls 'intrinsic distributions'.

Nathan points out that the principle of transformation groups is very costly.

> In each and every problem it would be necessary to examine all symmetries in order to ascertain consistency. Using Jaynes' method this is cumbersome enough with the simple circular target, and it becomes quite forbidding in more complicated cases.
>
> (1984:678)

We, however, are interested in Jaynes' principle in principle rather than in practice.

Nathan raises a number of worries about Jaynes' application of the principle. First, he rejects the idea that

> the imposition of symmetries.... [means]... that target symmetries affect line distribution... [because]... a replacement of the circle by a square must affect that distribution, which is absurd.
>
> (1984:678)

This thought seems to originate in the way Jaynes sets up Bertrand's question as analogous to tossing a sufficiently long straw onto a sufficiently small circle. Since the straw tossing as a random process amounts (roughly) to selecting a line in the plane, the distribution of such lines has nothing to do with their subsequent interactions with the circle in selecting chords. This is a shot at intrinsic distributions for chords of the circle or the square, since intrinsic distributions should be affected by their symmetries because symmetries are intrinsic properties.

He then moves on to suggesting that Jaynes' solution is in a certain sense trivial.

> if the distribution is independent of target shape then the problem cannot possibly be inconsistent.... We know, prima facie, exactly one distribution which ensures at one stroke all invariance properties and their consistency for any shape of target. It is, of course, the uniform and isotropic distribution of lines. Conversely, it seems evident and is easily proved that, given any finite target, the invariance of line intercept (or chord) distribution to arbitrary target displacements entails a uniform and isotropic distribution of lines.
>
> (1984:678–9)

Jaynes' probability density function *for the midpoints* of the chords (which is an instance of what Nathan means by an intrinsic distribution) just is what you get from an isotropic distribution of lines in the plane.

> It follows, then, that all he has demonstrated is the validity of the truism that, as far as Bertrand's Problem is concerned, a distribution of lines that is uniform and isotropic over a large area that includes the target entails a uniform and isotropic distribution within the area of the target.
>
> (1984:681)

Furthermore

> Jaynes' experiment is a good example of how easy it is to misinterpret experiments designed to corroborate a preconceived theory. His experimental results bear more than a superficial resemblance to the putative theory that every line that can be drawn is linear, confirmed by drawing lines and examining the curvature of a great number of small sections.
>
> (1984:679)

An isotropic distribution of lines is uniform in all directions, which, being defined in terms of parallel chords combined with his rotational symmetry point, is exactly what Bertrand's direction case is making use of—and doing so within the circle, which counts against Nathan's objection to intrinsic distributions.

In the rigorous treatment of the direction case in §3.12, this isotropy within the circle is fully represented by the use of the two-dimensional Lebesgue measure on $[-R, R] \times [0, \pi)$ for the normative measure. The reason is that that measure is identical to the product measure which is the product of the one-dimensional Lebesgue measures on each of $[-R, R]$ and $[0, \pi)$. Since $[0, \pi)$ defines the directions of chords and the Lebesgue measure being a uniform measure over $[0, \pi)$ gives equal measures to equal ranges of directions, it therefore represents the directions of chords being isotropic.[65] Since this is being done intrinsically to the chords, by which I mean nothing depends on the embedding of the circle in a larger space, I think Nathan's objection to intrinsic distributions falls and I won't discuss it further.

That point made, Nathan is saying that the application of Jaynes' principle of transformation groups amounts to assuming an isotropic distribution of straight lines, so it is hardly surprising if it gives a single answer, and one that agrees with Bertrand's direction case. Jaynes' solution is therefore unsatisfactory because

(1) Environmental distributions are independent of and cannot be inferred from symmetries, nor for that matter from any other properties of the demarcation of a target; neither can distributions of intercepts be so inferred although these are determined by target shape and environmental distribution and therefore reflect the symmetries; (2) nor can ambiguities be removed from probabilistic problems by the supposition that missing data must impose invariances.

(1984:684, my numbering)

In our terms, Nathan is saying that Jaynes has fallen for the Well-posing Danger: because other environmental distributions would give different probabilities, the principle of indifference does not here justify the application of Jaynes' principle of transformation groups to exclude them, and Jaynes' Invariance principle is false.

His first point, (1), is true but its force against Jaynes depends on the denial of intrinsic distributions. Granted that an empirical realization of the problem will amount to selecting a chord by an environmental distribution and grant that Jaynes tended to formulate his invariances in those terms, it is not clear that he must. One may read his points about

---

65 This can also be shown, with greater complication, in terms of the two-dimensional Lebesgue measure on $[-R, R] \times [0, \pi)$ without treating it as a product measure.

rotational, scale and translation invariance as standard points about the distinction between intrinsic and extrinsic features in geometry.[66] Speaking roughly, the latter may vary with the embedding in a space whereas the former do not and are therefore invariant with respect to embeddings. Rotations, scalings and translations of the circle are all relative to embeddings in a space (or a coordinate system) and hence invariance in those respects means eliminating the extrinsic features. Understood in that way, Jaynes' application of rotational, scale and translation invariance is, indeed, a matter of formulating the problem in terms of intrinsic features. Consequently, the mere fact that in general environmental distributions, which here means distributions of lines in the plane in which the circle is embedded, need not satisfy Jaynes' constraints only has force if intrinsic distributions are illicit. I have shown previously that they are not.

So we are left with (2), the rejection of Jaynes' Invariance principle. Nathan objects to being indifferent over problems:

> (a) a probabilistic problem is unambiguous precisely if (b) it defines a definite partitioning into events and assigns a probability to each and, conversely, (b) leads to (a).
>
> (1984:680)

Of course, if that is what it is to be a determinate probability problem we can abandon the principle of indifference altogether, since it is never needed. Here, then, Nathan has fallen into van Fraassen's trap, and does so because he is confusing having a determinate probability problem with having a problem which has a determinate solution. He continues:

> it is surely circular to claim that any indeterminacy determines a symmetry and subsequently to use the symmetry in order to remove the indeterminacy. If it were right there could not have been any indeterminacy to start with.
>
> (1984:680–1)

Well, yes, but that is why what Jaynes' is saying is that Bertrand's question merely *appears* to be ambiguous, to allow of distinct answers, due to a mistaken interpretation that fails to attend to the implicit constraint of invariance properties over what is left unknown in stating the problem. This is Jaynes' argument that the question poses a determinate problem.

---

66 This has a formal definition in differential geometry. Intrinsic properties are those that depend on the first fundamental form of a surface. See Pressley 2012:chapter 6.

Moreover, Jaynes' is not saying that *indeterminacy* determines a symmetry but that what is *unspecified* does so (more on this shortly).

Nathan next turns to a standard, but flawed, objection to the principle of indifference that we saw in Chapter 1 and applies it to Jaynes' principle of transformation groups.

> if what is meant is... that lack of specification in itself actually implies some correlated invariance, then the contention is simply erroneous. Lack of knowledge never entails knowledge, whatever form that knowledge may take.
>
> (1984:681)

Jaynes' actual statement is that

> every circumstance left unspecified in the statement of a problem defines an invariance property which the solution must have if there is to be any definite solution at all.
>
> (1973:488)

Grice's maxims of quantity and relation (give as much but no more information than is needed and be relevant Grice 1989:26-7) justify the assumption that the things that might have been specified but have not been specified won't change the problem. Jaynes is making a straightforward use of these maxims. Rotations, scalings and translations have been left unspecified so he is entitled to assume they make no difference, in other words that the problem is invariant with respect to them. So Jaynes can deny that the invariances were ever an indeterminacy because Grice's maxims mean that what is not said is not an indeterminacy but an irrelevance. That it was not said gives us knowledge of its irrelevance. So if these transformations are left unspecified then they are an irrelevance, which irrelevance implies their invariance, and because it was left unsaid we have knowledge of that irrelevance and thereby of their invariance.

In conclusion, whilst Nathan's discussion of Jaynes' is in places illuminating, insofar as it is a rebuttal of Jaynes' solution it fails. It fails because Nathan's rejections, of intrinsic distributions and of Jaynes' Invariance principle, fail.

6.5 Jaynes' fall to the Well-posing danger

van Fraassen seems to accept Jaynes' solution for Bertrand's paradox, but draws attention to Jaynes' concession that he doesn't see how to apply his approach to von Mise's water and wine problem. (van Fraassen 1989:315). Marinoff, on the other hand, thinks

> both Jaynes and van Fraassen erroneously claim that they are answering Bertrand's original question... which they both mistranslate as Q2.
>
> (1994:7)

Marinoff thinks this partly because he and Jaynes seem to agree on the answer to the empirical case of straw tossing. However, this is a significant misrepresentation of what Jaynes is doing. What Jaynes is doing is far more sophisticated, and strictly speaking, they do not agree on the answer to the empirical case.

Marinoff's Q2 is specified in terms of 'the random chord generated... by a procedure outside the circle' (1994:4). Marinoff's solution to Q2 (1994:7–11) is that the probability of the longer chord=the *limiting* probability as the distance of a point outside the circle from the centre of the circle tends to infinity, and that limit=½. If we are to understand Marinoff's Q2 as relevant to the empirical case, we must understand him as construing the procedure outside the circle as follows: the centre of the straw determines a point outside the circle. The length of the chord generated depends on the angle the straw makes with the extended diameter intersecting the centre of the straw. The angle is assumed to have a uniform probability density function, and straws which do not intersect the circle are ignored.

Marinoff is mistaken when he takes his limiting procedure to give the solution to the empirical case. Rather, it approximates cases where the straws are long relative to the circle diameter, so that the chance of the centre of the straw lying inside the circle is negligible. Furthermore, his solution method should not be taken to the limit for cases of specific relatively long straws. When it is not, for such relatively long straws of specific lengths his solution method will result in the probability of a longer chord being *strictly less* that ½, whereas Jaynes' solution to cases of finite straw length is precisely ½.

Marinoff's solution is a mere approximation in a restricted range of empirical cases because Marinoff's solution method to Q2 excludes straws whose centre lies within the circle, whereas Jaynes is exactly correct for an unrestricted range of empirical cases precisely because his solution does not exclude those straws. That the probabilities in Marinoff's solution converge quickly to ½ as straw length increases disguises this important distinction between their solutions.

Now for Marinoff, idealising to straws of infinite length gets rid of the problem of ignoring the straws whose centres are inside the circle. But that really amounts to abandoning the notion of a 'random chord generated...

by a procedure outside the circle'. Instead, it turns out that talk of points on extended diameters, uniform distributions over angles of lines through such points and taking limits as the distance of that point from the circle tends to infinity is an obscure way to obtain the answer for the probability of longer chords got from randomly selected lines (rather than line segments) in the plane. But put baldly like that, one now awaits a justification for why the former process should be taken as a solution to the latter problem.

Jaynes faces none of these problems, and in fact, only his approach can satisfactorily explain why the idealisation of infinite straws might be an answer to chords got from randomly selected lines in the plane. Jaynes' mathematics can apply to line segments (straws of specific lengths) but is independent of the finitude of such line segments. Consideration of the way he is applying the principle of indifference in terms of regions Γ and Γ' makes it clear that nothing *depends* on the relevant circles being close (as they have to be for finite straws), but allows them to be arbitrarily distant. In effect, his solution concerns itself with invariance of probability measure *given* infinite lines. This is significant, since it is the reason I think Jaynes' attempt at demonstrating the problem to be well posed fails.

The last point is also why Marinoff's criticism, although based on mistaking the relation of Jaynes' solution to Marinoff's Q2, is correct insofar as he convicts Jaynes of solving a particular version of Bertrand's question rather than the general question. Marinoff expresses it better in his conclusion when he says

> Poincaré's, Jaynes' and van Fraassen's appeals to invariance of probability density under the group of transformations in the Euclidean plane as the supreme arbiter of Bertrand's question merely suggest another version of [his question]: With what probability is Bertrand's condition fulfilled if we demand such invariance?... Bertrand himself made no such demand.
>
> (1994:22)

The problem is that if we don't accept the fully general mathematical extension of the empirical situation, then the problem Jaynes is considering is not Bertrand's, but a restriction of Bertrand's, not exactly Marinoff's Q2, but a restriction all the same, namely of a process of random choice relative to finite lines in the plane. If on the other hand we accept the full generalization, the situation is not improved. For quite clearly, what his application of the principle of indifference relies on is families of infinite lines which coordinate many regions Γ, Γ', Γ'', Γ''', ... in many circles.

Now this is indeed well motivated for the empirical problem of straw tossing but not for the general problem, since it still counts as specifying a particular way to select chords, namely, selecting them relative to infinite lines in the plane. So doing amounts to a way of determining a uniform distribution over lines in the plane, a point that Nathan also makes:

> all he has demonstrated is the validity of the truism that, as far as Bertrand's Problem is concerned, a distribution of lines that is uniform and isotropic over a large area that includes the target entails a uniform and isotropic distribution within the area of the target [i.e. the circle].
>
> (Nathan 1984:681)

Bertrand's question, however, is about any circle, not about any circle such that if this chord is selected in this circle, then that collinear chord is selected in that circle, and so on for all circles intersected by the extension of the chord in the first circle. Nothing about the problem as stated, nothing about the generality of circles spoken of, requires *this* coordination of events. It is rather the empirical situation of straw tossing that does so. That is to concede that Bertrand's general problem has not been solved, but only a particular problem, namely, the idealisation of the straw tossing variant. Jaynes claims that 'we do no violence to the problem if we suppose we are tossing straws' (Jaynes 1973:478), but it turns out that we do. Jaynes has fallen to the Well-posing Danger.

I have *not* proved here that the Symmetry requirement cannot underpin a successful attempt at the Well-posing strategy. I have only proved that Jaynes' specific use of the Symmetry requirement did not succeed. In a later chapter I will address the hope from symmetry in general.

6.6 Drory's critique of Jaynes

Drory sums up his critique of Jaynes:

> It is widely accepted that Jaynes proved that there is a unique solution that possesses rotational, scaling and translational invariance.... that is not the case... as with the principle of indifference, the application of the principle of transformation groups depends upon the method of selection of chords. The implementation of the symmetries turns out not to be unique, and leads to different mathematical requirements depending on the underlying process by which one imagines the random chords to be generated. In fact, contrary to Jaynes' assertion, each of the classical three solutions of Bertrand's problem... can be derived by the principle of transformation groups, using the exact same symmetries.
>
> (2015:441)

Drory's point is that what controls the specific way that Jaynes applies the symmetries in deriving his probability density function is the empirical method of throwing straws at a circle. Consequently

> We have the logical order of things backwards. The straw throwing experiment is not the empirical confirmation of an independent abstract analysis. Instead, the analysis is a mathematization of the straw throwing procedure. This means that the experimental procedure logically precedes the analysis.
>
> (2015:447)

Consequently

> if we had thought of a different experimental procedure in the first place, the principle of transformation groups would have yielded a different result.
>
> (2015:447)

Drory offers throwing a straw with a dart at its centre as an empirical selection corresponding to the midpoint case. He shows that when analysed with invariance under translation this gives probability density $f(r, \theta) = 1/\pi R^2$, giving the probability of longer $\frac{1}{4}$ (2015, 449). He then offers an empirical selection corresponding to the angle case. A spinner at the centre of the circle is spun once to pick an endpoint on the circumference and spun again to pick the direction of the chord with that endpoint. Analysed with rotational invariance gives probability density $f_1(\alpha, \beta) = 1/4\pi^2$, for $(\alpha, \beta) \in [0, 2\pi] \times [0, 2\pi]$, giving the probability of longer $\frac{1}{3}$ (2015:452) (see Figure 16).

Another example for the angle case uses a pointed stick balanced on its point on a random point on the circumference and is allowed to fall.

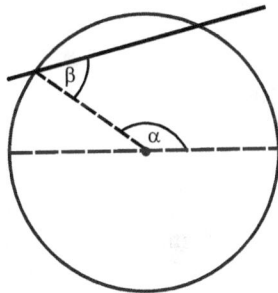

Figure 16 Angles as defined by Drory (2015:451).

Analysed with invariance gives probability density $f_2(\psi, \theta) = 1/\pi R^2$, for $\theta \in [0, 2\pi]$, giving the probability of longer ⅓ (2015:457) (angles $\psi = \alpha$ and $\theta = \beta$ as in Figure 16).[67]

The point of this second angle case is twofold. First, to meet the criticism of the first two that they are selections of points rather than chords (2015:454). In our terms, Drory doing the latter is the frailty of not using the POI on the chords, although it is also defensible in our terms by the Theorem of Induced Measure. Second

> More importantly... that when one applies the principle of transformation groups, one discovers that the symmetries function quite differently from the [spinner] case.
>
> (2015:455)

Unlike the spinner case, although rotational invariance applies to the initial point on the circumference, rotational invariance is insufficient to 'fix the PDF' (2015:455). Scale invariance also supplies no constraint.

What Drory then calls translation invariance fixes the probability density function. The invariance is among circles sharing the point on which the stick is balanced, rotated about that point, so this is really a second rotational invariance (as he appears to acknowledge in the rubric to his Figure 4). Therefore in both the spinner and pointed stick cases the probability density function is fixed by a pair of rotational invariances, which vitiates the claim of functioning differently. Indeed, it may appear that there is none.

The equations that express the invariances of the spinner and pointed stick angle cases are respectively

$$f_1(\alpha, \beta) = f_1(\alpha - \theta, \beta - \phi) \tag{15}$$

(2015:452)

$$\left[f_2(\psi, \theta) = \right] f_2(\theta) = f_2(\theta') = f_2(\theta - \phi) \tag{19}$$

(2015:457)

In the first α and β are angles defining a chord, but ψ and θ do that in the second. The difference in content is that in the first, θ and ϕ are the rotations for which invariance is required whereas in the second only ϕ is such a rotation.

---

67 In Drory's figure 4 he uses 'ψ' where he meant 'ϕ'.

Equation 19 omits the variable for the initial point on the circumference, ψ, because 'rotational symmetry implies that the p.d.f. is independent of ψ' (2015:455). But that point applies just as much to the spinner case, which would then omit the dependence on $\alpha-\theta$ and so also have been expressed in the same form, $f_1(\beta-\phi)$. Likewise, had that point not been applied to the pointed stick case its equation 19 expressing invariance would have had the form of Equation 15.

The difference Drory is really making use of is in the spinner case he sets it up for invariance over the choice of polar axis for each of α and β whereas in the pointed stick case he ignores that polar axis invariance altogether and considers only the rotation of the circle about the pointed stick point. In fact, we could have added polar axis invariance to the pointed stick case and would still have come up with an equation of the same form as Equation 15. Despite this, the difference in functioning of the symmetries between the two cases is really a matter of whether the needed invariances are dependent on or independent of the circle, in the sense I shall explain.

In the spinner case we need invariance over the polar axes because the length of the chord named by a specific $(\alpha, \beta) \in [0, 2\pi] \times [0, 2\pi]$ will be different for different axes and so the probability density function has variables for the rotations of the two axes, θ and ϕ. These axes and their rotations are *independent* of the circle in the sense that they can be defined without reference to the circle but only with reference to each other. Rotations of the circle about a point on the circumference, however, make no difference to the lengths of chords named by a specific $(\alpha, \beta) \in [0, 2\pi] \times [0, 2\pi]$ and so the probability density function is independent of those rotations and therefore need have no variable for those rotations.

In the pointed stick case, however, rotations of the circle about the point on the circumference on which the stick is balanced *do* make a difference to the length of a chord: a single fallen stick defines different length chords in the rotated circles. In this case, because ψ names the point on the circumference and θ is defined by the angle the fallen stick makes with the radius to ψ, any chord named by a specific $(\psi, \theta) \in [0, 2\pi] \times [0, 2\pi]$ will have the *same* length. *But*, the polar axis defining θ is not independent of the circles but varies with the rotation of the circle around ψ. Consequently, the invariance that produces the correspondence between chords of equal length can only be specified in terms of axes *defined by the point ψ on the circumference and some other point defined in terms of the circle* (the centre is the obvious point to use). So the probability density function is dependent on rotations about the point on the circumference and hence includes the variable ϕ for those rotations.

In conclusion, the big difference in the way the invariances function is this. For the spinner case the invariances, and hence the variables expressing those variances, can be defined entirely independently of the circles, whereas for the pointed stick case the definition of the invariance and its variable necessarily depends on the circles. To use some apposite metaphysical jargon, the former are only externally related to the circles whereas the latter are internally related. The difference between external and internal relations is regarded, at least by philosophers, as a fundamental difference.

All in all, then, Drory's argument is that Jaynes' way of applying the Symmetry requirement varies with the way of generating chords empirically. He suggests that Jaynes is therefore getting the logical order wrong when he claims to have given a general solution. Rather than the straw tossing offering empirical support for Jaynes' solution, it is the straw tossing that is being modelled by his solution. Since there are other empirical generators modelled with other transformation group invariances giving distinct probabilities of longer, Jaynes is misled in thinking his solution is general.

The suggestion could be challenged but I do not think that the challenge offers any help to Jaynes because I do not think the power of Drory's critique need depend on the logical order point. To my mind, it suffices to show, as he has done, that the invariances appealed to by Jaynes have different ways to be applied. In so doing, Drory has given us two independent arguments that Jaynes has fallen to the Well-posing Danger. The first argument is showing that all three of Bertrand's answers could be got from different implementations of the invariances. The second is showing two different implementations of the invariances give rise to the same answer. Having clarified just what that difference is in terms of internal and external relations, it seems to me that this is an additional and entirely independent argument whose import is to some extent left undeveloped by Drory. Yes, it supports his main point that that 'the implementation of the symmetries turns out not to be unique'(2015:441). It also shows that we may be misled into thinking we have a general solution just because two different implementations give the same result. It draws to our attention the need to show that *all* possible implementations give the same result.

Drory has amplified my original critique and shown that the problem of the Well-posing Danger is more dangerous than it might appear. Jaynes set off in one direction of applying the Symmetry requirement: had others set off in the other directions given by Drory they might equally well have thought that the Symmetry requirement solved the paradox in their way. Drory's exhibition of several such applications gave us two different argument for his main point and shows that the kind of use of the Symmetry requirement first instanced by Jaynes must fall to the Well-posing Danger.

6.7 Wang and Jackson's solution

Wang and Jackson don't explicitly announce a principle, but I think it is fair to enunciate it for them:

Wang and Jackson's principle:

> "Bertrand chords" and "random chords" mean the chords that are homogenously or uniformly distributed over the circle.
>
> (2011:77)

where

> 'Bertrand-chords'... represent the chords referred to in Bertrand's paradox.
>
> (2011:73)

Their argument for this is that

> everyone agrees on what Bertrand chords are like—they are homogeneously or uniformly distributed over the circle.
>
> (2011:76)

because

> we believe that Bertrand's problem has only one answer because we assume, implicitly, Bertrand-chords are truly random therefore homogeneously distributed over the circle. If, on the other hand, we did not assume homogeneity of Bertrand-chords, we would not insist on a unique solution.
>
> (2011:77–8)

But what is it that we are agreeing on?

> Random lines have been defined mathematically in the literature in terms of Poisson process or Poisson distribution. The essence of Poisson process is "pure randomness". Pure randomness leads to homogeneity. The essence of random lines, therefore, is homogeneity.[68]
>
> (2011:79)

---

68 For evidence that Wang and Jackson are in part thinking in terms of finite numbers of chords using Monte Carlo methods, see their figures and compare with those in Qin and Chen 2023. The last is not a claimed solution but drawing an analogy between Bertrand's paradox and fibres in face masks.

What this amounts to in the chord case is

> Definition-A. Chords in C are Bertrand-chords or homogeneously distributed chords if and only if their chord-directions are *uniformly distributed over range [0, 180)* and for any α between 0 and 180, the intersecting points of the chords in C_α with diameter $D_{\alpha\text{-normal}}$ are *uniformly distributed along $D_{\alpha\text{-normal}}$.*[69]
>
> (2011:80, my emphasis)

where

> Let $\phi(O, r)$ denote a circle with radius r and the center at O,...C represent a set of chords that are drawn at random on $\phi(O, r)$...C_α represents a set of parallel chords with chord-direction [α]...$D_{\alpha\text{-normal}}$ be the diameter $\phi(O, r)$ normal to the parallel chords in C_α
>
> (2011:79)

Evidently what they mean by homogeneous is isotropic. My earlier discussion of Nathan is therefore pertinent but I will confine myself here to addressing Wang and Jackson's use of isotropy. Then it is claimed that

> Since the chords defined in Definition-A are homogeneously spread over the circle, points in the circle must have same chance to be on a chord in C. So we have an alternative definition based on the homogeneity of the chords directly

> Definition-B. Let Δr denote an arbitrarily small but fixed amount. Chords in C are Bertrand-chords or homogeneously distributed if and only if their chord directions are uniformly distributed over range [0°, 180°) and for any two points, P and Q in circle $\phi(O, r)$, probability that circle $\phi(P, \Delta r)$ is on [i.e. intersects] a chord in C is same as probability that circle $\phi(Q, \Delta r)$ is on a chord in C.
>
> (2011:80)

There then follows some summary arguments in §6.4 with crucial premises that the midpoint case and angle case fail the final conjunct of Definition-B, whereas the direction case satisfies Definition-A. They conclude

> we... proved that two of the three alleged answers were not sound because of their false assumptions and that only one solution was correct. We also revealed... [what]... had persistently caused puzzles with regard to this paradox:... misunderstanding of the nature of homogeneously distributed chords.... The paradox is therefore no longer paradoxical.
>
> (2011:102)

---

69 In this section we follow their notation and note that their set of chords, C, is not clearly defined by them to be the same set as all the chords, C.

Thus we have their Well-posing strategy: Bertrand's question merely appears to allow of distinct answers, due to mistaken interpretation that permits non-homogeneous chords; absent this error, there is no paradox.

Wang and Jackson 2011:§5 is where we find the, at best incomplete, claimed proofs of the crucial premisses. I say at best incomplete because the technical aspects of the paper, whilst in some ways ingenious and interesting, are also a case study in the need for the rigour developed in Chapters 2 and 3.

From our earlier work we can immediately see that the direction case *trivially* satisfies Definition-A since that definition simply *defines* homogeneously distributed chords by the normalized product measure over our name space $[-R, R] \times [0,\pi)$. That product measure on $[-R, R] \times [0,\pi)$ is got from the normalized Lebesgue measures on each of $[-R, R]$ and $[0,\pi)$, which last is what is specified by the parts of Definition-A that I emphasised. As I explained earlier in discussing Nathan, the product measure on $[-R, R] \times [0,\pi)$ is the same measure as the $\mathbb{R}^2$ Lebesgue measure on $[-R, R] \times [0,\pi)$, which last is the name space I used for the direction case in §3.12.

Whether Definition-B is really equivalent to Definition-A is obscure. There is a way in which it is trivially implied by Definition-A but not in a way that helps them. The first conjunct follows immediately. The final conjunct is supposed to follow from and capture the claim that homogeneity entails that points in the circle must have the same chance to be on a chord in C. In the absence of any explicit definition of the event space, name space and application of POIS, the intuitive interpretation of that chance is the chance that *any* point is on any line in a bounded region, which is zero. On this basis, the claimed proofs for the midpoint and angle cases would have to be invalid.

The question of equivalence turns on the equivalence of the final conjuncts of the definitions. No argument is given for this equivalence. Furthermore, it seems that the first can be true whilst the second is false. Suppose $\Delta r = r$ and $P =$ the origin O and $Q =$ a point on the circumference. Once more, in the absence of explicit definitions, intuitively the probability that circle $\phi(P, \Delta r)$ is on a chord in C is 1 but for $\phi(Q, \Delta r)$ it is less than 1.

Intuition can take us only so far. When it comes to the proofs, there is a yawning gap between what is actually proved and what is claimed to be proved, a gap that the reader is helpless to fill due to the absence of explicit definition. For example, in the claimed proof that the midpoint case fails the final conjunct of Definition-B, what is proved is this: the midpoints of chords that all pass through the same point in the disk all lie on a circle[70] and if point X is closer to the origin than point Y, the

---

70 We have to count the origin as a degenerate circle to make this true.

circumference of that circle for X is smaller than that for Y. From this it is immediately inferred

> point X has a smaller chance to be on a chord in C_1 than point Y does (recall that 'chance a point P to be on a chord' means 'probability for $\phi(P, \Delta r)$ to be on a chord where Δr is an arbitrarily small but fixed amount'.

(2011:88)

Nothing more than this is done to connect up the difference in circumference lengths with the difference in probabilities for $\phi(X, \Delta r)$ and $\phi(Y, \Delta r)$ to be on a chord. I can see a variety of approaches, including taking limits as Δr tends to zero, but the connection still seems to come down to the chance of a point being on circle $\phi(X, \Delta r)$ or circle $\phi(Y, \Delta r)$. Absent explicit definition, that is zero for both because the measure of any circle in the disk is zero. There may be a way round this in terms of defining an event space of circles but some care would need to be taken since the obvious measure that gives non-zero measures to the circles[71] make it an infinite measure space, whereas we need a finite measure space for POIS to be used.

Similar points can be made about what is actually proved and what is claimed to be proved for the angle case. The issues in the proofs may be fixable using the resources of Chapters 2 and 3 and so having exhibited these lacunae as further support for the necessity of deploying those resources, I won't pursue this further. We leave this here because filling in the lacunae won't save their Well-posing attempt.

The game is given away in their later discussion:

> Two different concepts, homogeneously distributed Bertrand-chords and randomly drawn chords, have been mistakenly conflated by scholars who worked on this paradox.

(2011:93)

We can now see that the phrase "chords drawn at random" is an incorrect term for Bertrand-chords since randomly drawn chords are not necessarily homogeneously distributed and therefore not necessarily Bertrand-chords. No scholar who has written on Bertrand's paradox, including J. Bertrand himself, has ever pointed out the misrepresentation; and no one has ever tried to distinguish "a chord drawn at random" from "a random chord". Actually, the term "random chords" is an acceptable wording for Bertrand-chords if it means "truly random

---

71 I.e. the measure defined by their circumferences.

chords", because "truly random chords" would be homogeneously distributed.

(2011:94)

Distinguishing a chord drawn at random from a random chord, where the latter are *truly* random but the former are not, strikes me as a distinction without a difference. It gets any bite it might have from the identification of the truly random chords with homogeneous chords. This identification seems to be based on the thought that random processes selecting chords by endpoints or midpoints are not really random because they are biased, in the sense that Wang and Jackson's Monte Carlo method selection of finitely many gives an uneven output of chords (see their Figure 2 2011:77). But this is to make the mistake noted by Eagle:

> Those who would refuse to call biased chance processes 'random' are… letting views about the randomness of the product drive their opinion of randomness of the process.
>
> (Eagle 2016:442)

Recall that on their definition of homogenous chords, the chords of the direction case are trivially homogenous chords. Another way of putting that is to say that they have defined homogenous chords as the chords that are named by the direction case name space. In that sense, the definition of truly random is question-begging. I grant that in the background of that definition are their remarks about the Poisson process and their mention of literature that uses the Poisson process to define random lines *in the plane*. But this merely makes it clearer that the appeal to the truly random is not a distinction between the random and the not random. It is perhaps also worth noting that Ardakani and Wulff effectively *deny* the 'truly random' claim for the homogenous chords when they say about the direction case:

> In the random radius approach, the sample space is not defined properly for the problem. The major flaw for this approach is that chords are not drawn completely at random in the semicircle.
>
> (Ardakani and Wulff 2014:33)

What is really going on here is that with this distinction, Wang and Jackson have fallen for the Well-posing Danger.

> *Bertrand's problem was ill posed, because the wording incorrectly represented the original problem, rather than because of the vague wording.* In this sense, instead of ill posed, we would rather say Bertrand's problem was incorrectly posed.
>
> (2011:99 their emphasis)

The discussion of the determinacy of problems and of ill-posing in Chapter 4 and the defence of the simple generality of Bertrand's question in Chapter 5 shows this to be mistaken. When we ask, "What is the probability that a random chord of a circle is longer than the side of the inscribed equilateral triangle?", we have distinguished neither between a random chord and a chord drawn at random nor between a chord drawn at random and a chord that is truly random. We have simply asked the general question.

Consequently, in this remark they acknowledge that they are replacing Bertrand's question with what they think it ought to have been. And that amounts to substituting a restriction of the problem for the general problem rather than comprehending the general problem. They may have answered the question 'What is the probability that a homogeneously distributed Bertrand-chord of a circle is longer than the side of the inscribed equilateral triangle?'. In so doing, they have not answered Bertrand's question.

6.8 Rizza's premiss and principle

Rizza's paper is entitled 'A study of Mathematical Determination through Bertrand's Paradox'.

> Certain mathematical problems prove very hard to solve.... because certain intuitive features involved in their formulation cannot be codified by the mathematical apparatus canonically available to study them. In such cases what looks like an inherent difficulty of a given problem is best regarded as an effect of the fact that its intuitive content has not yet been resolved into mathematical determinations that can be relied upon in order to obtain a solution. This paper aims to explore and clarify this phenomenon with respect to one particular example, namely Bertrand's paradox.
>
> (2018:375)

Rizza is appealing to what I shall call his

Mathematical Determination principle: The mathematical apparatus applied to a problem must be adequate to represent its intuitive content.

Rizza's view is that Bertrand's question is answerable but unanswered because of what I shall call

Rizza's premiss: measure theory cannot supply a mathematical determination of the intuitive content of Bertrand's question.

We will later look further into what exactly Rizza means by a mathematical determination as it applies to Bertrand's question. To answer the question requires 'an expansion of the mathematical resources at hand' (2018:377) by Sergeyev's arithmetic of infinity, which gives us

> an elementary approach to Bertrand's paradox itself, motivated by the need to convert certain intuitive features of its geometrical setting into numerical determinations (more plainly, it is necessary numerically to specify the size of certain infinite collections of geometrical entities).
>
> (2018:376)

As we shall see, what Rizza means by numerically specifying the size of infinite sets here is being able to count chords using Sergeyev numbers.

Bertrand's question can be answered, then, and without paradox, because Sergeyev numbers satisfy the Mathematical Determination principle and prove

> *Rizza's Well-posing claim*: Sergeyev's arithmetic of infinity applied to the chords suffices for the principle of indifference to determine a unique probability of longer.

Thus does Rizza offer an instance of the Well-posing strategy.

6.9 Sergeyev's arithmetic of infinity applied to the chords

Sergeyev's arithmetic of infinity starts with an infinite integer called Grossone, which we symbolise by 'ϕ' and which he symbolises by a '1' in a circle as in Rizza's figures below. The basic idea behind it is that we have a numeral system for Sergeyev's numbers that includes infinite and infinitesimal numbers in terms of polynomials in the indeterminate ϕ for all positive and negative powers of ϕ (see Sergeyev 2009:181–3). Roughly, the arithmetic amounts to the arithmetic of a polynomial ring on ϕ over the reals.

> Inasmuch as it has been postulated that Grossone is a number, associative and commutative properties of multiplication and addition, distributive property of multiplication over addition, existence of inverse elements with respect to addition and multiplication hold for Grossone as for finite numbers.
>
> (Sergeyev 2009:181)

ϕ itself is the number of natural numbers but, unlike the cardinality $\aleph_0$, ϕ is not the number of even numbers, which is $\phi/2$. Similarly, adding and

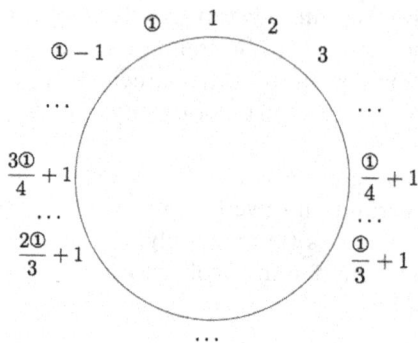

Figure 17 Rizza's figure 1. (2018:384).

subtracting finite numbers has an order effect not had on $\aleph_0$, for example $\phi-1<\phi<\phi+1$, and so on. The *Sergeyev numbers*, $\mathscr{S}$, as we shall call them, have real, infinite and infinitesimal parts,[72] and we can factorise and simplify as we do for polynomials in general (simple examples in the next section).

Rizza applies this to the chords in the following way

> The degree of accuracy selected to deal with Bertrand's problem is fixed as soon as it is declared, by means of a numerical specification, how many points on the boundary of the circle C can be discriminated.... An obvious, but fruitful, choice is to [discriminate] ϕ.
>
> (2018:383)

See Figure 17. What is going on here is that we have ϕ points on the circle dividing it into ϕ equal, infinitesimal, 'least discriminable arcs' (2018:384).

On this basis Rizza then works out the probability of longer by counting, using the Sergeyev numbers, all the chords, chords that are sides of an inscribed equilateral triangle and chords that are shorter. He divides the last two counts by the first and subtracts the result from 1 to give the probability of longer (2018:383–6).

I am not going to go through his calculations but will show what is going on here in our terms before we address his results. Rizza never defines a measure theory in terms of Sergeyev numbers but I will sketch it on his behalf. We define a *Sergeyev measure* to be a function that takes values in $\mathscr{S}$ with real parts in $\mathbb{R}^+$, a codomain we will call $\mathscr{S}^+$, which measure at least obeys the finitely additive measure axioms, and preferably the measure

---

72 E.g. In $3+2\phi+5\phi^{-1}$, 3 is the real part, 2ϕ is the infinite part and $5\phi^{-1}$ is the infinitesimal part.

axioms, up to infinitesimals.[73] The Sergeyev measure of a subset of chords is given by counting it with Sergeyev numbers. That counting Sergeyev measure is then used as the normative measure in POIS, giving the probability measure using corollary **9**.

Let[74]

$$\Psi = \{1, 2, \ldots \phi/4 \ldots \phi/3 \ldots \phi/2 - 1, \ \phi/2, \phi/2 + 1, \ldots 2\phi/3 \ldots 3\phi/4 \ldots \phi - 2, \ \phi - 1, \ \phi\}$$
$$= \{g \in \mathcal{G} : g = w + q\phi, \ w \in \mathbb{Z}, \ q \in \mathbb{Q} \cap [0,1] \text{ and } w < 1 \text{ if and only if } q > 0\}.$$

Rizza has implicitly used a naming of a subset of the circumference where Ψ is the codomain of the naming and is therefore the set of names. The arithmetic on these names of points on the circumference is modulo ϕ, since x and $x + \phi$ name the same point.

In naming the subset of points on the circumference Rizza has thereby defined a naming of a subset, Q, of the chords namely, $n: Q \rightarrow \Psi \times \Psi / \sim$, $n(c) = ([x, y])$ where x and y are the endpoints of c. Note the codomain, $\Psi \times \Psi / \sim$, which must be a quotient space for the same reason we saw in Chapter 3 when identifying chords by the angular component of the polar coordinates of their endpoints: for instance the chord with endpoints 1 and ϕ appears twice in $\Psi \times \Psi$. Here the equivalence relation defining the quotient is $(x, y) \sim (x', y')$ iff $(x, y) = (x', y')$ or $(x, y) = (y', x')$, giving us Figure 18. From hereon we name equivalence classes by either member as convenient.

The quotient space naming the chords is an infinite lattice with infinitesimal lattice spacing of $1/\phi$. We follow Rizza in excluding the degenerate chords and so each point in the lattice apart from the ϕ points on the diagonal is the name of a chord in Q. From hereon we speak of the names as the chords they name.

Rizza's view of what he is doing is this

the parametrisation of chords by pairs of distinct points on the boundary of C depends on a preliminary specification of the number of discriminable points. It is of the essence to realise that, when handling the computational instrumentality proposed by Sergeyev, there is no

---

73 A finitely additive measure is a function on an algebra of sets which is zero on the empty set and the measure of a finite union of sets is the sum of the individual measures. See Bruckner *et al.* 2008:26–7. One might ask Rizza to supply the proof that his use of Sergeyev's arithmetic behaves sufficiently like a measure for his purposes. For the sake of giving him his premises so we can pursue the discussion I do not take that absence further.

74 I do this because, although I suspect it would, I haven't proved that the von Neumann trick for the ordinals works properly for Sergeyev numbers. If it would then $\Psi = \phi + 1$.

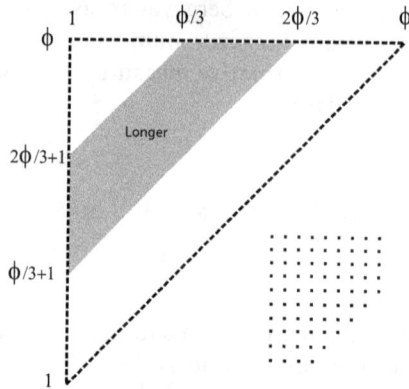

Figure 18 The lattice name space with infinite magnification of lattice on right.

question, in general, of obtaining exact numerical results: in what follows only approximate probability values or probability estimates are computed. This is, however, enough to restrict the range of inaccuracy to an infinitely small order of magnitude.

(2018:383)

In other words, what he is offering is a kind of approximate naming of the chords by the naming of the subset Q spoken of in terms of discriminability and the probabilities then derived are also approximations.

The function $f : \Psi \times \Psi/\sim \to \mathscr{G}^+$, $f([x, y]) = 1$, defines a counting Sergeyev measure, μ, on $\Psi \times \Psi/\sim$ thus: We assume we have an algebra or σ-algebra, $\mathscr{A}$, on $\Psi \times \Psi/\sim$ and then $\mu : \mathscr{A} \to \mathscr{G}^+$, $\mu(\alpha) = \Sigma_{[x,y] \in \alpha} f([x, y])$. This gives $\mu(\Psi \times \Psi/\sim) = (\phi^2 - \phi)/2$ and that is the Sergeyev number of chords in Q.[75] Rizza agrees

total number of discriminable chords is the infinitely large integer denoted by the term $(\phi^2 - \phi)/2$.

(2018:384)

The counting Sergeyev measure then plays the role of the normative measure in POIS, giving the probability measure on the name space $P = \mu/\mu(\Psi \times \Psi/\sim) = 2\mu/(\phi^2 - \phi)$.

75 The number of points in the lattice square $\Psi \times \Psi$ is ϕ^2 from which we subtract the ϕ in the diagonal of degenerate chords and then divide by two for the quotient space lattice points. Alternatively, and generally more useful for counting lattice points in a triangle is to use the formula $n(n+1)/2$ for the sum $1+2+3+...+n$. Since we exclude the ϕ in the diagonal then we are summing $1+2+3+...+\phi-1+\phi-\phi = 1+2+3+...+\phi-1 = (\phi-1)\phi/2$.

Chords, (x, y), are sides of inscribed equilateral triangles (from hereon, triangle chords) whenever $y = x + \phi/3$ or $y = x + 2\phi/3$ and therefore longer whenever $y > x + \phi/3$ or $y < x + 2\phi/3$, hence the turquoise area bounded by the triangle chord diagonals.

Rizza shows that there are ϕ triangle chords in $\Psi \times \Psi/\sim$,[76] giving the probability of triangle chords

$$P(e) = \frac{2\phi}{\phi^2 - \phi} = \frac{2}{\phi - 1}.$$

Similarly, he shows that there are $(\phi^2 - 3\phi)/3$ shorter chords giving the probability of shorter

$$P(s) = \frac{\phi^2 - 3\phi}{3} \frac{2}{\phi^2 - \phi} = \frac{2}{3} \frac{\phi - 3}{\phi - 1} = \frac{2}{3} - \frac{4}{3(\phi - 1)}.$$

Rizza then calculates the probability of longer. (2018:386)

$$P(l) = 1 - \left[P(s) + P(e) \right] = 1 - \left(\frac{2}{3} - \frac{4}{3(\phi - 1)} + \frac{2}{\phi - 1} \right) = \frac{1}{3} - \frac{2}{3(\phi - 1)}.$$

From the name space we can go more directly to the probability of longer via the number of lattice points in the longer area using the method in footnote 75, subtracting the number in the triangle bounded by and including the upper turquoise line from the number in the triangle bounded by and excluding the lower turquoise line:

$$\frac{1}{2}\left(\frac{2\phi}{3} - 1 \right)\frac{2\phi}{3} - \frac{1}{2}\frac{\phi}{3}\left(\frac{\phi}{3} + 1 \right) = \frac{\phi}{6}\left(\frac{4\phi}{3} - 2 - \frac{\phi}{3} - 1 \right) = \frac{\phi^2 - 3\phi}{6}$$

which is exactly half the number of shorter chords. Dividing by $(\phi^2 - \phi)/2$ gives the probability of longer.

$$P(l) = \frac{\phi^2 - 3\phi}{6} \frac{2}{\phi^2 - \phi} = \frac{1}{3}\frac{\phi - 3}{\phi - 1} = \frac{1}{3} - \frac{2}{3(\phi - 1)}$$

---

76 This is obvious from the name space. Being a lattice, there are as many points on the triangle chord diagonals bounding the longer chord area as on their projection on either axis.

6.10 Rizza's angle case

Rizza then turns to exhibiting Bertrand's procedures in his terms. Following Bertrand's treatment of the angle case, Rizza works out the probability of longer for chords sharing a single endpoint (see Figure 19).

This gives exactly the same result as before for the probability of longer, including the infinitesimal part. This is no surprise for us, since in doing this he is simply taking the $\lrcorner_1$ cross section of the naming space, as shown in Figure 20.

The number of chords in $\lrcorner_1$ is $\mu(\lrcorner_1)=\phi-1$ and likewise for any of the $\lrcorner$-cross sections,[77] which is $\mu(\Psi \times \Psi/\sim)$ scaled by $2/\phi$,[78] so these are uniform cross sections. As before, the shared endpoint is the degenerate chord named on the righthand diagonal edge of the quotient space. When we now take the Sergeyev measure μ restricted to $\lrcorner_1$ to be the normative measure in POIS, which is what Rizza is doing in his calculations, this explains why

> the... probability [for chords sharing the same endpoint] is the same as that obtained by the... method from the previous subsection... [because] under the given drawing procedure, the relative proportions of types of chords are preserved, although their numbers are scaled by the infinitesimal factor $2/\phi$.
>
> (2018:387)

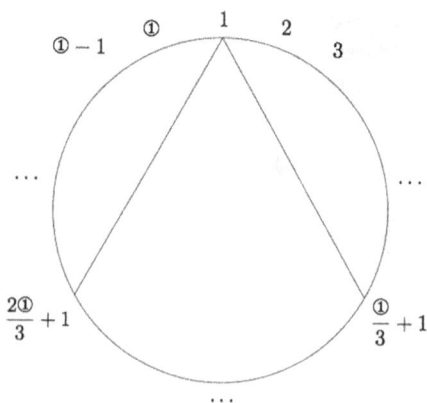

Figure 19 Rizza's figure 2. (2018:386).

---

77 The point $(1,1)$ at the bottom of $\lrcorner_1$ is not in the lattice since it names a degenerate chord. Likewise, $\lrcorner_x$ has x lattice points in its horizontal and $\phi-x$ in its vertical, giving ϕ in total, from which we subtract the degenerate chord (x, x).

78 There are ϕ cross sections with $\phi-1$ lattice points on each so $\phi(\phi-1)=2\mu(\Psi \times \Psi/\sim)$, double because a lattice point (x, y) gets counted twice, once on the on the vertical part of $\lrcorner_x$ and once on the horizontal part of $\lrcorner_y$.

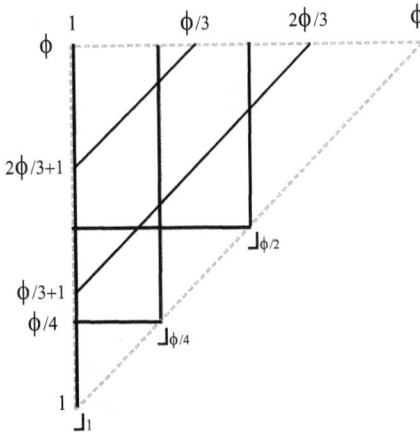

Figure 20 Angle cross sections in lattice name space.

As is evident from the cross section $\lrcorner_1$, we could go much more directly than does Rizza, who goes via counting the triangle chords and the shorter chords. The chords $(1, \phi/3+1)$ and $(1, 2\phi/3+1)$ are both triangle chords and hence the number of longer chords between them is the number of points between $\phi/3+1$ and $2\phi/3+1$ on the y-axis, which is $\phi/3-1$. Then

$$P(l) = \frac{1}{\phi-1}\frac{\phi-3}{3} = \frac{1}{3}\frac{\phi-3}{\phi-1} = \frac{1}{3}\left(1 - \frac{2}{\phi-1}\right) = \frac{1}{3} - \frac{2}{3(\phi-1)}.$$

The lattice name space shows why the answer comes out the same as angle case: if we replace ϕ by 2π on the axes we see the exactly similarity: essentially this lattice space is embedded in the complete name space on the chords in §3.11, Figure 10. The difference between them is the infinitesimal vertical and horizontal strips left out because the axes of the lattice space start at 1 whereas the axes of the naming space in §3.11 start at zero. The cross sections are essentially the same as those in Figure 11.

6.11 Rizza's direction case

Following Bertrand's treatment of the direction case, Rizza works out the probability of longer for chords sharing a single direction (see Figure 21).

In this case he gets a result for the probability of longer infinitesimally different from the other two results (2018:388–9):

$$P_2(l) + 1 - \left(P_2(s) + P_2(e)\right) = 1 - \left(\frac{2}{3} - \frac{8}{3(\phi-1)} + \frac{4}{\phi-1}\right) = \frac{1}{3} - \frac{4}{3(\phi-2)}.$$

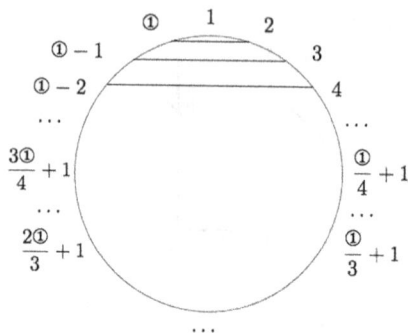

$$\text{①} \quad 1 \quad 2$$
$$\text{①} - 1 \qquad\qquad 3$$
$$\text{①} - 2 \qquad\qquad 4$$
$$\cdots \qquad\qquad \cdots$$
$$\frac{3\text{①}}{4} + 1 \qquad\qquad \frac{\text{①}}{4} + 1$$
$$\cdots \qquad\qquad \cdots$$
$$\frac{2\text{①}}{3} + 1 \qquad\qquad \frac{\text{①}}{3} + 1$$
$$\cdots$$

Figure 21 Rizza's figure 3 (2018:388).

We are now going to examine the cross sections of $\Psi \times \Psi/\sim$ defined by parallel chords and to do this I start by tabulating some points on those cross sections (see Table 1). The top row gives the $\Psi \times \Psi/\sim$ coordinates of a first chord of which the column below are parallel. Every chord, (x', y'), that is parallel to the first chord, (x, y), is defined by there existing $z \in \Psi$ such that $(x', y') = (x+z, y-z)$.[79] The first column is Rizza's cross section and the next two are mine. The difference between us is that mine start from a chord between consecutive points whereas he wanted to leave 1 at the top because he wants to talk of chords perpendicular to a diameter as Bertrand does. That doesn't really matter; what matters is that we have a cross section of parallel chords. As a result his sets of parallel chords have one fewer than mine, which we will eventually see makes a surprising difference. The last has more points calculated because the cross section is in two parts.

Table 1 Cross sections of $\Psi \times \Psi/\sim$ defined by parallel chords

Name and Colour of Cross Section in Figure 22	D_2 Pink Dot	D_1 Orange Solid	$D_{\phi/4}$, Purple Dash, in Two Parts
First chord	$(2, \phi)$	$(1, \phi)$	$(\phi/4, \phi/4+1)$ $(1, \phi/2)$
			$(\phi, \phi/2+1)$
Last chord	$(\phi/2, \phi/2+2)$	$(\phi/2, \phi/2+1)$	$(3\phi/4, 3\phi/4+1)$
Triangle chord	$(\phi/6+1, 5\phi/6+1)$	None	None
Triangle chord	$(\phi/3+1, 2\phi/3+1)$		

79 Recall $(x', y') \sim (y', x')$ because these are equivalence classes, and arithmetic on the names is modulo ϕ, facts that will be used in what follows without further mention.

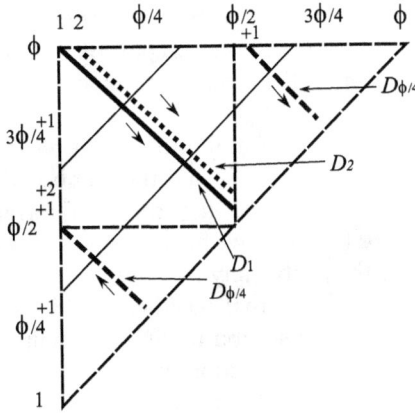

Figure 22 Parallel chord cross sections in lattice name space.

Now let us examine the lattice space in Figure 22. The little arrows show the progression from the first to the last in the parallel set names by the cross section. Although split in the diagram, the next chord in $D_{\phi/4}$ cross section after $(1, \phi/2)$ is $(1-1, \phi/2+1)=(\phi, \phi/2+1)$. So the split is an appearance of the representation: in the topology, they are joined.[80]

Being a lattice, there are as many points on the diagonal cross sections as on their projection on the axes. D_1 has $\phi/2$ chords whereas D_2 has $\phi/2-1$. This reduction by one corresponds to the number of points between the endpoints of the first chord in D_2 and the reduction carries on similarly for each extra point between the endpoints of the first chord of a cross section of parallel chords. All cross sections with $\phi/2$ chords must start or end with a lattice point on the diagonal line $y=x+1$ and apart from D_1 are in two parts like the example $D_{\phi/4}$. All cross sections with $\phi/2-1$ chords must start or end on the diagonal line $y=x+2$: in general for all $n\in\mathbb{N}$, all cross sections with $\phi/2-n$ chords must start or end on the diagonal line $y=x+n+1$. Let $D^n=\{$diagonal cross sections with $\phi/2-n$ chords$\}$. All members of D^n give the same probability of longer, which will differ only infinitesimally from the probability from any cross section in any D^m, $m\neq n$. D_1 and $D_{\phi/4}$ both belong to D^0 and we will calculate the probability of longer for D_1.

A triangle chord in D^0 on the cross section starting with chord $(b, b+1)$ is a chord (x, y) where $x=b+z$, $y=b+1-z$ and $x=y+\phi/3$ or $x=y+2\phi/3$.

---

80 What is going on is that this triangle lattice lives on a Möbius band with the diagonal including $(1, 1)$ forming its edge and the horizontal edge is joined to the vertical edge by a half twist, i.e. (ϕ, ϕ) to $(1, \phi)$ and $(1, \phi)$ to $(1, 1)$, with the distance along that half twist being the infinitesimal $1/\phi$. See picture in chapter appendix.

Solving these two sets of simultaneous equations gives $z = \frac{1}{2} + \phi/6$ or $z = \frac{1}{2} + \phi/3$. Since $b \in \Psi$, for these z the real part of $b + z$ is not an integer and therefore is not a member of Ψ. Hence no cross sections in D^0 have triangle chords.

$D_1 \in D^0$ so to calculate the probability of longer we need only count the number of chords that are longer, which is the number of lattice points with $1\frac{1}{2} + \phi/6 < x < 1\frac{1}{2} + \phi/3$, which is $\phi/6$. It is the same for any member of D^0, since starting with chord $(b, b+1)$, we have $b + \frac{1}{2} + \phi/6 < x < b + \frac{1}{2} + \phi/3$ giving the same result. Hence the probability of longer is $\phi/6 \div \phi/2 = \frac{1}{3}$. This is the surprising difference and shows that whether it was by luck or deliberate, Rizza had to choose his cross section to get the infinitesimal disagreement with Bertrand. In fact, any cross section in D^0 will give infinitesimal agreement with Bertrand's angle case and any cross section in D^n, $n \neq 0$, will not.

Rizza also treats the midpoint case with his name space. This gives a result infinitesimally different from the other cases. What he makes of that I will remark on next but apart from that I shall not address his treatment. We have now made clear enough of what he is doing to formulate my critique.

6.12 Rizza's premiss is false

Rizza's work is interesting and all very well so far as it goes. But it is no solution of the paradox. In this section I will show that his premiss is false. In the next two I shall demonstrate the failure of his solution in two ways. The first is that his solution works iff the formally identical solution works for the name space of §3.11, but the latter obviously doesn't work. The second will show that Rizza's Well-posing claim is false because I can demonstrate the exact recurrence of Bertrand's paradox in terms of Sergeyev's arithmetic.

Previously I gave

> *Rizza's premiss*: measure theory cannot supply a mathematical determination of the intuitive content of Bertrand's question.

Section 2 of his paper is where he explains this failure, drawing on the work of Rowbottom and Klyve. He tells us what he means by a mathematical determination by way of an example, whose features are to be extended to the paradox.

> consider... a fair die.... The probability model... adopted in this case is a uniform, [1] *discrete distribution* on the space of outcomes resulting from a throw. [2] *The totality of outcomes as well as the subset of relevant outcomes can be numerically specified and the numerical specifications can then be used to carry out computations of probability values.*

Bertrand's question about selecting a chord from the totality of all chords mirrors the character of the die problem in an infinite setting.... Following the template of the die model, such determinations should lead to the introduction of a uniform, [1] *discrete* distribution on the numerically specifiable totality of chords.

This approach is not viable if the canonical resources of probability theory are employed....[If] the totality of chords... is to be part of a workable probability model, a numerical estimate of its size, with which [3] *ordinary arithmetical computations* can be carried out, must be available. In other words, an intuitive feature of Bertrand's geometrical setup, i.e., the fact that a circle determines an infinite collection of chords, is to be assigned a mathematical determination, i.e., a [4] *numerical specification*, which cannot be offered in the canonical (i.e., measure-theoretic) context of probability theory.

(2018:377, my emphasis and numbering)

What is wrong with this is that in order to prove his premiss Rizza has helped himself to an unwarranted restriction on what it would be for measure theory to supply a mathematical determination of the intuitive content of Bertrand's question. I have numbered the critical elements on which I wish to comment.

It is initially difficult to see why he says that measure theory is unable to satisfy [3] and [4], since measure theory routinely assigns numbers to infinite sets. In the context, however, and recalling that Sergeyev numbers are spoken of as *integers*, what I understand by [4] is that a numerical specification assigns integers. This remains puzzling, since recalling TIM, there certainly will be ways of inducing a normative measure on the chords that assigns integers to at least some subsets. I think, however, that he means not only that integers are assigned but they are assigned by counting chords, in the way that applies to the case of the die and that we saw extended in his application of Sergeyev numbers to the chords.

With this in mind, we recognise [2] to be a very compressed version of the FPOIS and Theorem 7 and its corollaries with a certain modification and restriction of item 3 of FPOIS about the representation of events. Numerical specifications either are, or are defined by, a version of what we call *namings*, here being functions from the events to a mathematical space whose subsets can be assigned an integer by counting their contents. The counts can then define the probabilities similarly to Corollaries 8 or 9, that is dividing the count of the event of interest by the count of the totality of outcomes. Applied to our FPOIS as it stands, this amounts to a restriction on what measurable spaces are permitted for use in item 3 to those on which we have a counting measure. That this is what is intended is further supported by [1], the demand that we should end up with a discrete rather than continuous probability measure.

Now it is certainly true that with that restriction, the measure playing the normative measure role in POIS will be an infinite measure on the chords that therefore cannot be used in Corollaries 8 or 9 to produce a probability measure on the chords. So Rizza can have his premiss on these restrictions, but then his definition of mathematical determination is ten-dentious and irrelevant. Moreover, that restriction is arbitrary.

I granted Rizza's use of Sergeyev numbers not because they satisfy this restriction but only as a method of approximation. We need more than its success as such a method to justify the restriction that it satisfies. Prima facie, there is nothing about the intuitive content of Bertrand's question that justifies the restriction to counting measures for measures playing the normative measure role. Nor does that content justify the requirement that the probability measure on the chords should be a discrete measure. The only argument Rizza has given is the analogy with the die case, but that is a case in which the outcomes are countable whereas the chords are uncountable.

Rizza offers a different argument for his premiss that, in charity, I con-sider as independent of the restriction. He criticises Bertrand's square paradox:

> the sample space has changed from one scenario [choosing from 1 to 100] to the other [choosing from 1 to 10,000] and the question being answered is no longer the same (in the second case one is picking at random a number whose square root is greater than 50 and not a num-ber greater than 50). It is certainly possible to exchange a move to a different sample space with a move to a different distribution over the same sample space {1, 2, 3, ..., 100}, but the non-uniform distribution that gives rise to the probability value 3/4 has been manufactured out of the explicit consideration of a different problem.
>
> (2018:379–80)

Here he is pointing at a significant weakness in the square paradox, namely, that it works by covertly substituting a different event space.[81] He then directs our attention to the resource that undermines the paradox, namely,

> The problems in question here are easily distinguishable because suffi-cient numerical specifications are available to tell them apart.
>
> (2018:380)

---

81 A weakness that leads me in Chapter 13 to suggest the obvious improvement and which no doubt led van Fraassen to suggest a similar improvement that we will also see later.

Whether this really solves the square paradox can be questioned, since another way to run the square paradox is by pointing out that the correlation of numbers with their squares means we can rename the events in Bertrand's original case by their squares and we get distinct probabilities for the same events between their old and new names, contradicting Paris' Renaming principle.

Set that aside, however, since we are interested in what this example is supposed to show about Rizza's premiss. The numerical specification is that we have distinct integer name spaces for the distinct event spaces,[82] only one of which represents our equal ignorance of random choice. Hence his point about the manufacture by substitution of a different problem is made out.

By contrast

> This reduction is less straightforward in the context of Bertrand's geometrical problem because there are insufficient numerical resources to identify restrictions and effect discriminations.
>
> (2018:380)

It appears, then, that the lack in measure theory is that it fails to do for the chord paradox what can be done for the square paradox. It fails to distinguish, reveal or block a disguised elision of distinct event spaces.

My concern at this point is that Rizza may be running together two quite separate goals in what he wants out of a mathematical determination. The first is the accurate mathematical representation of the problem at hand, where the failure is

> certain intuitive features involved in [its] formulation cannot be codified by the mathematical apparatus canonically available.
>
> (2018:375)

The second is blocking paradoxical probabilities. It will not do to claim a failure at this second goal as a failure of mathematical determination for no other reason than that the mathematical apparatus doesn't provide a unique answer. That amounts to claiming uniqueness as a criterion of identity of the principle of indifference. That claim is a way of running the Irrelevance strategy and in Chapter 12 I will show what is wrong with it.

The contrast he draws for the success or failure of mathematical determination between the square paradox and the chord paradox is ambiguous between these two goals. It is ambiguous because in the square case the

---

82 In each case the naming is the identity, i.e. {1, 2, 3,...,100} names itself and {1,..., 10, 000} names itself.

claimed accurate representation 'reveals' the origin of the paradoxical probabilities to be an inaccurate representation via a renaming of the event space by the different event space of squares.

Yes, if our mathematical apparatus conflates the event space at issue with a different one, it is failing at accurate representation, which failure then has a further effect of producing paradoxical probabilities. In the square paradox, he is suggesting that the paradox makes that conflation but that, when applied correctly, the mathematical apparatus is capable of revealing and blocking that conflation.

He then offers Sergeyev's arithmetic of infinity as something doing the same job for the chord paradox, a job which the canonical tools of measure theory fail to do. In that case, what is supposed to be going on is that his application of Sergeyev's arithmetic of infinity gives *the* correct (approximate) naming of the chords and also determines the count measure which produces consistent probabilities of longer.[83]

So the failure of measure theory is supposed to be that, unlike using Sergeyev's arithmetic, it covertly elides distinct measure spaces and this is the origin of the paradoxical probabilities. Of course, there is a sense in which this is true, since as I analysed earlier, Bertrand's procedure derives the probabilities from measures on events spaces of properties of the chords rather than the chords themselves. So his angle, direction and midpoint spaces are indeed distinct event spaces and none of them is the event space of the chords, even though each is correlated in a certain way with the chords, just as the squares are correlated in a certain way with the their square roots.

I therefore recognise this as a restatement of frailty number 2 from §3.2. This was one of the reasons that I did the work on distinguishing event spaces and name spaces. In Chapter 3 I showed precisely how measure theory can resolve all the frailties of Bertrand's procedure and fully represent Bertrand's question. I cannot myself see what intuitive content of the question Chapter 3 has left without a mathematical determination. The vagueness of the latter means that having done the work of Chapter 3, I can only present it and challenge Rizza to state exactly what intuitive content of the question is without a mathematical determination.

Absent a reply, and on this unrestricted sense of mathematical determination, Rizza's premiss is false. That it is false means the paradox recurs in an interesting way, namely, the more than infinitesimal difference between his treatment of the direction case using approximation by Sergeyev's

---

83 Or at least, *a* correct approximate naming, without loss of generality. Obviously only infinitesimal differences in probabilities turn on whether we had used 2ϕ or ϕ^2 circumferential points instead, for example.

arithmetic and ours in Chapter 3 using standard measure theory: he gives the probability of longer as $\dfrac{1}{3} - \dfrac{4}{3(\phi - 2)}$ and I give ½.

6.13 A new source of Bertrand's paradox

I now do the same job that Rizza did for the direction case by determining the cross sections for parallel chords in the $[0, 2\pi] \times [0, 2\pi]/\sim$ name space used for the angle case. Recall that in this space I name the chords by the equivalence class of the pair of angular components of the polar coordinates of their endpoints, for example if a chord has endpoints (R, α), (R, β) it is named by $[(\alpha, \beta)]$, and once again I refer to equivalence classes by either member. For all $\alpha \in [0, \pi]$ the chords perpendicular to the diameter with endpoints $(\alpha, \alpha + \pi)$ have endpoints $(\alpha + \theta, \alpha - \theta)$ [mod 2π], $\theta \in [0, \pi]$. I now tabulate in Table 2 similar cases to those I did for Rizza's treatment and exhibit the cross sections in the space in Figure 23.

As before, longer chords are between the two turquoise lines. The cross sections of parallel chords are uniform, each having length $\sqrt{2}\pi$ and the length of the cross section between its triangle chords is $\sqrt{2}\pi/3$. So each such cross section gives the probability of longer for the direction case as 1/3, the same as for the angle case.

The point of this example is straightforward. This is not a solution to the paradox. Yet formally we have done nothing different from what Rizza did. His lattice is embedded (roughly speaking, since we would need to add hyper-reals) in our quotient space by replacing ϕ by 2π in purely rational members of Ψ.[84] To claim this as a solution we would be claiming that only the $[0, 2\pi] \times [0, 2\pi]/\sim$ name space is the only correct one. That, however, is sheer ad hoccery.

Table 2 Cross sections for parallel chords in the $[0, 2\pi] \times [0, 2\pi]/\sim$

Name and Colour of Cross Section in Figure 23	$D_{\pi/4}$ Pink Dot, in Two Parts	D_0 Orange Solid	$D_{\pi/2}$ Purple Dash, in Two Parts
First chord	$(\pi/4, \pi/4)$ $(0, \pi/2)$ $(\pi/2, 2\pi)$	$(0, 0)$ $(\pi/2, 3\pi/2)$	$(\pi/2, \pi/2)$ $(0, \pi)$ $(\pi, 2\pi)$
Last chord	$(5\pi/4, 5\pi/4)$	(π, π)	$(3\pi/2, 3\pi/2)$
Triangle chord	$(7\pi/12, 23\pi/12)$	$(\pi/3, 5\pi/3)$	$(7\pi/6, 11\pi/6)$
Triangle chord	$(11\pi/12, 19\pi/12)$	$(2\pi/3, 4\pi/3)$	$(\pi/6, 5\pi/6)$

84 I.e. we define an embedding by mapping all coordinates $(q\phi, r\phi) \in \Psi \times \Psi/\sim$: $q, r \in \mathbb{Q} \cap [0,1]$ to $(2q\pi, 2r\pi)$ and then fill in the infinitesimals between.

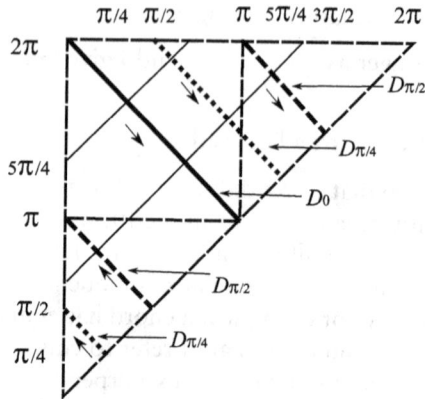

Figure 23 $[0, 2\pi] \times [0, 2\pi]/\sim$ name space with cross sections of parallel chords.

What justifies a name space as correct is that it represents equal igno-rance in a definable way. In this case it represents equal ignorance over the endpoints of the chords (and the further quotient space we developed from this quotient space represented equal ignorance over tangent angles). But does it also represent equal ignorance over directions, as does the $[-R, R] \times [0,\pi)$ name space of Figure 13?

If in fact $[0, 2\pi] \times [0, 2\pi]/\sim$ *does* represent equal ignorance over direc-tions, as Rizza's solution requires us to say, that would not improve mat-ters at all. Now we have two *different* name spaces[85] both representing the *same* equal ignorance and yet producing paradoxical probabilities of lon-ger. The paradox has turned out to be much stronger than we ever thought. This would not be merely a recurrence of the paradox but an entirely new source of Bertrand's paradox, since now *not even a single variety of equal ignorance*, namely equal ignorance over a single property, produces consis-tent probabilities.

6.14 Rizza's Well-posing claim is false

Rizza does nothing to show that his applications of Sergeyev's arithmetic of infinity represent any particular equal ignorance. He concerns himself only with naming a subset of the chords, which I have allowed to be an approximate naming of the chords, and producing from that naming an approximation of the probability of longer, which approximations are only infinitesimally different from the probability of longer for the angle case. He is merely assuming that his naming the chords with Sergeyev

---

85 I.e the names spaces $[-R, R] \times [0,\pi)$ and $[0, 2\pi] \times [0, 2\pi]/\sim$.

numbers is the only way to do so.[86] That is equivalent to me claiming that the naming by the $[0, 2\pi] \times [0, 2\pi]/\sim$ name space is the only correct way of using measure theory.

Rizza's name space, $\Psi \times \Psi/\sim$, and my $[0, 2\pi] \times [0, 2\pi]/\sim$ are both justifiable, since our equal ignorance over the chords means we are equally ignorant over their endpoints. But we are also equally ignorant over directions, which is what justified the $[-R, R] \times [0, \pi)$ name space. I shall now complete the refutation of Rizza's solution by showing that we can use Sergeyev numbers in the analogous way.

Doing so is straightforward. Recall that for the direction case we name the chords by the directional distance of their midpoint from the centre, which is the distance along the diameter to which they are perpendicular and the angle of that diameter. To produce the analogous approximation using Sergeyev numbers we divide each diameter into infinitesimal segments by $2\phi + 1$ points, named by Sergeyev numbers from $-\phi$ to ϕ and we divide the angle π into infinitesimal arcs named by Sergeyev numbers from 0 to ϕ. This gives the (approximate) name space $(-\phi, \phi) \times [0, \phi)$ for the chords in Figure 24.

The cross sections of parallel chords are uniform, each having $2\phi - 1$ points (following Rizza, we exclude the degenerate chords which are on the vertical cross sections $-\phi$ and ϕ). The triangle chords are on the vertical cross sections at $-\phi/2$ and $\phi/2$. The $I_{\phi/2}$ cross section, for example, has the horizontal chords in it, and between the triangle chords, $(-\phi/2, \phi/2)$ and $(\phi/2, \phi/2)$, it has $\phi - 1$ longer chords. We calculate the probability of longer for the cross section directly.

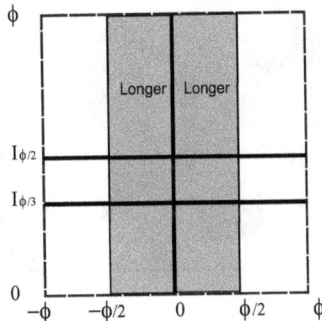

Figure 24 $[-\phi, \phi] \times [0, \phi)$ name space: cross sections of parallel chords in horizontal lines.

---

86 Without loss of generality up to multiples and powers of ϕ, as explained in footnote 84.

$$P(longer) = \frac{\phi - 1}{2\phi - 1} = \frac{\frac{1}{2}(2\phi - 1) - \frac{1}{2}}{2\phi - 1} = \frac{1}{2} - \frac{1}{4\phi - 1}$$

Calculating the probability of longer for the whole name space, we have $\phi(2\phi - 1)$ chords in total and $\phi(\phi - 1)$ longer chords, giving the probability of longer $= \phi(\phi - 1)/\phi(2\phi - 1)$, which gives the same answer as for the cross section because the ϕ cancels. This is infinitesimally different from ½ and is inconsistent with Rizza's probability that is infinitesimally different from ⅓.

Well-posing Requires a Principle, and Rizza has offered us his principle, that Sergeyev's arithmetic of infinity applied to the chords is a mathematical determination of the intuitive content of Bertrand's question, claiming that its application determines a unique probability of longer. We can apply Sergeyev numbers to give a name space with a counting Sergeyev measure that represents equal ignorance over direction just as well as we can apply them to give a name space with a counting Sergeyev measure that represents equal ignorance over endpoints. Doing so proves he hasn't avoided the paradoxical probabilities. Rizza's Well-posing claim is false and he has fallen for the *Well-posing Danger*. In applying his principle Rizza assumes that naming the endpoints is the only correct approximate naming using Sergeyev numbers. In doing so he covertly restricts Bertrand's question rather than comprehending the general problem.

Appendix

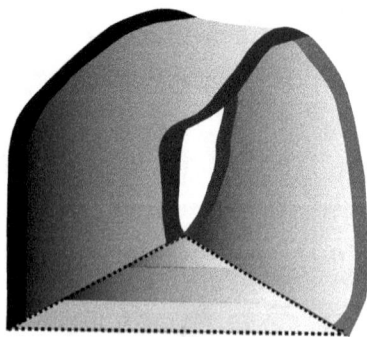

Figure 25 Name space lattice $\Psi \times \Psi/\sim$ lives on a Möbius band. The strip with the half-twist is infinitesimally long.

7 The Irrelevance Strategy

7.1 Introduction

Having looked at prominent instances of the Distinction and Well-posing strategies, I now turn to the Irrelevance strategy. I start with an examination of a proposal from Bangu, essentially that the error in thinking Bertrand's paradox can falsify the principle of indifference is the assumption that randomness transmits across entailment. I then turn to the current main contender for a successful instance of the Irrelevance strategy. It is also the second of the radical mathematical treatments that I examine. Gyenis and Rédei's classical interpretation of probability is expounded in Gyenis and Rédei 2015a, b, 2016, most completely in the first, 'Defusing Bertrand's Paradox', which will therefore be our focus. It is the most up to date and mathematically sophisticated development of classical probabilism and they claim that they show how to save the principle of indifference from refutation by Bertrand's paradox.

The authors formulate their classical probabilism in terms of compact topological groups and their principle of indifference in terms of the correlate group actions 'express[ing] epistemological indifference'. Supplementing their interpretation with a distinction between what they call labelling invariance and labelling irrelevance,[87] they claim thereby to show that the paradoxes that have troubled classical probabilism are defused.

According to labelling invariance, relabelling events doesn't change their probabilities. They claim that Bertrand's paradox is an example of, and hence a proof of, 'the violation of labelling invariance', but that under their interpretation this shows Bertrand's paradox should be diagnosed as nothing more than a Lebesgue measure case of a more general theorem from measure theory.

---

87 Their terminology is explained below as is its relation to Paris' Renaming principle.

DOI: 10.4324/9781003456308-7

They reject labelling invariance as a mistaken formulation of what really matters, which is labelling irrelevance. This distinction between what does and doesn't matter seems to open the possibility of solving the problems that paradoxes pose classical probabilism. We need only impose philosophically respectable conditions that rule out the relabellings that cause the trouble for labelling invariance whilst demonstrating that we can preserve labelling irrelevance.

They conclude that 'under the interpretation of Bertrand's paradox suggested in the paper, the paradox does not undermine either the principle of indifference or the classical interpretation' (2015a:349). Rather, their analysis diagnoses the common understanding of Bertrand's paradox to be a matter of mistaking a mathematical fact for a paradox. Once we are clear that it is labelling irrelevance rather than labelling invariance that matters, we can reject the paradox. Gyenis and Rédei thereby claim to have defused Bertrand's paradox and solved the philosophical problem that has undermined classical probabilism.

We recognise an instance of the irrelevance strategy. The error is mistaking the violation of labelling invariance to matter because we confuse labelling invariance, which doesn't matter, with labelling irrelevance, which does. The paradox is an instance of the violation of labelling invariance and its appearance of refuting the principle is due to the error. The inference to line 5 of the Rigorous argument, that the answer to Bertrand's question is single, rests on that error. Removing the error both blocks the inference and shows that the paradox is irrelevant to the truth of the principle.

Gyenis and Rédei's proposal is an elegant deployment of some elegant mathematics. It is the most interesting proposal that has been made in aid of classical probabilism for a long time. To have shown that Bertrand's paradox fails in its philosophical import would amount to offering us a consistent classical probabilism for continua of events. So if successful, Gyenis and Rédei have a big achievement on their hands and their classical interpretation could fairly be claimed to be the current state of the art for applying probability to continua.

In this chapter, I show that Gyenis and Rédei's classical probabilism does not succeed in defusing Bertrand's paradox. Having outlined their classical probabilism (§§7.3–7.5) I turn to what Gyenis and Rédei call an interpretation of Bertrand's paradox, their General Bertrand's paradox. The latter's relation to the square paradox is apparent but its relation to the chord paradox is not explained and left obscure, leaving it unclear how any proposed solution of General Bertrand's paradox is relevant (§7.6). Nevertheless, given the failure of others to solve Bertrand's paradox and given the intrinsic interest in Gyenis and Rédei's proposal, the question arises of whether their classical probabilism succeeds in its own terms. Furthermore, if it does succeed in its own terms I discern a route to extend that success to cover Bertrand's chord paradox, and, despite them giving no explanation or even mention of that extension, that may have been their intention all along. So I examine their interpretation of Bertrand's paradox and their diagnosis and defusal (§§7.7–7.8). I then show that the use of their distinction incurs the significant philosophical cost of rejecting Paris' Renaming principle (§7.10) and, more importantly, their crucial distinction between labelling invariance and labelling irrelevance collapses (§7.11). As a result, their entire diagnosis and defusal of Bertrand's paradox collapses (§7.12). All in all, Gyenis and Rédei's classical interpretation of probability fails because it is internally inconsistent with their declared criterion of consistency, and they do not show the Rigorous argument from the paradox to the falsity of the principle of indifference (§4.3) to be unsound. Rather, their attempt at the Irrelevance strategy by defusal fails and leaves Bertrand's paradox ready to detonate.

7.2 Bangu on transmission of randomness failure

Bangu claims that Bertrand's paradox requires an assumption about the transmission of randomness across a certain kind of entailment. Bangu only discusses the square paradox and his approach does not apply to the chord paradox. I will lay out his approach in some detail and then indicate briefly why it doesn't apply and how his point might be extended to the chord paradox.

For the square paradox, the assumption is that choosing at random in $[0, 100]$ is choosing at random in $[0, 10,000]$, just because if $x \in [0, 100]$ then $x^2 \in [0, 10,000]$. On this assumption, and applying the principle of indifference, the probability of $y \in [0, 100]$, $y \geq 50$ should be the same whether we calculate it from choosing it in $[0, 100]$ or choosing its square in $[0, 10,000]$. The principle gives a uniform distribution across each, giving probability of ½ in the first case and ¾ in the second and so it is inconsistent. But, says Bangu, randomness need not transmit across the conditional. Yes, $x \in [0, 100]$ entails $x^2 \in [0, 10,000]$ but that doesn't mean that if x was chosen randomly then x^2 was chosen randomly. In fact, to assume that it does is an error.

The paradoxical probabilities that are taken to refute the principle of indifference only arise if we make that assumption. So yes, Bertrand's question, here for the square paradox, produces his paradox but the appearance of falsifying the principle of indifference is founded on this erroneous assumption about the transmission of randomness. Due to this error, the inference to line 13 of the Rigorous argument (modified for the square paradox) is invalid. Thus is Bangu's proposal an instance of the Irrelevance strategy.

Having started from the square paradox Bangu treats the question in general terms:

> One begins with a variable x ... and then one considers a scaling transformation θ such that $x' = \theta(x)$.

(2010:31)

Without loss of generality, we will assume that θ is monotonic increasing. Running the argument against the principle of indifference in these terms we have

1. Suppose the principle of indifference is true.
2. $[x,y] \subset [a,b]$ entails $[\theta(x),\theta(y)] \subset [\theta(a),\theta(b)]$.
3. Randomness transmits across event entailment.
4. Therefore choosing at random in the interval $[a,b]$ is choosing at random in the interval $[\theta(a),\theta(b)]$. (2, 3)
5. Therefore the probability of the event $[x,y] \subset [a,b]$ is the same as the probability of the event $[\theta(x),\theta(y)] \subset [\theta(a),\theta(b)]$. (4)
6. By the principle of indifference

$$P([x,y]) = \frac{y-x}{b-a} \qquad P\big([\theta(x)],[\theta(y)]\big) = \frac{\theta(y)-\theta(x)}{\theta(b)-\theta(a)} \qquad (1)$$

7. The probabilities of line 6 are inconsistent iff θ is a non-linear function (from Corollary **13**, see below).
8. θ in the square paradox is the non-linear squaring function.
9. The probabilities of line 6 are inconsistent. (7, 8)
10. Therefore the principle of indifference is false. (1, 9, RAA)

Proof of line 7

The events are named by intervals in $\mathbb{R}$ and θ is a renaming. We have taken the normative measures in POIS to be the Lebesgue measure on $\mathbb{R}$ restricted to the intervals, μ on $[a,b]$ and ν on $[\theta(a),\theta(b)]$. Let a single event e be named by both $[x,y]$ and $[\theta(x),\theta(y)]$. θ defines the function f between measures μ and ν for Corollary **13** thus: $f(\mu(e)) = \theta(y) - \theta(x) = \nu(e)$. f is linear iff θ is linear because for constant c, $\nu(e) = c\mu(e)$ iff $\theta(y) - \theta(x) = c(y-x) = cy - cx$ iff for all z, $\theta(z) = cz$.[88]

---

[88] For the reason given in the proof of Corollary **13**, for linear function $f(z) = cz + b$, $b = 0$ because μ and ν are measures.

Bangu explicitly rejects line 4, and therefore Premiss 3, because it assumes that

> R: If the argument of a scaling function is random in an interval, then the scaled value is random as well (in the scaled interval).
>
> (2010:33)

Of course, there is a trivial case under which R is false: if θ is a constant function. Bangu, however, needs it to be false no matter what kind of scaling transformation θ is.

Bangu does not address the chord paradox but only those that rely on Corollary 13 in the way just shown. The analysis conducted in Chapter 13 will show that the chord paradox does *not* work on the basis of Corollary 13. This is why I gave a more general premiss 3 from which the assumption Bangu rejects is entailed.

We return to Bertrand's own procedures to illustrate how Bangu might extend his critique to the chord paradox. We might say, for example, that just because the event of the angle with the tangent being in a certain range entails the event of the distance from the centre being in a certain range, we cannot assume that randomness transmits across that entailment. But Bertrand's procedure (and the Rigorous argument) assumes that it does when it finds fault in the disagreement between probabilities for the angle case and the direction case. Hence does the apparent falsification of the principle by the chord paradox rely on this erroneous assumption. I shall not take this further, since we can address the rejection as Bangu has formulated it against line 4 and the generalization to a defence of Premiss 3 will then be obvious.

Bangu takes unpredictability 'to be the standard conception of randomness' and argues against R:

> Suppose a machine, a random number generator, picks a value x in the interval [a,b].... The machine records it, but does not communicate it to us. We now ask whether the transformed value is also random in the interval [a',b'].... One might say that this is not so, as there is a crucial difference between the value of x and the transformed value. While we cannot predict the value of x, we can predict the value of the transformed: we find what value has been recorded, and we scale it. So, is the transformed value random in [a',b'] after all? Or, is the sense of 'prediction' not sufficiently well defined to be useful here?
>
> (2010:33)

Bangu's argument here rests on a premiss that is patently false. In a perfectly straightforward sense (namely prior to the machine selecting), the transformed value is no more predictable than the original. What he calls the transformed value being predictable is the situation of being given the value the machine picked and using that to predict, for example, the square by squaring it. But of course, in that sense of predictable the value the machine picked is precisely as predictable! Take the value the machine picked as the prediction of the value the machine picked.

Grant that randomness should be understood in terms of unpredictability. When we are considering the question of the truth of R, what is at issue is whether if x is not predictable, $\theta(x)$ is not predictable. Bangu's argument seems directed, instead, at whether if x is not predictable but we know x, $\theta(x)$ is not predictable. Obviously, the unpredictability of x will not carry over to $\theta(x)$ once we know x. But that is irrelevant.

The truth of R can be illustrated by an elementary example. Let X be the random variable that takes its values from the roll of a fair die. Let $\Phi(X)$ be the random variable that is zero in the event that X is not 1, and one in the event that X is 1. Φ is known. Even though the image of Φ is smaller than its domain, $\Phi(X)$ is no more predictable than X. In order to see this, consider a sequence of values of X from such a die roll and the corresponding values of $\Phi(X)$:

$$S_X \quad 1,2,1,1,5,3,3,1,4,1,6,4,2,\ldots x_n$$
$$S_\Phi \quad 1,0,1,1,0,0,0,1,0,1,0,0,0,\ldots \Phi\left(x_n\right)$$

If either of these sequences is predictable then knowing some number of the previous members of the sequence suffices for knowing the next one. Suppose that S_X is unpredictable. That means that for all n, knowing $\langle x_1,\ldots, x_n\rangle$ does not suffice for knowing x_{n+1}. Now we consider the predictability of S_Φ. We can suppose that we know $\langle x_1, \ldots, x_n\rangle$ and we know $\langle \Phi(x_1),\ldots, \Phi(x_n)\rangle$, and because S_X is unpredictable we do not know x_{n+1}. So for all we know x_{n+1} could be 1 or 6 and so $\Phi(x_{n+1})$ could be 1 or 0 and hence we do not know $\Phi(x_{n+1})$. So knowing $\langle \Phi(x_1), \ldots, \Phi(x_n)\rangle$ does not suffice for knowing $\Phi(x_{n+1})$ and so S_Φ is unpredictable. Hence if X is unpredictable, $\Phi(X)$ is unpredictable, whence if X is random, $\Phi(X)$ is random.

In short, knowledge of how to map S_X onto S_Φ does not render S_Φ any more predictable than S_X, and that is all that is required for the truth of R in this example. This argument is in fact general. It neither depends on the discreteness of the event spaces nor on Φ. It applies to a wide range of functions. Its conclusion is equivalent to the assumption that Bangu rejects.

This argument generalizes immediately to cover Premiss 3. The relation of event entailment between X and $\Phi(X)$ is defined by a map. The argument given proves the transmission of randomness from any map's argument to the value taken by the map for that argument, provided only that the map is not a constant map. Consequently, since any kind of event entailment defines a map from the entailing events to the entailed events, the argument proves that randomness transmits across event entailment, provided only that the event entailment is not constant.

7.3 Gyenis and Rédei's classical interpretation

I now turn to the current main contender for a successful instance of the Irrelevance strategy. In summary, Gyenis and Rédei's classical interpretation of probability (GR-classical probabilism hereafter) and its defence from paradox consists of:

1. their definitions of their classical interpretation and their principle of indifference;
2. their definitions of 'labellings', 'relabellings', 'labelling invariance' and 'labelling irrelevance';
3. their proposal that labelling irrelevance is what matters and that it is not equivalent to labelling invariance;
4. their 'encoding' of labelling irrelevance, their application to probabilistic modelling and their 'non-trivial strategy';
5. their diagnosis and defusal of Bertrand's paradox.

The items in (4) are not immediately relevant to the concerns of this chapter but will come back into view in Chapter 13, so we will be examining only the other four.

Gyenis and Rédei define thus their

> General classical interpretation: If X is a compact topological group, then the probabilities of the events are given by the Haar measure p_H on (the Borel sets of) X, and (frequency link:) the numbers $p_u(A)$ [sic][89] will be (approximately) equal to the relative frequency of A occurring in a series of trials producing elementary random events from X.

> General principle of indifference: If X is a compact topological group and if the group action expresses epistemological indifference about the elementary random events in X, then the general classical interpretation is correct.

> (2015a:356)

Taken on its face, this is not a principle of indifference at all. It is, rather, a statement in the philosophy of probability stating a (rather peculiar) sufficient condition for the truth of the general classical interpretation as

---

89 Should be $p_H(A)$.

defined. The antecedent is plainly insufficient for the truth of the frequency link but this is irrelevant to our concerns so we can grant that for the sake of argument and omit it from hereon.[90]

Together the definitions entail a proposition closer to a principle of indifference as normally formulated and in charity we take this entailment to be the:

> *GR-principle of indifference*: If X is a compact topological group and if the group action expresses epistemological indifference about the elementary random events in X, then the probabilities of the events are given by the Haar measure on X.

Gyenis and Rédei tend to elide the distinction between the events and the elements in the topology, X, which label those events, in part because we can sometimes take the events to *be* elements in X which can therefore label themselves. Whenever I speak of events or elements in X in what follows, I am exploiting the same ambiguity as do they, that is speaking of elementary random events *in or labelled by X*. Where it matters, I will draw the distinction.

Gyenis and Rédei don't explain why this amounts to a principle of indifference. To make it comprehensible, I am going to give a quick summary of how it works before I explain the technicalities. What is essentially going on here is that we assume we are epistemologically indifferent between the elementary random events in X, that is in our terms, we have no reason to discriminate between the elements of X. The group action *expresses* that epistemological indifference by *mapping* elements of X between which we are epistemologically indifferent to each other. The Haar measures of a compact topological group are defined by the group and give equal measures to the elementary events, so giving equal measures to elementary events between which we have no reason to discriminate. The virtue of the Haar measures of a compact topological group is that they are *unique up to a multiplicative constant*. It is this special feature that creates the promise of consistent probabilities for continua. When used to play the normative measure role in the POIS they all produce the same probability measure, which probability measure is what Gyenis and Rédei mean by '*the* Haar measure on X'. Note that in so speaking they have not noticed the *Need-a-Measure-to-Get-a-Measure* principle and conflated the two roles.

---

90 See chapter appendix for their concern with this link.

Technically, each member of the group, G, of the compact topology, X,[91] is a bijection mapping X to itself, that is, is a permutation of X. The group action is the 'sum' of those permutations in the sense of being a function $G \times X \to X$, $(g,x) \to g(x)$. So for each permutation, g, and for each element, $x \in X$, the group action maps that element to its value $g(x)$ under the permutation.

The central idea behind 'the group action express[ing] epistemological indifference about the elementary random events in X', is that if we have no reason to discriminate between elementary events then we have no reason to discriminate between a permutation of them. Hence permutations preserve epistemological indifference about elementary random events. Consequently, the group action of compact topological group X expressing epistemological indifference about elementary random events amounts to this:

GR-Representation Lemma 30

Given a labelling of a set of elementary random events, E, by a compact topological group, X, we are epistemologically indifferent between a pair of elementary random events iff their labels in X are mapped to one another by the group action on X.

Given a topology on X, the σ-algebra generated by the open sets of that topology is called its Borel algebra, named S by Gyenis and Rédei. Haar (1933) showed that if X is a compact topological group,[92] then there are certain measures on X, now called Haar measures, that are determined by the group and are multiples of one another. Consequently, if we normalize these measures, we end up with one and the same probability measure, the Haar measure on X of which Gyenis and Rédei speak.

So in this way Gyenis and Rédei have made use of the special normalized measure spaces, (X, S, p_H), that satisfy Kolmogorov's axioms and that are defined by normalizing any of the Haar measures on a compact topological group X. The group action that represents epistemic indifference, in determining the Haar measure, provides their principle of indifference to satisfy Salmon's Ascertainability criterion.

---

91 For Gyenis and Rédei, X is itself a group acting on itself, but here I want to keep track of the group and what the group acts on. For a fuller explanation of the relevant group theory see §13.7.

92 The weaker condition of being locally compact suffices, but Gyenis and Rédei need their Haar measure to be finite on the whole space, so in a locally compact group they will have to confine their space to a compact subset.

7.4 A strength and a weakness of GR-classical probabilism

Gyenis and Rédei call their interpretation classical in part because groups
are the mathematical way of treating symmetry. They could, if they chose,
give up the claim to be offering an interpretation of probability, in the
standard sense of telling us what probabilities are. The theory as just explained could be equally well adopted by non-classical interpretations as
the correct one to make use of when dealing with continua among whose
atomic events we have no epistemic reason to distinguish, when that equality of epistemic status can be represented by a compact topological group
and its Haar measures.

The power of the proposal is the combination of two things. The uniqueness of the normalized Haar measure for a given compact topological
group apparently blocks paradoxical probabilities, provided using the
Haar measure can be justified as the right measure rather than being ad
hoc.[93] That the Haar measure is defined by, and therefore internally related
to, the group is what justifies it as the right measure.

For example, let us replace Bertrand's square paradox with the random
square machine producing random squares with areas up to 4 square feet.
What is the probability of getting a random square of 1 square foot? Equal
ignorance over the side lengths gives the answer ½ but over the areas gives
answer ¼. The Haar measure seems to give a solution. If the area measure
is a Haar measure of squares then that justifies the area measure as the correct measure, giving a unique probability of ¼.

If Gyenis and Rédei's interpretation is confined to compact topological
groups, their proposal would be too narrow a satisfaction of Salmon's
Ascertainability criterion to be of much use. Although the spaces $\mathbb{R}^n$ are
locally compact topological groups, they are not compact, their Haar measures are infinite measures and so useless for defining probabilities over $\mathbb{R}^n$.
The spaces of most common interest for probabilities, intervals of $\mathbb{R}^n$
(products of n intervals of $\mathbb{R}$), are *not* topological *groups*.

One can induce a measure on an interval from the Haar Measure space
of a compact topological group using TIM, which Gyenis and Rédei do in
their appendix.[94] So doing exhibits a satisfying consilience. For intervals,
the equal ignorance for events is represented by the Lebesgue measures of
those intervals. The normalized measures induced by TIM on intervals
from compact topological groups are identical with the normalized
Lebesgue measures. For example, the measure Gyenis and Rédei induce
from the Haar measure space of a compact topological group on an

93 See van Fraassen briefly moot this virtue of Haar measures. (van Fraassen 1989:316).
94 Despite knowing my earlier paper in which I first introduced TIM, they appear not to
  realize they could simplify their proof by using it. See chapter appendix.

interval [0,1) is its Lebesgue measure, which is also the standard normal measure for that interval, equivalent to the uniform probability distribution on [0,1). So the proposal is consilient with our expectation of what the principle of indifference applied to [0,1) gives.

Nevertheless, it has to be said that we have at this point met a weakness of their proposal, a weakness that Gyenis and Rédei obscure when they say that

> the normalized restrictions of the Lebesgue measure on [intervals in] $\mathbb{R}^n$ can also be regarded as Haar measures... to see how the Lebesgue measure on [a,b) is a Haar measure in its own right, it is enough to see how the (normalized) Lebesgue measure... on the interval [0,1), emerges as a Haar measure.
>
> (2015a:370)

What they offer as proof of their claim is a function $f:[0,1) \rightarrow S^1$, where S^1 is the unit circle (a compact topological group) and a proof that f is a measure isomorphism between the Lebesgue measure on [0,1) and the Haar measure of S^1.[95] But being measure isomorphic doesn't mean the Lebesgue measure *is* a Haar measure, for the simple reason that the interval is not a compact topological group.

So regarding the Lebesgue measure as a Haar measure no more makes it a Haar measure than regarding a giraffe as a hedgehog makes it a hedgehog. Now here is the problem. Since the measure is *not* a Haar measure *we have lost the justification for the Lebesgue measure being the unique correct measure*. For example, the mooted solution of the square factory paradox fails because the area is *not* a Haar measure, even if it can be induced from the Haar measure of $S^1 \times S^1$.

One might attempt to rescue this by claiming that homeomorphism retains the internality of relation and hence retains the justification. The claim would have to be that the square is homeomorphic to $S^1 \times S^1$. This is false, but nearly true, since $S^1 \times S^1$ is homeomorphic to a quotient space of the square (we saw the homeomorphism in §3.11). So one can define a group on the quotient space of the square in terms of the group of $S^1 \times S^1$. The doubt that then arises is over whether the group so defined can really count as internally related to the *square* when the homeomorphism is to its quotient. If not, the resultant Haar measure is only externally related to the square, and so the justification remains lost.

If Gyenis and Rédei's classical probabilism is not to be hobbled by narrowness of application we need some warrant for why a measure induced

---

95 What this also proves is that the measure induced on [0,1) by using f and the Haar measure on S^1 in TIM is the Lebesgue measure on [0,1).

from a Haar measure inherits the Haar measure's justification for being the correct measure. In the chapter appendix I go into greater depth on why I think Gyenis and Rédei's construction fails at that warrant. I then offer a construction that does a better job, at least in my opinion, because the compact topological groups from which measures are induced on intervals in my construction are internally related to those intervals in a way that Gyenis and Rédei's groups are not. Whether it is a good enough job can still be questioned.

I will not take this discussion further. For now I grant the premiss that a measure induced from a Haar measure inherits the Haar measure's justification of being the correct measure. We are interested in whether, granted that premiss, they have solved the chord paradox.

7.5 Labelling, labelling irrelevance, relabelling and labelling invariance

Labellings are not explicitly defined by Gyenis and Rédei. At the beginning of their discussion of probability modelling they say

> the elements in the pair of sets $(X, S,)$ label the (elementary, respectively, general) random events of some random phenomenon.

> (2015a:366)

Labellings that are obviously presupposed in Gyenis and Rédei's paper are compact topological groups labelling themselves with the identity function. Whatever they do mean by a labelling isn't guaranteed to be the same as a naming as defined in §2.6, since there we defined the σ-algebra by the σ-algebra of events but here they take the σ-algebra to be the Borel algebra on X. Of course, when they are taking the events *to be* members of the Borel algebra and using the identify function then their labelling would be a naming. Beyond that we cannot say.

Labelling irrelevance 'states that labelling of random events does not affect their probabilities' (2015a:357). So labelling irrelevance is not a definition of the mathematical representation of anything but a proposition in the philosophy of probability stating a criterion of probabilistic consistency on any labelling of events.

Gyenis and Rédei define a *relabelling* to be

> if (X, S, p_H) and (X', S', p'_H) are two probability spaces describing the same phenomenon, then the map $h{:}X{\rightarrow}X'$ is called a relabelling if it is a bijection ... and both h and its inverse h^{-1} are measurable, i.e....:

$$h[A] \in S' \text{ for all } A \in S \tag{4}$$

$$h^{-1}[B] \in S \text{ for all } B \in S' \tag{5}$$

(2015a:357)

They define

> *Labelling invariance*... probabilities... are invariant with respect to re-naming; that is to say, if (X, S, p_H) and (X', S', p'_H) are two probability spaces and h is a relabelling between X and X', then it holds that
>
> $$p'_H(h[A]) = p_H(A) \text{ for all } A \in S \tag{6}$$
>
> $$p_H(h^{-1}[A']) = p'_H(A') \text{ for all } A' \in S'. \tag{7}$$

(2015a:358)

The conditions of lines (6) and (7) mean that
Labelling invariance can therefore be expressed compactly by saying

> *Labelling invariance*: Any relabelling between probability spaces (X, S, p_H) and (X', S', p'_H) is an isomorphism between these probability spaces.

(2015a:358)

Here, by 'isomorphism' they mean a probability isomorphism.

Whenever I use any of this terminology (labelling, labelling irrelevance, relabelling and labelling invariance) and especially when I give arguments in these terms, I abjure any intuitive meanings and mean only and precisely what Gyenis and Rédei have defined them to mean.

7.6 Misrepresenting Bertrand's paradox

Gyenis and Rédei offer what they call 'a new interpretation of Bertrand's paradox':

> General Bertrand's paradox: Let (X, S, p_H) and (X', S', p'_H) be probability spaces with compact topological groups X and X' having an infinite number of elements and p_H, p'_H being the respective Haar measures on the Borel σ-algebras S and S' of X, and X'.

(2015a:359)

170 *The Irrelevance Strategy*

This quotation is not incomplete but is the entirety of their *definition* of their General Bertrand's paradox. They then state the existence of paradoxical probabilities

Then Labelling Invariance does not hold for (X, S, p_H) and (X', S', p'_H) in the sense that

- either there is no re-labelling between X and X';
- or, if there is a re-labelling between X and X', then there also exists a re-labelling that violates Labelling Invariance.[96]

(2015a:359)

We will examine the consequence in the next section. For now, we concern ourselves with their 'interpretation' of Bertrand's paradox.

On its face, General Bertrand's paradox has nothing to do with Bertrand's chord paradox. Nowhere in the paper has it been explained how this 'interpretation' of Bertrand's paradox is supposed to interpret the paradox, and, indeed, nowhere in the paper is the set of chords even mentioned. The closest we get to some relevant remark is

Both the original Bertrand's paradox and all of the simplified versions of it take the normalized restriction of the Lebesgue measure to some bounded, compact sets in R^n ... as the measure that expresses the principle of indifference.

(2015a:356)

I think this manifests a failure to see the difference between Bertrand's chord paradox and his square paradox, in which we do at least have two Lebesgue-derived probability spaces on events.[97] It appears that Gyenis and Rédei think they have covered the chord paradox because Bertrand's treatment uses Lebesgue-derived probability measures for the angle, direction and midpoint property spaces, and so those probability measures can play that role in their General Bertrand's paradox. Nevertheless, Bertrand's spaces are not compact topological groups and so do not have Haar measures. Consequently, Gyenis and Rédei would have to rely on the Theorem of Induced Measure.[98]

---

96 In their later paper the definition is united with inconsistent probabilities in 'Proposition 1 (General Bertrand Paradox)' (Gyenis and Rédei 2015b:268).
97 $([0,100], \mathscr{L} \cap [0,100], \lambda|_{[0,100]}/100)$ and $([0,10,000], \mathscr{L} \cap [0,10,000], \lambda|_{[0,10,000]}/10,000)$.
98 The functions in TIM would be compositions of the functions α, ρ and κ in Chapter 3 with quotienting functions and then functions to compact topological groups. See chapter appendix.

It must be conceded that Gyenis and Rédei are not alone in classifying the chord and square paradoxes together, and Bertrand is perhaps to blame for this, since it requires seeing the significance of the frailties of his procedure to see the necessity of distinguishing them. Given that necessity, it is evident that using functions from the chord properties to compact topological groups, whilst in line with their actual methods, is not what their classical probabilism promises, which requires a compact topological group *of chords*.

Gyenis and Rédei do not attempt to apply any Haar measures to Bertrand's chord paradox. Insofar as we can identify how they might do so, there seem to be only two ways. The first, which I think best fits their methods and their General Bertrand's paradox, would be to treat it entirely in terms of relabellings by appropriate compact topological groups of the measure spaces from Bertrand's procedure.[99] Relabellings are possible since these spaces have the same cardinality and therefore bijections between them, and they would have to ignore the third frailty, that none of Bertrand's procedure uses a naming. In this way we turn the probabilistic inconsistency for the events of Bertrand's paradox into an instance of the violation of labelling invariance, thereby fitting with their remark:

> the original Bertrand's paradox... claimed that there exist Haar measures and re-labellings that violate labelling invariance.
>
> (2015a:360)

The other way GR-classical probabilism can be applied to Bertrand's paradox is by determining the compact topological group for which the set of chords is the base set. $\mathbb{C}$ is not a topology, however. Yes, we can define a topology on $\mathbb{C}$. For example,

$U \subseteq \mathbb{C}$ is open in $\mathbb{C}$ iff $\{(x,y) \in R^2 : \exists c \in U : (x,y)$ is the centre point of chord $c\}$ is open in R^2.

Then the Haar measure on $\mathbb{C}$ will be defined by the Haar measure on the closed disk, induced on it as in the appendix if necessary, which is the same as the Lebesgue measure on that disk.[100] Evidently this is just a complicated a way of using the Theorem of Induced Measure, and it gives Bertrand's treatment of the midpoint case. All well and good, so far as it goes, but the problem is that each of Bertrand's *other* ways of determining probabilities can be used to induce a different compact topological group on the chords, resulting in

---

99 See chapter appendix.
100 To make the disk a group and get a Haar measure we might have to take the quotient of the disk by its boundary, which quotient is homeomorphic to the sphere, S^2.

different probabilities for the longer chord.[101] Consequently, not only have Gyenis and Rédei failed to address Bertrand's chord paradox, but their classical probabilism cannot solve it *by the Well-posing strategy*.

Since Gyenis and Rédei have shown no relation between their General Bertrand's paradox and Bertrand's chord paradox, at this point it might be thought we need go no further. Perhaps, but we shall see in the next section that their General Bertrand's paradox is at least a general version of paradoxes like Bertrand's square paradox (such as the cube paradox and the water/wine paradox—more on these in a later chapter) and their instance of the *Irrelevance strategy* is the first new approach in a long time. A classical probabilism that can produce probabilities for continua without falling foul of these paradoxes would be a significant restoration in its own right. We must therefore consider whether GR-classical probabilism succeeds in its own terms. In so doing we will then see, in charity, a different route by which their Irrelevance strategy might apply to Bertrand's chord paradox.

7.7 The General Bertrand's paradox and labelling invariance violation

Having defined their General Bertrand's paradox, Gyenis and Rédei then make use of this theorem:

Proposition 1

(van Douwen 1984; Rudin 1993) If X is an infinite, compact topological group with the Haar measure, p_H, on the Borel σ algebra S of X, then there exists an autohomeomorphism θ of X and an open set E in S such that $p_h(\theta[E]) \neq p_h(E)$.

(2015a:359)

Applying it to the General Bertrand's paradox, they demonstrate that

if there is at all a relabelling between two probability spaces (X, S, p_H) and (X', S', p'_H) with infinite X and X', then there is also a relabelling between these spaces that violates labelling invariance, and for any space (X, S, p_H) with an infinite X, there exists a space (namely itself) and a self-relabelling of (X, S, p_H) that violates labelling invariance.

(2015a:360)

---

101 For Bertrand's angle case we can define a topology on C by the topology on [0, π) but that would not be a topological group. Instead we would have to use a function C→[0, π)→S¹ and use the topology and group of to S¹ define the topology and group on C. Likewise for the direction case.

Labelling invariance is violated by one and the same event under an original labelling and its relabelling having different Haar measures under each and hence inconsistent probabilities, contradicting line (6) of their definition of labelling invariance (above and 2015a:358). This is what they mean by the claim

> Thus Bertrand's 1888 paradox can be viewed as the specific 'Lebesgue measure case' of a mathematical theorem that was only proved in full generality in 1993.
>
> (2015a:360)

Earlier, Gyenis and Rédei tell us what they claim to show about Bertrand's paradox:

> The interpretation proposed here should make clear that Bertrand's paradox cannot be 'resolved'—not because it is an unresolvable genuine paradox but because there is nothing to be resolved: the 'paradox' simply states a provable nontrivial mathematical fact, a fact that is perfectly in line both with the correct intuition about how probability theory should be used to model phenomena and with how probability theory is applied in the sciences.
>
> (2015a:350)

So this is an error theory of the paradox: we misunderstand it and think we must solve it but we can instead reject it altogether because it is a mere restatement of a fact 'perfectly in line' with the relevant issues in the philosophy of probability. This quotation makes clear that they are pursuing the Irrelevance strategy.

At this point I have also brought into view the different route I mooted previously, the route by which they may intend to reject rather than resolve Bertrand's chord paradox. As they present it, they do not intend to confine the import of the error theory to the General Bertrand's paradox but intend it to be understood to apply equally well to Bertrand's chord paradox. If so, they leave this extension of the error theory entirely unexplained and unargued.

7.8 Gyenis and Rédei's diagnosis and defusal of Bertrand's paradox

Now we turn to the proposed diagnosis of Bertrand's paradox, the diagnosis on which basis GR-classical probabilism is supposed to defuse the danger the General Bertrand's paradox poses to the principle of indifference, thereby leaving in place a consistent theory, namely, GR-classical probabilism.

The essential point of their diagnosis is the distinction between labelling invariance and labelling irrelevance. It is *labelling irrelevance* that has 'conceptual importance' because

> its violation would entail a radical ambiguity and arbitrariness in assigning probabilities to random events... and... is obviously incompatible with any interpretation of probability that treats probability as an objective feature of the world. Furthermore, violation of labelling irrelevance would make subjective degrees of belief vulnerable to a Dutch book.
>
> (2015a:357)

Labelling irrelevance captures what they called earlier 'the correct intuition about how probability theory should be used' (2015a:350) and so must be held to rather than violated if GR-classical probabilism is to be consistent. Then, because

> labelling invariance entails [labelling irrelevance], one expects labelling invariance to hold for the general classical interpretation as well.
>
> (2015a:357)

This expectation is *the* mistake that is made in inferring the inconsistency of classical probabilism from the General Bertrand's paradox. Yes, were labelling irrelevance to entail labelling invariance, the violation of the latter by the paradox would entail the violation of the former, and hence entail the inconsistency of GR-classical probabilism. BUT

> *labelling invariance and labelling irrelevance are not equivalent.*
>
> (2015a:357 my emphasis)

Mistakenly thinking they are equivalent leads us into error:

> because of this conflation of labelling invariance and labelling irrelevance ... violation of labelling invariance in the category of Haar measure spaces (i.e. General Bertrand's paradox) appears paradoxical.
>
> (2015a:366)

It is a mere appearance, however. Yes,

> Bertrand's paradox can be viewed as a proof that labelling invariance does not hold in the general classical interpretation.
>
> (2015a:357)

But that doesn't matter because

> It should be emphasized... that this [violation of labelling invariance] does not entail that labelling irrelevance cannot be maintained in general, nor does it follow from violation of labelling invariance that the classical interpretation based on the principle of indifference is inconsistent.
>
> (2015a:357)

So we do not need to solve the General Bertrand's paradox. We can acknowledge its truth as a mathematical theorem but because it does not threaten what matters we simply reject it as irrelevant to the question of the consistency of GR-classical probabilism and the truth of the GR-principle of indifference.

So the diagnosis and defusal is this:

Diagnosis: Labelling irrelevance is the relevant criterion for the consistency of probabilisms with respect to labellings. Labelling invariance is not a criterion of consistency. The General Bertrand's paradox proves that GR-classical probabilism violates labelling invariance. We mistakenly conflate labelling irrelevance and labelling invariance. From this error springs the further error of mistaking the violation of labelling invariance as sufficient to prove the inconsistency of GR-classical probabilism and the falsity of the GR-principle of indifference. Rather, since labelling irrelevance, not labelling invariance, is the criterion, what is required for that proof is for labelling irrelevance to entail labelling invariance. Only then would the violation of the latter entail the violation of the former. But there is no such entailment so the proof of inconsistency and of the falsity of the principle is blocked.

Defusal: Since labelling irrelevance does not entail labelling invariance, the General Bertrand's paradox does not violate the relevant criterion of consistency. Consequently the General Bertrand's paradox is irrelevant to the question of the consistency of GR-classical probabilism and can therefore be rejected rather than resolved.

General Bertrand's paradox thus defused is the basis on which it is claimed that Bertrand's paradox is defused. Although they do not analyse the application to Bertrand's chord paradox, on their behalf we can sketch how it might go. We treat each of Bertrand's original cases as relabellings of each other. Bijections are available because each of $[0, \pi)$, $(-R, R)$ and D have the same cardinality. Such relabelling is how the General Bertrand's paradox works, but since we have shown that the consequent violation of relabelling invariance is irrelevant to the GR-principle of indifference, so too is the violation by Bertrand's paradox—and indeed, so are all the paradoxes because they can all be treated in terms of relabellings (we proved this for renamings in the Paradoxical Probabilities Equivalence theorem **12**). Consequently we can reject rather than solve Bertrand's chord paradox.

7.9 Gyenis and Rédei's Irrelevance strategy

Putting this all together, Gyenis and Rédei start with about as strong an assertion that Bertrand's question produces his paradox as is possible: namely, that General Bertrand's paradox produces paradoxical probabilities for their classical probabilism is in fact a theorem of measure theory! It is the theorem that relabelling invariance is violated, a corollary of van Douwen's theorem. However, the claim that the paradoxical probabilities refute the GR-principle of indifference is founded on an error, namely, the assumption that what matters, labelling irrelevance, entails relabelling invariance. That assumption is false and consequently the violation of relabelling invariance does not imply the violation of what matters. Therefore, although Bertrand's question produces his paradox, the appearance of falsifying the principle of indifference is founded on the error of confusing the consistency criterion that matters with one that doesn't. This mistake means that the inferences from Premisses 2 and 10 to lines 5 and 13 of the Rigorous argument are invalid. The paradox is therefore irrelevant to the truth of the principle and so can be rejected. Thus do we have Gyenis and Rédei's instance of the Irrelevance strategy.

7.10 Committed to rejecting Paris' Renaming principle

I now turn to demonstrating what is going wrong. The first thing is to show that rejecting labelling invariance amounts to rejecting Paris' Renaming principle. To do this I show that Paris' Renaming principle, when applied to events named by topological groups, implies Gyenis and Rédei's labelling invariance.

We must remember that Paris' renamings and Gyenis and Rédei's relabellings are of one and the same events, not changes in naming or labelling due to changes in the events of interest. That is why Gyenis and Rédei's definition of relabelling says '(X, S, p_H) and (X', S', p'_H) are two probability spaces describing *the same phenomenon*' (2015a:357 my emphasis). So we are not talking about reassigning names from a representation of, for example, poker hands to a representation of bridge hands, but reassigning a representation of poker hands by X to a representation of poker hands by X'.

To avoid confusion with my definition of a renaming in Chapter 1, I shall speak of a Paris-renaming. I have to do this because we cannot assume that Gyenis and Rédei's labellings are namings and therefore cannot assume that a relabelling is a renaming.

To express Paris-renaming when topological groups are used for labelling events we consider two probability spaces on topological groups (X, S, p) and (X', S', p'), where the first already labels the events. We Paris-rename by replacing the labels from the first topological group with labels from the second. The labels of the elementary events are from the topological group itself, that is are labelled by $x \in X$ or by $\{x\} \in S$, and labels of the general events are from the σ-algebra, that is are labelled by $s \in S$. If (X', S', p') is to represent the same phenomenon then the new labels of the elementary events come from X' and those of the general events come from S'.

We therefore define a Paris-renaming by a function $g: X \rightarrow X'$. Assuming labels must be unique, g must be an injection. The image of X under g, $g(X)$, is a subset of X' and if it is a proper subset we can confine our second space to the image. So without loss of generality, we can assume $g(X) = X'$ and hence Paris-renamings are defined by bijections $g: X \rightarrow X'$.[102]

The Paris-renaming mustn't change probabilities so in particular it mustn't result in events lacking probabilities. Hence

$$g(s) \in S' \text{ for all } s \in S \tag{4'}$$

The image of S under g, $g(S)$, is a σ-algebra on X' and if it is proper subset of S' we can confine our second space accordingly. So without loss of generality, we can assume $g(S) = S'$.[103] Since g is a bijection

$$g^{-1}(s') \in S \text{ for all } s' \in S' \tag{5'}$$

Comparing the numbered lines with the similar numbered lines of Gyenis and Rédei's definition of relabelling (above and 2015a:357) shows that any Paris-renaming between topological groups describing the same phenomenon is a Gyenis and Rédei relabelling, and vice versa.

So a Gyenis and Rédei relabelling is what you get if and only if you apply a Paris-renaming to the topological groups of GR-classical probabilism. Paris' Renaming principle, in requiring that a Paris-renaming does not change probabilities of events, entails that a Paris-renaming preserves probabilities, which is to say that a Paris-renaming is a probability isomorphism

---

102 This is so *a forteriori* if '(X, S, p_H) and (X', S', p'_H) are two probability spaces describing the same phenomenon' as in Gyenis and Rédei's definition of relabelling.

103 Once again, *a forteriori* if as in the preceding footnote.

(Renaming Theorem 4). But a Paris-renaming of topological groups is equivalent to a Gyenis and Rédei relabelling. Therefore, Paris' Renaming principle entails that any Gyenis and Rédei relabelling is a probability isomorphism, which last just is what Gyenis and Rédei call the compact expression of labelling invariance (above and 2015a:358). Hence Paris' Renaming principle implies Gyenis and Rédei's labelling invariance.

Consequently, it follows that in saying 'labelling invariance does not hold in the general classical interpretation' (2015a:357), Gyenis and Rédei are thereby conceding that GR-classical probabilism contravenes Paris' Renaming principle. Gyenis and Rédei have not given us any reason to reject Paris' Renaming principle except that if we reject it the problem General Bertrand's paradox poses GR-classical probabilism goes away; i.e. the rejection is ad hoc. Nor will the claim that it is only labelling irrelevance that matters help, since that is question begging if taken in defence of *this* rejection. We can agree that labelling irrelevance matters, but why does *only* labelling irrelevance matter?

Worse, Paris' Renaming principle is an entailment of what we might call the *Naming Irrelevance principle*: that naming shouldn't affect probabilities. Obviously, the last is awfully close to labelling irrelevance. I won't take this further here, since in the next section the looming difficulty emerges entirely in Gyenis and Rédei's own terms.

So merely rejecting the conflation of labelling invariance and labelling irrelevance is not enough. Some reason must be given for rejecting Paris' Renaming principle. The Renaming principle, however, is *hard* to reject: it looks as if it expresses an internal relation between probabilities of events and names of events, which is to say, it is another *a priori* principle about the nature of probability. So their unprincipled rejection is a significant philosophical cost of GR-classical probabilism.

7.11 Distinction between labelling irrelevance and invariance collapses

Gyenis and Rédei note that 'labelling invariance... entails... labelling irrelevance' (2015a:357) and that entailment is what is supposed to *mislead* us into thinking the converse entailment true. It is absolutely critical to the philosophical project of rejecting the General Bertrand's paradoxes, and thereby all the others, that the converse should *not* hold. For if labelling irrelevance *should* entail labelling invariance, then the violation of labelling invariance proved by their Proposition 1 entails the violation of labelling irrelevance, and then their defusal fails entirely. I shall now prove that labelling irrelevance entails labelling invariance.

Suppose Gyenis and Rédei labelling irrelevance is true, that is 'labelling of random events does not affect their probabilities' (2015a:357). By the definition of a relabelling, each of the probability spaces involved already describes the same random events, which have therefore already been labelled by the two topological groups. By labelling irrelevance, those original labellings do not affect the probabilities, so the same event receives the same probability under each labelling. A relabelling is just a function between the original labels and hence it produces a new labelling of the random event by the second topological group (which might be, but need not be, the same labelling as the original labelling by the second topological group). By labelling irrelevance, the new labelling doesn't affect the probabilities either, since it is just another labelling. So no relabelling affects the probabilities of an event. That relabellings of the same events does not affect their probabilities means that relabellings preserve probabilities (since not preserving probabilities is affecting the probabilities). The definition of labelling invariance is that relabellings preserve probabilities.[104] So labelling irrelevance entails labelling invariance.

Here is an example (purely to give a feel for what is going on—proof follows) using the finite event space of 6 sides of a die: We can label each side by a unique label in $\{0, 1, 2, 3, 4, 5\}$ or the same sides in the same order by $\{0, 12, 24, 36, 48, 60\}$.[105] The function $x \rightarrow 12x$ relabels the first by the second and produces a new labelling that is in the same order as the original labelling by the second. By contrast, $x \rightarrow 12(5 - x)$ produces a new labelling in reverse order from the original. By labelling irrelevance, neither the original nor the new labellings affect the probabilities. But if the probability of one side of the die is different under the new labelling by $\{0, 12, 24, 36, 48, 60\}$ from its probability under the original labelling by $\{0, 12, 24, 36, 48, 60\}$, then the labelling would have affected the probability. That is ruled out by labelling irrelevance so the relabelling preserves probabilities.

An objection I received to a presentation of the argument of the penultimate paragraph is that it is merely intuitive, based on intuitions about 'labelling invariance' and 'labelling irrelevance', whereas Gyenis and Rédei are explicitly challenging any such intuitive relation. This is difficult to see when the argument uses the terms as defined by Gyenis and Rédei (see §7.5), supplemented only by a phrase that names the conditions in labelling invariance (see Footnote 104). I gave the aforementioned argument because it is perhaps more illuminating than the mathematical proof, which to still any reservations I now give.

---

104 The phrase 'preserving probabilities' is the shorthand for the conditions (c) and (d) introduced in the proof of Renaming theorem 4, which are the same as Conditions (6) and (7) in Gyenis and Rédei's definition of labelling invariance (2015a:358).

105 Compact groups under addition modulo 6 and 72 respectively.

Labelling Irrelevance Entails Labelling Invariance Theorem 31

If 'labelling of random events does not affect their probabilities' (2015a:357) then any Gyenis and Rédei relabelling satisfies Gyenis and Rédei's definition of labelling invariance.

Proof: Let $\mathscr{E}$ be the set of all events of the phenomenon of interest, f a labelling to a probability space (X, S, p_H) and f' a second labelling to a probability space (X',S',p'_H). Consider any event, $e \in \mathscr{E}$, whether elementary or general, and its labels $f(e)$ and $f'(e)$.

According to Gyenis and Rédei labelling irrelevance, labelling of e does not affect the probability of e. Suppose $p'_H(f'(e)) \neq p_H(f(e))$. That would mean that the second labelling f' affected the probability of e, contradicting labelling irrelevance. Hence for all events e in E and all labellings, f, f', of $\mathscr{E}$

$$p'_H\big(f'(e)\big) = p_H\big(f(e)\big) \tag{1}$$

Now consider any Gyenis and Rédei relabelling $h{:}X{\to}X'$. Since h is a bijection the function $h{\circ}f{:}\mathscr{E}{\to}X'$ is just another labelling to which (1) applies, so letting $f'=h{\circ}f$ we have

$$p'_H\Big(h\big(f(e)\big)\Big) = p'_H\big(f'(e)\big) = p_H\big(f(e)\big) \tag{2}$$

Likewise, $h^{-1}{\circ}f'{:}\mathscr{E}{\to}X$ is a labelling, namely, the labelling $h^{-1}{\circ}f'=h^{-1}{\circ}h{\circ}f=f$, and so again

$$p_H\Big(h^{-1}\big(f'(e)\big)\Big) = p_H\big(f(e)\big) = p'_H\big(f'(e)\big) \tag{3}$$

Finally, since labellings are injections for all $A \in S$ and $A' \in S'$ there exist unique events d and e respectively such that $A=f(d)$ and $A'=f'(e)$. Consequently

$$p'_H\big(h(A)\big) = p'_H\Big(h\big(f(d)\big)\Big) = p_H\big(f(d)\big) = p_H\big(A\big) \tag{6'}$$

$$p_H\big(h^{-1}(A')\big) = p_H\Big(h^{-1}\big(f'(e)\big)\Big) = p'_H\big(f'(e)\big) = p'_H\big(A'\big) \tag{7'}$$

using (2) and (3) for the middle equality of (6') and (7') respectively. The latter are numbered for ease of comparison with Conditions (6) and (7) of Gyenis and Rédei's definition of labelling invariance (above and 2015a:358). By inspection, Conditions (6') and (6), and Conditions (7') and (7), are identical. QED.

Remark: Note that although I used the notation for Gyenis and Rédei's Haar measure spaces so that inspection shows the identity of conditions, the proof applies to all probability spaces. This is what we should expect, since Gyenis and Rédei's labelling irrelevance is evidently a general principle in the philosophy of probability applying to all probability measures and Gyenis and

Rédei's definitions of relabelling and labelling invariance applies to any probability spaces.

Finally, having proved this, it is difficult to see why Gyenis and Rédei aren't additionally committed to this entailment by their claim that

> labelling irrelevance is encoded in the notion of isomorphism of probability spaces.
>
> (2015a:366)

The proof just given proves that labelling irrelevance entails relabellings are probability isomorphisms. It is difficult to see what adjustments they could make to what they mean by labelling irrelevance being 'encoded' by probability isomorphism that could simultaneously evade the proof just given whilst supporting the encoding claim.

7.12 A defence?

A defence I received to a presentation of the preceding section is that, since labelling irrelevance is equivalent to labelling invariance, which last has been shown to false by the theorem in measure theory (*Proposition 1* above and 2015a:359), labelling irrelevance is also false. But I must be committed to the truth of labelling irrelevance. So if I don't want what I have shown to have that undesirable consequence I must deny labelling invariance fails and this means that I have to show that the aforementioned mathematical theorem is false. But obviously if I have to do that something must have gone badly wrong, since the theorem is proved.

The measure theoretic theorem, *Proposition 1*, shows that an auto-homeomorphism (=a relabelling) doesn't preserve the Haar measure (= failure of labelling invariance). That doesn't matter within mathematics because mathematics alone does not require autohomeomorphisms of Haar measures to be measure isomorphisms. It is only when, in GR-classical probabilism, we take the Haar measure to represent *probabilities*, understood to be that quantitative feature, whatever it is, that we are talking about when we say events or propositions are probable, that the theorem causes problems. But it doesn't cause them for mathematics: it causes them for GR-classical probabilism.

There are principles about probability, principles grounded in philosophical considerations about the nature of probability, which GR-classical probabilism must satisfy. Those principles therefore impose additional mathematical constraints that the Haar measure should satisfy insofar as it is to represent probability successfully. I showed in §7.10 that labelling invariance is for Gyenis and Rédei the correct formal representation of

Paris' Renaming principle. In rejecting labelling invariance, GR-classical probabilism rejects Paris' Renaming principle, holding instead that labelling irrelevance is the true principle.

It is only now, from the definition of GR-classical probabilism, that *Proposition 1*, in entailing some homeomorphisms don't preserve the Haar measure, entails that some relabellings of the General Bertrand's paradox don't preserve probabilities, thereby violating labelling invariance. Consequently GR-classical probabilism, in violating labelling invariance, violates labelling irrelevance, thereby violating one of its own commitments. My work is done at this point and it does not depend on my taking a position on labelling irrelevance, only on the fact that GR-classical probabilism does.

Why, then, is there an appearance of something going wrong? Because the defence offered is misrepresenting the dialectical situation. Yes, if labelling irrelevance is true, the violation of labelling irrelevance means we have to give up something, but the defence is supposing that there are only two premises that might be given up (labelling irrelevance and the measure theoretic theorem) when in fact there are three: the measure theoretic theorem, labelling irrelevance *and* GR-classical probabilism. If labelling irrelevance is true, its combination with the measure theoretic theorem shows GR-classical probabilism to be false. The defence, therefore, assumes the truth of GR-classical probabilism and thereby begs the question. I no more have to show the measure theoretic theorem false than Bertrand had to show ½ ≠ ¼ false when he originally put forward his chord paradox.

7.13 Not defusal but detonation

Gyenis and Rédei sum up their diagnosis and defusal of GR-classical probabilism at the end of their paper:

> Bertrand's paradox shows violation of labelling invariance. That is, relabellings are not necessarily [probability] isomorphisms in the category of Haar probability measure spaces with infinite random events. But the violation of labelling invariance does not undermine the classical interpretation of probability understood with the principle of indifference.... violation of labelling invariance does not entail violation of labelling irrelevance.

(2015a:369)

We have now seen that the final sentence is false (§7.11). This is more serious than it may appear. It refutes the entire diagnosis and defusal.

According to Gyenis and Rédei, labelling irrelevance is the criterion of consistency that matters and hence is a normative commitment of GR-classical probabilism. Since labelling irrelevance entails labelling invariance,

GR-classical probabilism, having been proved to violate labelling invariance, violates labelling irrelevance as well. Hence Gyenis and Rédei's classical probabilism contradicts its own normative commitment and is therefore internally inconsistent.

Gyenis and Rédei's diagnosis depended on it being an error to think the General Bertrand's paradox violates labelling irrelevance because it violates labelling invariance—but it is not an error. The defusal depended on the General Bertrand's paradox being irrelevant to the consistency of GR-classical probabilism and the GR-principle of indifference, and therefore fit for rejection, because it doesn't violate the relevant consistency criterion (labelling irrelevance)—but it does violate the relevant criterion and so is not rejectable. Nothing is left of the diagnosis and defusal and hence it cannot be extended to Bertrand's chord paradox either. Consequently they have no basis left for their claim that 'Bertrand's paradox ... does not undermine the classical interpretation' (2015a:349). Gyenis and Rédei's Irrelevance strategy has failed. Their attempt at defusal leaves Bertrand's paradox ready to detonate.

7.14 Appendix

7.14.1 Gyenis and Rédei's remarks on relative frequency seem to be about clarifying their commitment to the ontology of probabilities.

> Without such an interpretive link, the classical interpretation is not an interpretation of probability at all: the numbers... are just pure, simple mathematical relations. There are two standard interpretive links: the frequency link and the degree of belief link. We restrict the discussion to the frequency link because the classical interpretation emerged historically and was formulated on the basis of this link.
>
> (2015a:353)

It may appear they intend their classical probabilism to be a variety of frequentism. They deny this, saying that the future tense 'will' in their frequency link statement means that

> it is this reference for future random trials that distinguishes the classical interpretation (with the frequency link) from the frequency interpretation, in which the ensemble of elementary random events determining A's relative frequency must be specified before one can talk about probabilities.
>
> (2015a:353)

There is some obscurity in thinking the future tense can achieve this: after all, for long-run frequentists much of what determines the probability is in the future. Nevertheless, I take it that they are suggesting that they could, if they so choose, advance their classical probabilism as being ontologically objective but distinct from frequentism because they take the opposite direction of explanation over the relation of the laws of large numbers and relative frequencies approximating probabilities. Having said this much, I leave aside this question since it is not germane to us here.

7.14.2 In their appendix Gyenis and Rédei define probabilities on intervals of $\mathbb{R}^n$ by the Haar measure of the $T^n = S^1 \times \ldots \times S^1$ product topology. They claim that they thereby show 'the Lebesgue measure... [to be] a Haar measure in its own right'(2015a:370). In fact, when their classical probabilism is applied to spaces other than compact topological groups, TIM is not simply unavoidable but *strongly* unavoidable. This is because what they want is a probability measure on the space and what they have is a normal Haar measure on a *different* space so they *must* use TIM to induce the measure they want from the measure they've got. For brevity we are going to discuss their treatment of the unit interval in $\mathbb{R}$ because whatever they achieve for the unit interval generalizes to intervals of $\mathbb{R}^n$, so we omit the merely algebraic complications necessary to cover the latter.

Intervals are compact but they are not groups in the reals and so they do not have Haar measures. Gyenis and Rédei's attribution of Haar measures to them is therefore doubtful. The basis of the attribution is the proof that the bijection $f:[0,1) \to S^1$, $f(x) = e^{2\pi i x}$ is a measure isomorphism between the Lebesgue measure on $[0,1)$ and the normalized Haar measure of S^1. A proof of this can also supplied by more straightforwardly using their function f in the Theorem of Induced Measure. We take the measure space in that theorem to be $(S^1, \mathscr{S}, \nu)$ where $\mathscr{S}$ is the Borel algebra on S^1 and ν is the normalized Haar measure. The induced measure defined in the theorem by $\mu(\sigma) = \nu(f(\sigma))$ is the Lebesgue measure on $[0,1)$. By Corollary 19 the inducing Haar measure and the induced Lebesgue measure are measure isomorphic.

In doing only that much, however, Gyenis and Rédei, have not shown that the Lebesgue measure *is* a Haar measure on $[0,1)$. Neither their proof nor the proof using TIM entails that the induced Lebesgue measure *is* a Haar measure, any more than using TIM to induce a measure on the chords from the chord tangent angles shows that the Lebesgue measure on $[0, \pi/2]$ *is* a measure on the chords.[106] This point applies equally well to what they then build on this starting point for defining measures on intervals in $\mathbb{R}^n$ using the exponential function above for a component-wise function to T^n.

What is needed to even approach the claim that the Lebesgue measure on $[0,1)$ is a Haar measure is for $[0,1)$ to be homeomorphic to the compact

---

106 The error of assuming the identity is the first in the list of the frailties of Bertrand's procedure in §3.2.

group S^1. Gyenis and Rédei say that 'f... is a continuous and continuously invertible bijection' (2015a:370) and if that were true then they would be homeomorphic. But it isn't true and they are not homeomorphic.[107] The simplest proof uses connectivity to show the failure: remove a point from the interval and it is no longer connected but S^1 less a point is connected, so they are not homeomorphic.

A further problem is that, once we are clear that $[0,1)$ is not a compact topological group and that we are using the Haar measure on a compact topological group and TIM to induce a normative measure on $[0,1)$, it appears that we have now not just a square paradox but the even simpler line paradox. Different compact topological groups can give distinct resulting probabilities for the event $[0,1/2)$ and who is to say which is the correct one? Presumably Gyenis and Rédei will regard this as a mere violation of labelling invariance, which they intend their defusal to address.

In place of their construction I offer the following construction on their behalf. It is defensible as showing the uniquely correct compact topological groups to use for intervals in $\mathbb{R}^n$ because of the way it is internally related to those intervals and would therefore avoid the just mooted problem.

Although intervals are not groups in the reals and so they do not have their own Haar measures, certain quotient spaces of intervals are compact groups. Instead of unit intervals such as $[0,1)$ we should use the quotient space $X=[0,1]/\{\{0\},\{1\}\}$. The quotient space is homeomorphic to S^1 by $f:[0,1]/\{\{0\},\{1\}\} \to S^1, f([x])=e^{2\pi i x}$, hence is itself a compact topological group and so has a Haar measure of its own, i.e. that is internally related to it. TIM induces the Lebesgue measure on $[0,1]$ by $g:[0,1] \to [0,1]/\{\{0\},\{1\}\}$, $g(x)=[x]$. The same can be done for n-dimensional cubes in $\mathbb{R}^n$, $\prod^n[0,1]$, by taking its quotient $\prod^n([0,1]/\{\{0\}, \{1\}\})$ giving a product topological group which is also compact (Willard 2004:134) and homeomorphic with T^n. The extension to intervals in general is straightforward.[108]

With this construction understood, we can see that wherever we speak of the Haar measure of intervals, we should instead speak of the Haar measure of their quotient spaces. The important fact that licences the Haar measure of the quotient group of intervals for determining the probabilities of intervals in $[0, 1]$ is that the equivalence class of an interval in $(0, 1)$ contains only that very interval. And since the probabilities of singletons in

---

107 The function f is not continuously invertible. $[0,1/2)$ is open in the interval. The pre-image of $[0,1/2)$ under the inverse, f^{-1}, is the part of S^1 with positive y coordinates plus the point $(1,0)$. But that pre-image is not open in S^1, precisely because there is no open set in S^1 containing the point $(1,0)$ that is contained in the pre-image.

108 To treat Bertrand's spaces $[0, \pi)$ and $(-R, R)$ we would need to take quotients of their closures and homeomorphisms to S^1. Treating D is less clear since it is not an interval of $\mathbb{R}^2$. The quotient by its boundary is homeomorphic to the sphere S^2 not the torus $T^2 = S^1 \times S^1$.

continuum spaces are zero,[109] they have no effect on probabilities and hence identifying the probability of any interval in [0, 1] with that of the corresponding interval without 0 or 1 makes no difference to the probability. This seems to me a better warrant for why a measure induced from a Haar measure inherits the Haar measure's justification for being the correct measure than the warrant supplied by Gyenis and Rédei.

That being said, the paradox might still recur. Instead of the aforementioned quotient we can take the quotient of $\Pi^n[0, 1]$ by its boundaries, which quotients are homeomorphic with S^n and also with $D^n/\partial D^n$ (Crossley 2010:79), whence proving that these quotients are also compact topological groups and hence have their own Haar measures. The recurrence would be if the Haar measure of an interval is different under these Haar measures than of those in the penultimate paragraph, which might be the case since T^n is not homeomorphic with S^n.

109 Or infinitesimals, if we can use hyperreals.

8 The Maximum Entropy Principle

8.1 Introduction

This chapter addresses the third radical mathematical treatment: the application of a principle proposed by Jaynes as a replacement for the principle of indifference, namely the Maximum Entropy principle. Somewhat surprisingly, his paper considered in Chapter 6 makes no use of entropy!

The Maximum Entropy principle can be considered as a refinement and generalization of the principle of indifference: they agree on cases to which the latter principle applies whilst the former can yet be used where the latter has no purchase. Its application to the paradox is an instance of the Well-posing strategy. Since it constitutes a way of evaluating all possible probability measures for how well they respect our equal ignorance of the chords and then choosing one that does so maximally, we count it as an instance of the Well-posing strategy pursued by use of meta-indifference (more on this counting in Chapter 10).

Apart from the paper upon which this chapter draws (Shackel and Rowbottom 2020), there is no literature examining whether the Maximum Entropy principle is able to resolve Bertrand's paradox. Indeed, there is no literature even showing how the Maximum Entropy principle can get a purchase on the paradox. In this chapter, I show that even under the most favourable assumptions allowing for that purchase, Bertrand's chord paradox undermines the Maximum Entropy principle. Additionally, the course of the analysis brings to light a new paradox, a revenge paradox of the chords, that is unique to the Maximum Entropy principle.

8.2 The Maximum Entropy Principle

Objective Bayesians such as Jaynes (2003), Rosenkrantz (1977) and Williamson (2010) have suggested that the Maximum Entropy principle can replace the principle of indifference.

DOI: 10.4324/9781003456308-8

The Maximum Entropy Principle... the prior probability assignment should be the one with the maximum entropy consistent with the prior knowledge.

(Jaynes 1968:229)

Jaynes, the originator of this proposal, proposes this because the maximum entropy probability measure is 'maximally non-committal with regard to missing information' (Jaynes 1957a:620) and 'the most unbiased representation of our knowledge' (Jaynes 1957b:171). Jaynes reiterates this thought with slight variation through his papers (Jaynes 1957a, b, 1963, 1968, 1985). In his book (2003) he explicitly joins it up with Shannon's development of entropy in information theory.

The reason entropy helps Jaynes here is the thought that bias is a kind of unwarranted lack of uncertainty, so being maximally unbiased would be maximising uncertainty, and Shannon's suggestion (1948) that entropy can be understood as a measure of uncertainty.[110] Here are some platitudes deploying our pre-theoretical concepts of uncertainty and probability:

1. our uncertainty depends on how likely events are and is entirely absent if all but one have no chance at all,
2. we are most uncertain when possibilities are equally likely,
3. our uncertainty does not change if we entertain additional events with no chance of happening and
4. entertaining additional events with some chance of happening makes us more uncertain in a way that depends on the uncertainty of the events added.

The use of 'chance' here is colloquial rather than a deployment of the term when stipulated to mean objective probability.

Shannon gave initial proofs that his entropy satisfies mathematical requirements we can see as capturing much of these platitudes (1948:10 & 28). Khinchin shows more rigorously (Khinchin 1957:9 Theorem 1) that Shannon entropy satisfies Khinchin's three axioms for uncertainty,[111] axioms that we might reasonably take to be the mathematical equivalents of

---

110 Nothing turns on whether we formulate this in terms of certainty instead of uncertainty. It is done this way because Shannon entropy increases with increasing uncertainty, i.e. increases with decreasing certainty. I reject Knight's terminology (Knight 1921) under which uncertainty means without probability. For why, see Shackel 2022, n.d.-c.

111 '1. For given n and for $\sum_{k=1}^{n} p_k = 1$, the function $H(p_1, p_2,...p_n)$ takes its largest value for $p_k = 1/n$ $(k=1, 2, ...n)$. 2. $H(AB) = H(A) + H(B)$. 3. $H(p_1, p_2,...p_n, 0) = H(p_1, p_2,...p_n)$.' (Khinchin 1957:9) $H(p_1, p_2,...p_n)$ is defined to be 'a quantity that measures the amount of uncertainty' (Khinchin 1957:3).

the first three platitudes above (Khinchin 1957:9 and Beck and Schlögl 1993:47).

So Shannon entropy is a reasonable measure of our intuitive notion of uncertainty. Taking the probability measure compatible with information possessed that otherwise *maximizes* uncertainty is a way to *minimize* bias in representing our knowledge. Hence using the Maximum Entropy principle seems to be a way of satisfying the Ascertainability criterion that fits with Jaynes' claims for the principle whilst satisfying the ignorance and non-presumptuous requirements.

In capturing the second platitude the Maximum Entropy principle can be regarded as a generalisation of the principle of indifference. Furthermore, the second platitude requires an equal ignorance probability measure and, by that platitude, an equal ignorance probability measure has maximal uncertainty and therefore has maximum entropy. In fact, Khinchin's axiom that captures the second platitude is equivalent to the proposition that entropy is maximal iff the probability measure is an equal ignorance one.

The Maximum Entropy principle has been applied to some of the paradoxes that trouble the principle of indifference and it has been claimed to resolve some of them. Paris and Vencovská 1997, for example, claim that correct application of the Maximum Entropy principle defeats the book paradox we saw in §1.4.[112] So it may well appear that the Maximum Entropy principle is a good prospect for solving Bertrand's paradox (indeed, there is an impression that Jaynes (1973) did exactly that, see more following). In fact, and perhaps surprisingly given the extensive literature on Bertrand's paradox and the principle of indifference, nowhere in the literature prior to Shackel and Rowbottom 2020 has there been an analysis of applying the Maximum Entropy principle to Bertrand's paradox. By an analysis, I mean showing whether the mathematics of entropy, when applied to Bertrand's paradox, determines a unique probability measure with maximum entropy. I demonstrate here that it does not.

Although this result may have been suspected by some, it has not previously been proved. To put it crudely, absent that paper, and now this chapter, for all we know the Maximum Entropy principle might yet solve Bertrand's chord paradox. This chapter proves it does not. Moreover, the course of the analysis exhibits a further problem for the Maximum Entropy principle: I present a new paradox of the chords that arises only for the Maximum Entropy principle.

Having proved all this, the work of the chapter is done. Before doing that work, I need to dispel the impression that prior literature has analysed the application of the Maximum Entropy principle to Bertrand's paradox. The source for this impression is Jaynes' paper on Bertrand's paradox that

---

112 For a recent summary of Paris' claimed solutions see Paris 2014.

we looked at in Chapter 6. Rosenkrantz explicitly treats Bertrand's para-dox in his book (1977:73), but whilst the context may give the impression that he is applying the Maximum Entropy principle, in fact he is expounding Jaynes' paper. Insofar as Williamson discusses Bertrand's paradox and the Maximum Entropy principle (2010:152), he merely mentions it, mentions Jaynes' 1973 paper and perhaps endorses it. He makes no attempt to apply the Maximum Entropy principle to Bertrand's paradox nor does he make any statement on whether doing so would result in unique probabilities.

Jaynes' abstract for his paper is misleading: it speaks of the Maximum Entropy principle but the paper itself does not apply the Maximum Entropy principle to Bertrand's paradox. Instead, and as we saw earlier, it applies his principle of transformation groups and the principle of indifference. Jaynes' view of the Maximum Entropy principle as a generalization of the principle of indifference means Jaynes' paper may seem to apply the Maximum Entropy principle to Bertrand's paradox by applying the prin-ciple of indifference to Bertrand's paradox. This thought is tempting[113] but too quick when we are talking about infinite event spaces, although that will only be evident after we have plowed through the work herein. What I can do for now is to show that this thought is a distraction.

First, if Jaynes was in some sense applying the Maximum Entropy prin-ciple in his paper, the refutation in Chapter 6 means that it was not done successfully. Second, the mathematics of entropy are not actually used in Jaynes' paper.[114] No equations for the entropies of any probability measures are given and nowhere does the paper make any attempt to show that the probability measure derived is the one with maximum entropy. Neither en-tropy itself nor the criterion of having maximum entropy do any work in the paper. Consequently, whilst Chapter 6's refutation of Jaynes' paper shows that this special sense of applying the Maximum Entropy principle is unsuccessful, it does not show that the Maximum Entropy principle itself cannot solve Bertrand's paradox. For all we know, then, were we to actually use the mathematics of entropy, the Maximum Entropy principle might yet solve Bertrand's paradox. It is only the work herein that proves that it does not.

Finally, the interaction of the Maximum Entropy principle and Bertrand's paradox has not been previously examined by philosophers perhaps in part because entropy can involve highly abstract mathematics. On the other hand, some proponents of the Maximum Entropy principle are

---

113 Tempting because of the second platitude above and because for finite event spaces it is easy to prove that the uniform probability measure is the function with maximum entropy.

114 An examination of the appendix to this chapter will show what would have been neces-sary for the mathematics of entropy to have been used.

mathematicians and physicists who are uninterested in or intolerant of the philosophical issues, and consequently, some of its specialist literature does the maths whilst discarding awkward philosophical problems as mere matters of convention, or alternatively, burying them in the maths. This is off-putting for philosophers and I have sought to do the opposite. It turns out that although we need to take the output of the relevant mathematics as premises, the central philosophical issues do not depend much on why those premises are true. For this reason, whilst I have included the mathematics sufficient to the latter question in an appendix, I have been able to articulate the philosophical import of the problems without requiring the reader to plough through that mathematics.

8.3 The Maximum Entropy Principle applied to Bertrand's paradox

The Deluge Theorem 24 shows there are an infinity of measures we might induce on the chords by the Theorem of Induced Measure and we have yet to see a principled way of picking among them. The apparent virtue of applying the Maximum Entropy principle is that, rather than *ignoring* the infinitude of inducible probability measures, applying the principle amounts to attempting to evaluate the members of that infinitude. Taking the one with maximum entropy is a way of choosing the maximally ignorant one, the least biased of that infinitude.

For this to work requires that

1. there is a correct measure of entropy and it can be applied to Bertrand's paradox
2. each probability measure in the set of probability measures inducible on $\mathbb{C}$ by the Theorem of Induced Measure has a unique entropy,
3. we can identify at least one function with maximum entropy,
4. and if there is more than one function with maximum entropy they all agree on the probability of the longer chord.

There are some significant technical difficulties that stand in the way of the first two of these, and some of those difficulties, if not resolved, do reinstate the paradox. However, I have traced a path through the difficulties, making use of Shannon's differential entropy, that a proponent of the Maximum Entropy principle may reasonably take. For that reason, and to avoid obstructing the flow of argument, I have described those difficulties and a path through them in the appendix to this chapter.

As I shall now show, Bertrand's paradox exhibits serious obstructions that lie in the way of fulfilling the other conditions (3 and 4 of our list). I now demonstrate two obstructions, each of which, if correct, shows that the Maximum Entropy principle fails to determine a unique probability for being longer. I then give a third obstruction in a coda which is unavoidable if the first two are avoided.

8.4 Unknowable maximum entropy probability measures are not acceptable

Bertrand's paradox brings into view the danger that certain spaces pose to knowing the maximum entropy probability measure. If there are cases in which we cannot know, then the principle fails to satisfy Salmon's Ascertainability criterion. We do not know the maximum entropy probability measure over the set of chords, and as I shall now show, we have good reason to think we cannot know it.

Since the set of chords, C, is continuum sized, the entropy of a probability measure of C is defined by an integral (see chapter appendix). Many integrals are not analytically solvable—call these obstinate integrals.[115] If the integrals are obstinate, we don't know if the integrals are numerically tractable (approximately solvable in reasonable time using numerical methods), and we do know there are serious difficulties in evaluating continuous entropies even just for $\mathbb{R}^n$.[116] This is a neglected issue because physicists are used to finding methodological tweaks to get round obstinacy and their empirical focus means they do not attend to the fully general problem that concerns us as philosophers of probability.

One route to knowing the maximum entropy probability measure would be to evaluate entropies. Unfortunately, this is blocked by obstinate integrals leading to numerical intractability. The necessity of using the Theorem of Induced Measure means we know that there will be functions from C, onto codomains with measures giving rise to obstinate integrals, that induce measures on C resulting in obstinate integrals for defining entropy for probability measures over C. Since we cannot solve analytically for their entropy, solving for the probability measure with maximum entropy over C may require numerical methods. However, many such will not be numerically tractable and that would mean we won't be able to know the maximum entropy probability measure in this way.

Numerical intractability, it might be objected, poses a merely practical problem. In principle, numerical evaluation is possible and the

---

115 Roughly, they lack a solution expressible as a formula not involving differential or integral equations, power series or limits. If a solution is expressible in closed form it is possible to evaluate in finitely many steps and is for that reason called tractable. Analytically solvable is slightly wider than closed form solvable by allowing some special functions such as the gamma function and some infinite expressions.
116 See Pearl (1988:chapter 9) who objects to the Maximum Entropy principle on the grounds of computational intractability.

intractability is just a matter of it taking a very long time to do the figuring. This, it must be conceded, is certainly true of each individual obstinate integral. Unfortunately, since there are infinitely many such integrals, this does not offer an in principle defence of knowability. The problem is that to determine the Maximum Entropy principle in this way would require performing infinitely many numerical evaluations and this we cannot do.

The other route we might take to knowing the maximum entropy probability measure would be having a constructive existence proof of which one it is that does not depend on evaluating entropies. The necessity of using TIM means the set of probability measures over which we seek the one with maximum entropy, PM_C, is both enormous[117] and essentially unrestricted.[118] The size of PM_C means, absent relevant proofs, there is no reason to expect such a constructive proof would be available. Since it is unrestricted, that there are some sets of probability measures for which such constructive proofs are available, which sets will be subsets of PM_C, tells us very little. There are, after all, also sets of probability measures for which there are no such constructive proofs and they, too, will be subsets of PM_C. So consider the union of a set for which there is a constructive proof and a set for which there isn't, and for which the entropies of the first are all less than those of the second. There will be no constructive proof for such a union and the union will be a subset of PM_C. This counts against such a constructive proof being available for PM_C.

It is true that there may be specific empirical situations whose modelling sufficiently restricts the set of probability measures induced on $\mathbb{C}$ to one for which no entropy is defined by an obstinate integral or one for which there is a non-evaluative constructive proof. But that is not the general case.

Although these several points do not constitute a conclusive proof that (if there is one) the unique maximum entropy probability measure for Bertrand's paradox is unknowable, they are good reason to think that it is

---

117 Take a probability space (C, Σ, P) where P is an induced probability measure on $\mathbb{C}$ for which the subset of the σ-algebra, Σ, having non-zero probabilities is continuum sized, with continuum many distinct probabilities. Any permutation of Σ corresponds to a distinct probability measure on $\mathbb{C}$ provided it produces a difference in probability for a single member of Σ. Any such distinct probability measure, P', on $\mathbb{C}$, can be induced on $\mathbb{C}$ using the Theorem of Induced Measure (take any continuum sized set, X, and use a bijection $f:X \to C$ to induce from the given P' a measure on X, (X, T, M), and now, using f^{-1} to induce a probability measure on $\mathbb{C}$ from (X, T, M) will induce the given probability measure, P'). The cardinality of the set of such permutations $= 2^{|\Sigma|} = \text{Beth}_2 > $ cardinality of the continuum $=\text{Beth}_1$. Hence the set of all probability measures inducible on $\mathbb{C}$ is at least as big. (Relation to Aleph cardinals: $\text{Beth}_n \geq \text{Aleph}_n$. The general continuum hypothesis implies equality.)

118 Since $|\mathbb{C}| = |\text{continuum}|$, for any continuum sized space, X, there is a bijection from it to $\mathbb{C}$ and so for any bounded measure on X that bijection can be used with that measure in the Theorem of Induced Measure.

not. We need to be clear why not knowing or being unable to know amounts to a failure. After all, many problems in real analysis are intractable. The nature of the problem here depends on whether the Maximum Entropy principle is taken to be a purely metaphysical principle or whether it is in part or wholly an epistemic principle.[119]

If the Maximum Entropy principle is taken to be a purely metaphysical principle, in the sense that it *determines* a single correct probability measure for Bertrand's paradox, whether or not we are able to *know* which it is, then perhaps this is not a failure. There just has to *be* a unique function with maximum entropy. Nevertheless, in this case the proponents of the principle owe us a general existence proof since otherwise they are merely claiming that the principle will solve the paradox. No such proof has ever been offered for Bertrand's paradox. Shortly I show that there is a function for Bertrand's paradox with maximum entropy, but not in a way that will help proponents.

If, however, the principle is in part or purely an epistemic principle then this is a serious problem. The role of the principle in, for example, objective Bayesianism is epistemic. Indeed, among Jaynes' philosophical reasons for advancing the principle was to solve the problem of justifying applying statistical mechanics to a deterministic physical system. Jaynes considered that standard explanations of statistical mechanics involved objectionable obscurities whereas:

> [the] independence [of Jaynes' derivation of the Boltzmann distribution] from difficult and dubious physical assumptions like ergodicity, and its avoidance of fictions like 'virtual ensemble' have impressed many students of statistical mechanics and inductive logic.
>
> (Shimony 1985:38)

Jaynes gains this independence because he bases the derivation on an application of the Maximum Entropy Principle. This allows probabilistic properties of statistical mechanics to be understood in epistemic terms rather than requiring them to be physically real. Consequently, statistical mechanics is not a purely physical theory but is instead an epistemic theory of our knowledge of thermodynamic systems. Clearly all this requires taking the Maximum Entropy Principle to be an essentially epistemic principle.

Finally, the very fact that the principle is supposed to address the Ascertainability criterion shows it to be at least in part an epistemic principle. The whole point of the principle is to produce determinate probabilities when we do not know, without restriction on the extent of our

---

119 For discussion on the same point as it arises for the principle of indifference, see Shackel 2007:161–2.

ignorance. Even if the principle succeeds in determining that there are such probabilities, if we cannot know what they are because we cannot know which probability measure has maximum entropy, then it has failed at the point of that point, namely, to be the method by which we can satisfy Salmon's Ascertainability criterion.

Some Bayesians countenance the axioms of probability as regulative ideals that we cannot know whether we meet because our finitude entails lack of logical omniscience and probabilistic completeness. It might then be argued that the Maximum Entropy Principle could function similarly despite the unknowability of the function with maximum entropy (provided we had an existence proof). The question then would be how well it could function as such. It might give some local direction when one function had higher entropy than another, but we shall be seeing later that in general there will be too many functions with the same entropy and sometimes unbounded entropies for that to help. I show in the following that Bertrand's paradox is a case in point and hence the principle does not solve the paradox even when taken as a regulative ideal.

8.5 Either Bertrand's paradox returns or we have a revenge paradox of the chords

There is an important difference between entropy for discrete and continuous event spaces: entropy can be unbounded for continuous distributions (more loosely put, can be infinite). We seek the maximum entropy over *all* probability measures that could be induced on ℂ. Since entropy can be unbounded for continuous distributions there will be probability measures of that type that can be induced on ℂ. Indeed, there will be infinitely many probability measures of that type that can be induced on ℂ. The fact that they might not be well motivated geometrically is beside the point of the strategy of going for the maximum entropy probability measure. Therefore there will be no single maximum entropy probability measure. So for the Maximum Entropy principle to solve Bertrand's paradox would require all the probability measures for which entropy is unbounded to agree on the probability of longer. Provided there is just one such for which the probability of longer $>\frac{1}{2}$ there will be another that is its dual giving the probability of longer $<\frac{1}{2}$ (see the Duality theorem **38** in chapter appendix). That would immediately give rise to Bertrand's paradox. So for this strategy to work it would require that all the probability measures, P, for which entropy is unbounded give the probability of longer, $P(L) = \frac{1}{2}$.

Now consider the following events whose disjunction is the same as *not being longer: being less than half the length of the inscribed equilateral triangle; being between half the length and the length of the inscribed equilateral triangle*. The first disjunct partitions ℂ into sets H and $\neg H$ and the second disjunct into G and $\neg G$. The Duality theorem applies to these sets also and thus so does the reasoning just given about the event of *being longer*. Hence, on pain of the immediate recurrence of Bertrand's paradox, all the probability measures, P, for which entropy is unbounded must give $P(H) = \frac{1}{2}$ and $P(G) = \frac{1}{2}$. Now we have a new paradox of the chords:

$$
\begin{aligned}
\frac{1}{2} &= P(L) \\
&= P(\neg L) \\
&= P(H \text{ or } G) \\
&= P(H) + P(G) \ \text{(by additivity of measures, since } H \cap G = \varnothing) \\
&= \frac{1}{2} + \frac{1}{2} = 1
\end{aligned}
$$

So, on the assumption that, despite unbounded entropies, there is a solution to Bertrand's paradox furnished by the Maximum Entropy principle, a revenge paradox of the chords appears.

The only way out of this is to exclude probability measures with unbounded entropies. Yet the reason grounding the Maximum Entropy principle is that by rejecting those with less than maximum entropy we *reject* all probability measures that are more biased than is warranted by our state of knowledge. If instead we now *exclude* probability measures with unbounded entropies we are *confining* ourselves to probability measures that are more biased than is warranted by our state of knowledge. Consequently it is very difficult to see how we could justify excluding such functions.

I am therefore confident that this challenge is correct and conclusive. There is no unique maximum entropy probability measure for the set of chords, but infinitely many which have unbounded entropies and are therefore equally maximal. Either they do not agree on a unique probability of longer or they do. If they don't we have Bertrand's paradox again, but if they do then we end up in the revenge paradox.

8.6 Coda: collapse into the original Bertrand's paradox

Since I see no prospect for a principled exclusion of probability measures with unbounded entropies, I believe that at this point I have established the result of the chapter. I continue with a coda only for the sake of demonstrating that *even if* someone were to come up with such a principled exclusion, its effect would not save the Maximum Entropy principle from Bertrand's paradox but result in it collapsing into a version of the original chord paradox.

So now we exclude the induced probability measures on $\mathbb{C}$ with unbounded entropies. With this exclusion another problem comes into view:

that although all probability measures left have finite entropies, the set of entropies need not have a maximum. So we shall also have to assume that problem away. We shall now see quite directly that there will still be more than one probability measure with maximum entropy and those probability measures will disagree on the probability of the longer chord. Furthermore, despite the apparent simplicity of the route now taken, the tactics applied by proponents of the Maximum Entropy principle to simpler paradoxes cannot get a grip just because of the richness of Bertrand's paradox. Any such attempt will amount to a covert restriction rather than comprehension of the case and be thereby an evasion of the general problem posed by Bertrand's paradox.

Shannon entropy satisfies Khinchin's axiom, which axiom, recall, is equivalent to the proposition that entropy is maximal iff the probability measure is an equal ignorance one.[120] This is no accident. Khinchin's second axiom effectively captures one of the central simple truths about uncertainty, that we are most uncertain when possibilities are equally likely. Consequently, if a measure of entropy does *not* satisfy Khinchin's second axiom then that brings into question the use of that measure for applying the Maximum Entropy principle. It means that the argument I gave earlier defending the Maximum Entropy principle on the grounds of platitudes about uncertainty would not apply to that measure. It would mean that the measure fails to satisfy Williamson's equivocation norm (Williamson 2010:28). It would be contrary to what Jaynes requires when he says the Maximum Entropy principle reduces to 'the principle of insufficient reason... in case no information is given except enumeration of the possibilities' (Jaynes 1957a:623), since to do this the principle must give an equal ignorance probability measure in the case of complete ignorance.

From this it follows (under the assumptions made at the beginning of this section) that the maximum entropy probability measures on $\mathbb{C}$ are equal ignorance probability measures when not constrained by further information. In Bertrand's paradox we are not constrained. Equal ignorance probability measures on $\mathbb{C}$ are defined, via the Theorem of Induced Measure, by equal ignorance functions and equal ignorance measures on the inducing space. From the original version of Bertrand's paradox we know of two inducing spaces whose equal ignorance measures induce two equal ignorance probability measures on $\mathbb{C}$, P_α from the angle case (§3.11) and P_ρ from the direction case (§3.12). By the entropy measure satisfying Khinchin's second axiom, entropies of both P_α and P_ρ are equal and maximum. P_α and P_ρ give contrary probabilities for the longer chord. Therefore Bertrand's paradox recurs under the Maximum Entropy principle.

---

120 Shannon asserts the truth of this when the underlying event space has bounded volume but does not address the completely general case. Shannon 1948:35.

8.7 Conclusion

I have shown that applying the Maximum Entropy principle to Bertrand's paradox faces grave difficulties. The first obstruction is damaging because the Bertrand's chord set shows a significant narrowing of the extent to which the maximum entropy function is knowable. This has been neglected because physicists have applied the Maximum Entropy principle to event spaces with sufficient empirical constraints to be within that narrowed extent. Of course, when you have suitable further information you can avoid the problem here: this is why I drew attention to the generality of the problem that a claimed solution to Bertrand's paradox faces. The analysis of Bertrand's question in Chapter 5 showed that the question is answerable, not unanswerable, and hence the existence of restricted questions does not remove the determinate general problem his question poses. The fact that some empirically constrained cases do not fall to this obstruction is not a resolution of it.

The second obstruction, which I regard as conclusive, shows that the Maximum Entropy principle falls to Bertrand's paradox or faces a new paradox, the revenge paradox of the chords. The third, a coda based on the supposition that the previous obstructions can be avoided in a principled way, shows that the Maximum Entropy Principle collapses into the original Bertrand's paradox.

The relation of the principle of indifference and Bertrand's paradox has given rise to an extensive analytical literature. Prior to Shackel and Rowbottom 2020, there is not a single paper that has conducted an analysis of the relation of the Maximum Entropy Principle and Bertrand's paradox. In this chapter I have conducted the needed analysis and proved that the Maximum Entropy Principle cannot solve Bertrand's chord paradox. In so doing, I have proved that Bertrand's paradox remains the exemplar it has always been. The problem for the principle is not confined to probabilities for chords. The obstructions I have shown are a lurking threat for the application of the Maximum Entropy Principle to *any* continuum sized event space.

8.8 Appendix

I include this appendix to show that I haven't concealed philosophically significant moves in the presupposed mathematical results and to demonstrate the status of those presupposed results, including where there are mathematical burdens of proof that a proponent of the Maximum Entropy principle faces. Despite some difficulties I mention about such burdens, the appendix is not intended as a further line of argument against the Maximum

Entropy principle. In earlier work on these burdens, in the appendix to Shackel and Rowbottom 2020, I fulfilled some but only sketched a fulfilment of a threatening difficulty of non-unique entropies, and I now think that material was inadequate. Consequently from Subsection 8.8.3 onwards I have significantly reformulated all the material. I think I have now solved all the difficulties, including the threatening difficulty, and satisfied the burdens of proof fully on behalf of the proponents of the Maximum Entropy principle.

8.8.1 We need a definition of continuous entropy

Shannon entropy is defined only for discrete spaces, spaces that are finite or countable. But $\mathbb{C}$ is continuum sized and since Shannon entropy is not defined for it we need a definition of continuous entropy. Unfortunately there is not a simple definition of continuous entropy that can be considered *the* limiting case of discrete entropy.[121] Consequently, scientists have found a variety of definitions of continuous entropy worth using depending on the empirical situation. There is no reason to expect those definitions to agree on the function with maximum entropy, nor to expect the various functions that have maximum entropy under the various definitions to agree on the probability of the longer chord, and hence Bertrand's paradox recurs immediately.

Nevertheless, there is a definition of continuous entropy, a definition that Shannon offered in an appendix of his original paper (1948:35), which has some claim to be considered standard, namely, differential entropy. It is at least possible that good grounds could be found for this being the uniquely correct definition. For example, if we follow Williamson in seeking the function with minimum Kullback-Leibler divergence from the equivocator function, in finite state spaces this is provably the maximum differential entropy function (Williamson 2010:28–9). Possibly something similar could be proved for continuous state spaces.[122]

---

121 See Jaynes 1968:235. Nor will Kullback's principle of minimum discrimination information, using Kullback-Leibler divergence, help here. Although it is sometimes advanced as a natural extension of the Maximum Entropy principle to continuous spaces, it only tells us to use the nearest posterior distribution given new information to a *prior* probability distribution, whereas our problem is that we have no prior to start with.

122 Williamson addresses Kullback-Leibler divergence for continuous spaces at Williamson 2010:154 but does not consider to what extent the result for the finite spaces may carry over.

8.8.2 *The entropy measure depends on a prior definition of the probability density function*

Applying the definition of differential entropy to $\mathbb{C}$ gives

$$entropy\left(\mathbb{C}_R\right) = -\int_{\mathbb{C}} p\log\circ p\,d\mu$$

where $\mathbb{C}_R$ is a random variable taking values in $\mathbb{C}$, p is not a probability measure over $\mathbb{C}$ but a probability density function, $\log\circ p$ is the composition of the logarithm function and p,[123] and μ is a measure on $\mathbb{C}$. The entropy measure for $\mathbb{C}$ therefore depends on the prior definition of a probability density function p. Given a probability measure, P, on $\mathbb{C}$, we define the probability density function by

$$\int_{c\in\sigma} p(c)d\mu = p(\sigma)$$

for all $\sigma\in\Sigma$. So these two integrals together are what are required in order to define the entropy of a probability measure induced on $\mathbb{C}$ by an application of the Theorem of Induced Measure.

8.8.3 *The integrals need to exist*

We need to check whether the integrals defining the entropy exist. I shall now show that they exist using standard measure theory. For these integrals to be defined the functions must be μ-measurable. Since their domains are $\mathbb{C}$, which is not a subset of $\mathbb{R}^2$ but of $P(\mathbb{R}^2)$, μ cannot be a Lebesgue measure.

Here we are going to exploit the TIM Is Unavoidable theorem **23** in order to apply the Radon-Nikodym theorem. A more basic approach from first principles that does not rely on Theorem **23** is given in the appendix to Shackel and Rowbottom 2020. I now regard that approach as not taking us far enough for what is needed, whereas we shall see that the use of the Radon-Nikodym theorem eventually gives us what we need to solve what the earlier approach, I now say, did not.

For a technical reason that will become clear I have had to introduce a restriction on the probability spaces, $(\mathbb{C}, \Sigma, P)$ to which the following theorems defining entropy apply. If there are any uncountable $\sigma\in\Sigma$ with $P(\sigma)=0$ then $(\mathbb{C}, \Sigma, P)$ is excluded. However, this doesn't mean entropy is left

---

123 I.e. $\log\circ p(c)=\log(p(c))$ for all $c\in\mathbb{C}$. The base doesn't matter to us.

undefined for such excluded spaces. It is defined by the entropy of the sub-space we get from the original space by removing all the uncountable events with probability zero.

Let $(\mathbb{C}, \Sigma, P)$ be an excluded space and $W \subset \mathbb{C}$ be the union of all such uncountable $\sigma \in \Sigma$ with $P(\sigma) = 0$. Then $P(W) = 0$ and consequently to sat-isfy the relevant version for continua of Khinchin's 3rd axiom (and to satisfy our third platitude on uncertainty) $P(W)$ must make no difference to the entropy of P. Hence we define the entropy of P in $(\mathbb{C}, \Sigma, P)$ to equal the entropy of $P|_{\Sigma \cap \mathbb{P}(C-W)}$ in $(\mathbb{C}-W, \Sigma \cap \mathbb{P}(\mathbb{C}-W), P|_{\Sigma \cap \mathbb{P}(C-W)})$, that is to equal the entropy of the probability measure of the subspace we get by removing W from $\mathbb{C}$. The entropy of $P|_{\Sigma \cap \mathbb{P}(C-W)}$ is defined by what follows, just be-cause $(\mathbb{C}-W, \Sigma \cap \mathbb{P}(\mathbb{C}-W), P|_{\Sigma \cap \mathbb{P}(C-W)})$ has no uncountable $\sigma \in \Sigma \cap \mathbb{P}(C-W)$ with $P|_{\Sigma \cap \mathbb{P}(C-W)}(\sigma) = 0$. Thus do I define entropy for all probability mea-sures on $\mathbb{C}$.

Lemma 32

Given a probability space $(\mathbb{C}, \Sigma, P)$ where P is positive for any uncountable members of Σ, there exists a measure space $(\mathbb{C}, \Sigma, \mu)$ and a μ-measurable positive function $p:\mathbb{C} \to (0,\infty)$ such that

$$\int_\sigma p \, d\mu = P(\sigma) \text{ for all } \sigma \in \Sigma$$

and p is a probability density function.

Proof: From TIM Is Unavoidable theorem 23 we have a measure space $(Y, \mathscr{S}, v)$ and a function $f:\mathbb{C} \to Y$ such that $v(f(\mathbb{C})) < \infty$ and $P = \mu/\mu(\mathbb{C})$ where μ is the measure induced on $\mathbb{C}$ by TIM from $(Y, \mathscr{S}, v)$ and f. From TIM, $\mu(\mathbb{C}) = v(f(\mathbb{C})) < \infty$ and $P(\mathbb{C}) = 1$ so both are finite measures spaces and there-fore σ-finite measures spaces. P is absolutely continuous with respect to μ, since for any $\sigma \in \Sigma$ such that $\mu(\sigma) = 0$, $P(\sigma) = \mu(\sigma)/\mu(\mathbb{C}) = 0$. Therefore by the Radon-Nikodym theorem (Bruckner *et al.* 2008:220, Theorem 5.29) there exists a μ-measurable function $g:\mathbb{C} \to \mathbb{R}$ such that

$$\int_\sigma g \, d\mu = P(\sigma) \text{ for all } \sigma \in \Sigma$$

Since μ and P are non-negative measures and positive for any uncountable members of Σ, g is positive except on a null set, Z (i.e. a set with μ-measure zero). Since two functions differing only on a null set have the same integral (Rudin 1987:27), define $p:\mathbb{C} \to (0,\infty)$ by

$$p(c) = \begin{cases} g(c) & \text{if } c \in \mathbb{C}-Z \\ 1 & \text{if } c \in Z \end{cases}$$

and then p is a μ-measurable positive function for which the integral result follows. p is a probability density function because p is non-negative and

$$\int_c pd\mu = P(\mathbb{C}) = 1. \text{ QED}$$

Lemma 33

Given μ-measurable probability density functions, p, q, got from (C, Σ, μ) as in Lemma 32, then $p=q$ almost everywhere (i.e. $p=q$ except on a μ-null set).

Proof: We are given that for all $\sigma \in \Sigma$, $\int_\sigma pd\mu = P(\sigma) = \int_\sigma qd\mu$. Applying Capinski and Kopp 2004:90 Theorem 4.22 gives $p=q$ almost everywhere. QED

Remark: What the lemma doesn't tell us is that if p is μ-measurable but q is *m*-measurable, from a measure space (C, Σ, m) that *also* induces P on $\mathbb{C}$ by TIM, $p=q$ almost everywhere. There is a problem here that I will address in the next subsection.

Lemma 34

Given the probability space (C, Σ, P), let (C, Σ, μ) and a probability density function p be as derived in Lemma 32. Then $p \times (\log \circ p):C \to \mathbb{R}$ is a μ-measurable function.

Proof: Log is defined on $(0, \infty)$ but not on $[0, \infty)$ and this is why I defined the codomain of p in Lemma 32 to be $(0, \infty)$. Since log is continuous and p is a μ-measurable function, the compound function $\log \circ p:C \to \mathbb{R}$ is a μ-measurable function, whence the product $p \times (\log \circ p)$ is a μ-measurable function (Bruckner *et al.* 2008: 165, Theorem 4.8(iii) and (iv)). QED

Theorem 35

Given the probability space (C, Σ, P), a μ-relative differential entropy, H_μ, for $\mathbb{C}$ exists.

Proof: Take the measure space (C, Σ, μ) and probability density function, p, derived from (C, Σ, P) by Lemma 32. By Lemma 34 $p \times (\log \circ p):C \to \mathbb{R}$ is a μ-measurable function and therefore the integral

$$H_\mu(\mathbb{C}_R) = -\int_{c \in C} p \times \log \circ \, pd\mu$$

exists by the definition of the integral (Bruckner *et al.* 2008:201). QED

Remark: This establishes nothing more than the existence of differential entropy, H_μ, where the subscript does significant work, since this entropy is relative to the measure μ. The remark to Lemma 33 alerts us to why we haven't yet proved entropy for $\mathbb{C}$ to be well-defined. We need the further proof that H_μ is unique.

8.8.4 A threat of non-unique entropies

To get the entropy we first require a probability density function to be integrated over, which in turn is implicitly defined by an integral relating it to the probability measure, which last is logically prior and has been induced on ℂ by the Theorem of Induced Measure. Absent a uniqueness proof, there is a danger that the probability density function is not *uniquely* defined (almost everywhere) for each probability measure. This would arise if two measures on ℂ induced the same probability measure whilst producing distinct probability density functions. This possibility was explained in the remark to Lemma **32**.

An illustration of the kind of thing I mean is this. Take the state space $[0, 2]$ and the σ-algebra $\{\varnothing, [0, 1), [1, 2], [0, 2]\}$. Let $p(x) = ½$ for x in $[0, 2]$ and $p'(x) = 1$ for x in $[1/2, 3/2]$ (both zero elsewhere). Then both p and p' agree on the probabilities for events in the σ-algebra but they are distinct probability density functions. The problem is that they give distinct entropies, for p it is log 2 and for p' it is 0. Although in this illustration the state space is discrete and finite, the relevant parallel is when our interest is in a discrete set of events that are determined by an underlying continuous space. Because {longer chords} and $[1, 2]$, and {not-longer chords] and $[0, 1)$, are equinumerous, there are bijections that we can compound to give a bijection from ℂ to $[0, 2]$ to use in the Theorem of Induced Measure.[124] This will then give a probability measure for ℂ with two different entropies (although perhaps with a somewhat restricted, if still infinite, algebra of events for ℂ, see potential significance of this below). The example is very simple but it makes clear the general danger. *If* non-uniqueness of this kind obtains for *any* functions on *any* measure space that is the same size as ℂ, then because of the Theorem of Induced Measure, there will be non-uniqueness of this kind for Bertrand's paradox.

Just to be clear of the dangers here, non-uniqueness of probability density functions can entail accompanying non-unique entropy for the corresponding probability measure. Recalling that the entropy is supposed to measure the presumption or bias of a probability measure, non-uniqueness in that measure is itself intolerable for the philosophical reasons motivating the Maximum Entropy principle in the first place. This amounts to a subtle, interesting and logically prior occurrence of the kind of non-uniqueness

---

124 If g is a bijection from {not-longer chords} to $[0,1)$ and h is a bijection from {longer chords} to $[1,2]$, define $f:ℂ\rightarrow[0,2]$ by $f|_{[0,1)}=g$ and $f|_{[1,2]}=h$. A similar construction for $p'(x)=1$ for x in $[1/2, 3/2]$ and $p''(x)=1$ for x in $[1/4, 5/4]$ foreshadows the return of Bertrand's paradox itself. These probability density functions give rise by the same indirect route just articulated to two distinct probability measures on ℂ, P' and P'', having the same entropy, and with P'(longer)=¼ and P''(longer)=½. This specific problem is forestalled by these entropies not being maximal, which is why the arguments in the body of the chapter do not rest on this simple example.

that Bertrand's paradox poses. Bertrand's paradox usually undermines by non-unique probabilities: here it undermines earlier by non-unique entropies themselves being objectionable and by them blocking the route to the Maximum Entropy principle furnishing probabilities at all.

8.8.5 Solving the uniqueness problem

In my earlier work (Shackel and Rowbottom 2020) I thought that what is here Lemma 33 essentially removed this problem. In so doing I was attending to the fact that the codomain of the function $p \times (\log \circ p):C \to \mathbb{R}$ is $\mathbb{R}$ and so treating it as if we could use the Lebesgue measure on $\mathbb{R}$ for the integral defining the entropy for probability measures on C. If that were correct, Lemma 33 would suffice.

However, I now think I neglected to address clearly enough the question of what the μ-measure in Subsection 8.8.2 is.[125] We want to define entropy by the integral

$$entropy(\mathbb{C}_R) = -\int_{\mathbb{C}} p \log \circ p \, d\mu$$

By the definition of the integral, $p \times (\log \circ p)$ must be μ-measurable and that means that μ must be a measure not of the codomain but of the domain of the functions being integrated: here, that means the μ-measure must be a measure on the set of chords.

The work of Chapter 3, especially the TIM Is Unavoidable theorem, shows both that there is no difficulty in having such a measure, but also that the differential entropy thereby defined is unavoidably μ-relative and so Lemma 33 does too little work. We need to give a uniqueness proof for the μ-relative differential entropy of Theorem 35, that is, we need to prove

Conjectured Theorem 36

Let measures spaces $\{C, \Sigma, \mu\}$, $\{C, \Sigma, m\}$ induce probability measure space $\{C, \Sigma, P\}$ by TIM, let p be the μ-measurable probability density function and q the m-measurable probability density function got from Lemma 32, AND [a presently unknown hypothesis]. Let H_μ be the μ-relative differential entropy and H_m be the m-relative differential entropy from Theorem 35. Then $H_m = H_\mu$.

We could prove this immediately from the lemma

---

125 In the original paper 'v' was used to name the measure named by 'μ' in Subsection 8.8.2 above.

Conjectured Lemma 37

Let μ and m be measures on Σ and for all $\sigma \in \Sigma$, $\int_\sigma p d\mu = \int_\sigma q dm$ AND [a presently unknown hypothesis]. Then $\int_C p \log \circ p d\mu = \int_C q \log \circ q dm$

The conjectured theorem and lemma as given lack a needed hypothesis, since we know from the previous example of Subsection 8.8.4 that it is possible for the μ-relative differential entropy of Theorem 35 to be not unique.

What gave us the non-uniqueness was a highly restricted algebra of events on $\mathbb{C}$. It is worth noting that with the algebra of events we used in the example, we could have used Shannon's discrete entropy rather than our differential entropy. Had we done that, both probability densities, despite being distinct, give the same probabilities to the same events and so the Shannon discrete entropy is unique. So if there is a way out of this it is that the algebra of events induced on $\mathbb{C}$ by TIM in these problematic cases is objectionably restricted.

Consequently, a good candidate hypothesis would be to rule out algebras of events on $\mathbb{C}$ to which Shannon's discrete entropy could be applied. The obvious way to do this is to require Σ to be the σ-algebra generated by the set of singleton sets of chords. This would immediately block our example since $p \neq q$ on $[0,1/2)$, which is not a null set, so their integrals would not be the same and they couldn't both give the same probability measure on Σ.

What needs to be done, therefore, is state what are philosophically objectionable restrictions, formulate that condition mathematically as hypotheses for conjectures 36 and 37 and prove the above lemma. The rejection of the discrete entropy algebra seems defensible, since the interest we have in the differential entropy is the problem of entropy for continuum sized state spaces. The proposal that it be the σ-algebra generated by the set of singleton sets of chords might be defensible on two grounds. First, that every chord must be counted so all the singleton sets should be in the σ-algebra. Second, the generated σ-algebra is, by definition, the *smallest* σ-algebra containing all the singleton sets, so it is the least presumptuous way of including them. So that is my proposal for the needed hypothesis.

Nevertheless, there remains a significant worry here. The motivating power of applying the Maximum Entropy principle to solve the paradox of the chords is that it is a way of being meta-indifferent over all the possible probability measures, that is over all $P \in PM_C$. We have now excluded any probability measures that happen, through the way they can be induced by TIM, to have a discrete, countable or finite, σ-algebra.

8.8.6 A better solution

An alternative solution is to attempt to avoid the problem of measure relativity by using the probability measure P to be the measure defining the integrals and therefore to define the probability density function and entropy.

$(\mathbb{C}, \Sigma, P)$ can itself play the role of the other measure space in Lemma **32**, giving us a P-measurable probability density function for which

$$\int_\sigma p dP = P(\sigma) \text{ for all } \sigma \in \Sigma$$

Then Theorem **35** gives us a P-relative entropy

$$H_P(\mathbb{C}_R) = - \int_{c \in \mathbb{C}} p \times \log\circ p dP$$

If we apply this to our problematic example, we get

$$\int_\varnothing p dP = 0$$

$$\int_{[0,1)} p dP = 1/2$$

$$\int_{[1,2)} p dP = 1/2$$

$$\int_{[0,2)} p dP = 1$$

where functions, p, satisfying these can only differ on a null set. Consequently they must all be almost everywhere equal to ½ for x in [0, 2], giving the unique P-relative entropy of log 2.

So taking this line of approach we need no longer worry about hypotheses restricting the algebras and we have thereby avoided the restriction of probability measures over which the Maximum Entropy principle is used as a form of meta-indifference, solving the worry at the end of the last subsection. There remains the worry about uniqueness.

The move to make at this point is say that *this* measure-relativity doesn't matter. What we were worried about was that different inducing measures giving *the same probability measure* on the chords gave different relative entropies *for the same probability measure*.

Arguably, however, defining the entropy for the probability measure by an integral *defined directly in terms of that measure itself* is the *correct* measure-relative entropy to use. Indeed, when what we want is an entropy for each probability measure, P, on the chords, what we are wanting is *precisely* a *P-relative entropy!* And since, for each probability measure, P, on $\mathbb{C}$, H_p is a *P*-relative entropy, what we wanted is precisely what we have now got! Since H_p is unique for each P, we have solved the uniqueness problem. We now have a unique entropy for each probability measure. I am satisfied that this is a conclusive solution of the unique entropy problem.

8.8.7 Revenge paradox theorem

Duality Theorem 38

Suppose $(X, \mathscr{S}, \nu)$ and $f{:}\mathbb{C}{\rightarrow}X$ give TIM begotten measure space $(\mathbb{C}, \Sigma, \mu)$ which begets probability measure P_f. Then there exists a dual function, $g{:}\mathbb{C}{\rightarrow}X$, with a dual TIM begotten $(\mathbb{C}, \Sigma, \kappa)$ and dual probability measure P_g such that H_{P_f} is unbounded iff H_{P_g} is unbounded and for $L \in \Sigma$, $P_g(L)=P_f(\neg L)$ and $P_g(L)<\tfrac{1}{2}$ iff $P_f(L)>\tfrac{1}{2}$.

Proof: L and $\neg L=\mathbb{C}-L$ partition $\mathbb{C}$. Take a permutation $h{:}\mathbb{C}{\rightarrow}\mathbb{C}$ for which $h(L)=\neg L$ and let $g=f{\circ}h$, when $g(L)=f(\neg L)$. Then

$$P_g(L) = \frac{\kappa(L)}{\kappa(\mathbb{C})} = \frac{\nu(g(L))}{\nu(g(\mathbb{C}))} = \frac{\nu(f\circ h(L))}{\nu(f\circ h(\mathbb{C}))} = \frac{\nu(f(\neg L))}{\nu(f(\mathbb{C}))} = P_f(\neg L).$$

Consequently $P_g(L)<\tfrac{1}{2}$ iff $P_f(\neg L)<\tfrac{1}{2}$ iff $P_f(L)>\tfrac{1}{2}$. The unbounded entropy correlation follows from the definitions above and because σ-algebras are closed under complements. The difference between g and f, and therefore between measures μ and κ, is from the permutation h. Since h permutes L to its complement, it leaves Σ intact (by design) and swaps the measures of some elements in Σ with the measure of their complements, which makes no difference to the entropy. QED

9 The Universal Average

9.1 Introduction

This chapter addresses the fourth radical mathematical treatment, the proposal from Aerts and Sassoli de Bianchi (2014). They address my earlier paper (Shackel 2007), and say that they do not fall under either the Well-posing or Distinction strategies as identified therein, but instead offer a measure theoretic meta-indifference they call the universal average, which is defined by an approach similar to that used in defining the Weiner measure. Examining their solution, we find that the universal average is applied to distinct questions, each addressing a *kind* of random process rather than an *instance* of a random process for choosing chords. In doing this they do not fall into van Fraassen's trap in the way that Marinoff does: they fall into it in a different way. Their universal average does not average over all the kinds of random processes, which is what is required to answer Bertrand's question by an instance of Well-posing by meta-indifferent average, but is taken only over each kind separately. Nevertheless, I can on their behalf avoid van Fraassen's trap by showing how to generalize the application of their universal average to Bertrand's question. I then conduct an analysis of the extent to which the universal average is really a meta-indifferent average and give an independent argument for the recurrence of Bertrand's paradox.

9.2 Aerts and Sassoli de Bianchi's distinction strategy

There is a tendency of physicists in particular to interpret the paradox in terms of tossing ideal sticks, or other things, onto ideal circles. Although no doubt relevant to considering specific empirical situations, this can be misleading when analysing the paradox. In fact, it is one of the lures in van Fraassen's trap. This way of thinking leads Aerts and Sassoli de Bianchi to conclude that what is needed is to solve the paradox is to

DOI: 10.4324/9781003456308-9

disambiguate Bertrand's question by specifying the nature of the randomized entity.

(2014:15)

Doing that is solving what they call the easy problem of Bertrand's paradox, the 'modelization problem', defining an experimental realization of Bertrand's question by giving

a sufficiently precise definition of the entity which is subjected to the randomization.

(2014:5)

When this thought is brought to bear on Bertrand's cases, their point is that experimental realizations of them show each to be, in fact, different two step procedures,

[the angle case] consists in a procedure where one selects, in sequence, two points on the circle, to which only subsequently one associates a chord. This is what one would naturally do by tossing two pebbles... [the direction case] describes a two-step procedure where an orientation is first chosen for the chord, and then an orthogonal displacement for it is produced. This is what one would naturally do (although usually not in sequence) when tossing a stick.

(2014:5)

What they call the hard problem of Bertrand's paradox, the 'randomization problem' (2014:5), is producing a probability measure that meets the requirement to

understand what... maximum lack of knowledge really implies, in relation to an entity which has been... unambiguously defined.

(Aerts and Sassoli de Bianchi 2014:5)

We will be looking at their solution to the hard problem later. On their view, Bertrand's approach to the angle case is answering

Modified Bertrand's Question (BQ2): We draw at random two points onto a circle... What is the probability that the chord passing through these two points is longer than the side of the inscribed equilateral triangle?

(2014:5)

Evidently this is not equivalent to Bertrand's question, but

> if we uncritically mix non-equivalent questions (erroneously considering them as equivalent) into a single fuzzy question, by doing so we will artificially create a paradox.
>
> (2014:5)

Bertrand's question is fuzzy because it muddles up the easy and hard problems. Their disambiguation is distinct from Marinoff's because

> it is not the term "at random" which is [ambiguous] but the nature of the entity which is randomized. Is such entity a straight line (a stick), a couple of points (two pebbles), or a point and a straight line (a pebble and a stick)? Only once the above question is clearly answered, by specifying the nature of the entity which is subjected to the random process, one obtains a well-posed problem, which can then be solved.
>
> (2014:14)

> It could be objected that in our approach Bertrand's paradox recurs in a different form, as we have obtained three different averages... which correspond to the three different values proposed by Bertrand. It is crucial however to observe that these three averages are now the solutions of three very different problems, referring to three very different entities.
>
> (2014:15)

This is obviously an articulation of the diagnosis of the Distinction strategy—Bertrand's question is unanswerable *as it stands*, being a clumsy way of raising whilst confounding distinct *well-posed* questions with unique answers. And yet, Aerts and Sassoli de Bianchi say.

> Shackel pointed out... two different strategies [Distinction and Well-posing]... The approach we have here presented, however, cannot be reduced to one of them.
>
> (2014:14)

It is something of a mystery to me why they think they are not using the Distinction strategy merely because their method of finding probability measures for their distinct well-posed questions, in other words solving the hard problem, is (as we shall see) a kind of meta-indifference.

Their distinguishing modelization and randomization may appear to be a further critique of the rigour of Bertrand's procedures, and insofar as Bertrand ever thought in these terms they have a point. I doubt he did think in these terms and I think the terms of my rigorous construction of his cases are a better representation of what lies behind his quite brief exposition of the paradox. That being said, and whatever is true about what was in his mind, the rigorous construction I conducted of Bertrand's cases shows that Bertrand is not remotely committed to the ground of this critique.

Part of the rigorous construction was the explanation of how using the Theorem of Induced Measure produced a measure representing equal ignorance over the chords. That explanation has nothing to do with procedures for selecting a chord by their endpoints or direction. Instead, it is entirely about properties of, or facts about, the chords—angles with the tangent, distances of parallel chords from the centre, midpoint locations—and about our equal ignorance over all those facts or properties. If all we know of a chord is that it was chosen randomly, and we know nothing of any procedure by which it was chosen, we are equally ignorant over the angle it makes with the tangent, its distance from the center and the location of its midpoint. So yes, Aerts and Bianchi are quite correct to raise the point about understanding 'what a situation of maximum lack of knowledge really implies'. Our rigorous construction of Bertrand's cases not only does precisely that, but it shows what a measure representing that must do for any property of the chord of which we are equally ignorant, and does so without being in any way vulnerable to Aerts and Sassoli de Bianchi's claim that Bertrand's question is a fuzzy uncritical mix of non-equivalent questions.

Consequently modelization, which is their distinction of questions for their Distinction strategy, and a natural point for physicists to make, is entirely irrelevant to the procedures Bertrand is using, or at least, irrelevant to my rigorous reconstruction of those procedures. In particular, those procedures have nothing to do with procedures for selecting the chord but are instead procedures articulating the various equal ignorances over the various properties had by a chord chosen at random.

9.3 Aerts and Sassoli de Bianchi's meta-indifference

Why, then, are we looking further at Aerts and Sassoli de Bianchi's work? Because they say

> BQ can be consistently answered by applying what Shackel calls the *principle of meta-indifference*…. We think that our analysis precisely

avoids the arguments presented by Shackel [against that principle], as it does not lead to the vicious regress he indicates. This because, on one hand, we have disambiguated Bertrand's question by specifying the nature of the randomized entity, and on the other hand we have identified where and how meta-indifference should be applied in the problem, giving rise to a universal average which can be calculated by a discretization procedure, much in the spirit of what is done in mathematics with the so-called Wiener process.

(2014:15, their emphasis)

What they are saying is that having chosen, for example, the modelization of tossing a stick, they offer a way of being meta-indifferent over all randomizations of that model by a way of being meta-indifferent over the probability measures of such randomizations.

I am not here going to worry about the fact that resorting to distinct models falls into van Fraassen's trap. I am interested in the possibility mooted by their kind of meta-indifference and for the sake of the interest in that, although what I will initially look at are their two cases, having done so I will then indicate where and how the meta-indifference might be further generalized and then discuss what hope their kind of measure-theoretic meta-indifference gives for being an instance of the Well-posing strategy.

9.4 The stick-tossing case

The case Aerts and Sassoli de Bianchi address in detail is their modelization of tossing a stick. They offer a way of being meta-indifferent over all randomizations of that model by a way of being meta-indifferent over the probability measures of such randomizations, where the probability measures are defined by probability densities for their model.

A simple parameterization, well adapted to the problem, consists in describing the different lines (sticks) in terms of the angle $\theta \in [0, \pi]$ they make with respect to a given axis, and of their distance $x \in [0, 2]$ from the bottom of the circle in a direction perpendicular to their orientation.

(2014:5)

See Figure 26.

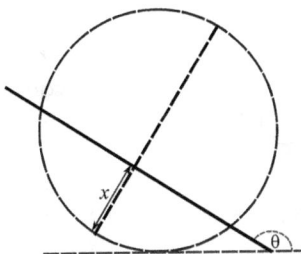

Figure 26 Reprinted from Aerts and Sassoli de Bianchi Figure 2 (2014:6). Chord solid, diameter dashed.

Aerts and Sassoli de Bianchi

> modelize the different possible "ways of tossing a stick".... by means of different probability densities $\rho(x,\theta) \geq 0$... if $A \subset [0, 2]$, and $B \subset [0, \pi]$, then[126]

$$P(x \in A, \theta \in B \mid \rho) = \int_0^2 dx \int_0^\pi d\theta \chi_A(x) \chi_B(\theta) \rho(x,\theta)$$

(2014:6)

So if we consider the set of all such probability densities, we have a family of the probabilities for all the different ways of tossing sticks. Evidently, the stick-tossing treatment is Bertrand's direction case.

> considering that the length of a chord only depends on x, and not on θ, the probability of obtaining a chord of length L greater than $\sqrt{3}$, given that the stick is tossed in the ρ-way, is

$$P(L > \sqrt{3} \mid \rho) = P(x \in \left[\frac{1}{2}, \frac{3}{2}\right] \mid \rho) = \int_0^2 dx \chi_{\left[\frac{1}{2}, \frac{3}{2}\right]}(x) \rho(x) \text{ where}$$

$$\rho(x) = \int_0^\pi d\theta \rho(x,\theta).$$

(2014:6)

---

126 Fubini's theorem justifies this nested integral from the product measure integral $P(E|\rho) = \int_E \rho \, d(\lambda|_{[0,2]} \times \lambda|_{[0,\pi]})$ for $E \subset [0,2] \times [0,\pi]$ that actually defines the probability measure from ρ.

So we have a set of probability densities, each of which gives us the probability measure of a way of randomizing stick tossing and for each member of that set we have the probability of longer conditional on that member. Aerts and Sassoli de Bianchi then address

> The problem with which we are confronted is that of defining and calculating a uniform average over all possible ρ(x), which we will call a *universal average*.

(2014:7)

They define their universal average in terms of step functions. A step function is defined to take a finite number of values on a finite number of intervals of its domain. For example, the function on [0,2] that takes the value zero on [0,1] and one on (1,2] is a step function.

Aerts and Sassoli de Bianchi divide the interval [0,2] into n intervals, or *cells*, as they call them, of length $2/n$ and define $\rho_n(x)$ to be a step function on [0,2] taking one of at most two values on each of the cells. Their example of a step function on 20 cells takes either 0 or 10/9 on each interval 1/10 long (2014:8). I shall call their step functions *bivalent densities*.

They then define the *Average Probability* (capitalizing their name) of these step functions

$$P(x \in [x_1, x_2] \mid n) = \frac{1}{2^n - 1} \sum_{\rho_n} P(x \in [x_1, x_2] \mid \rho_n) \qquad (15)$$

where the sum runs over all the possible $2^n - 1$ (non-zero) cellular probability densities $\rho_n(x)$, made of n elementary cells.

(2014:9)

and then define their meta-indifference, the Universal Average (capitalizing their name), from that average.

$$\langle P(x \in [x_1, x_2] \mid \rho) \rangle^{uinv} = \lim_{n \to \infty} P(x \in [x_1, x_2] \mid n). \qquad (16)$$

(2014:9)

Then we meet the *Crucial Theorem* of their solution:

Theorem

The universal average, *over all possible probability densities* $\rho(x)$, on the interval $[0, 2]$, produces the same probabilities as those calculated by means of a uniform probability density $\rho_u(x) = \frac{1}{2}$, in the sense of the equality

$$\langle P(x \in [x_1, x_2] \mid \rho) \rangle^{uinv} = P(x \in [x_1, x_2] \mid \rho_u) = \frac{x_2 - x_1}{2}. \tag{17}$$

(2014:9 my emphasis)

and consequently the probability of longer given by the Universal Average is $\frac{1}{2}$.

9.5 The pebble-tossing case

Having addressed their stick-tossing case Aerts and Sassoli de Bianchi move on to considering their pebble-tossing case, where a chord is selected by tossing two pebbles onto the circumference.

we need to choose some convenient parameters to univocally identify two points intersecting the unit circle (representing the two pebbles in EQ2). A natural and simple way to do this is by means of two angles $(\alpha, \beta) \in [0, 2 \geq \pi] \times [0, 2\pi]$; see Figure 5.

(2014:12)

The angles in question are the angular component in the polar coordinates of the endpoints. Their Figure 5 is essentially the square in my Figure 9 of Chapter 3. Once again, the ways of tossing two pebbles are formulated as density functions

$$P\big((\alpha, \beta) \in R \mid \rho\big) = \int_0^{2\pi} d\alpha \int_0^{\pi} d\beta \chi_R(\alpha, \beta) \rho(\alpha, \beta)$$

(2014:12)

To define the Average Probability and the Universal Average, two-dimensional step functions on square cells are needed and mentioned. Although they don't give the extension of their method to treating the two-dimensional space in the paper, they give a reference for it (Aerts and Sassoli de Bianchi 2015a, b) and one can see how it would go in broad outline. They then report

a similar proof as per above applies, and one can show that, for any region $R \subset [0, 2\pi] \times [0, 2\pi]$

$$\langle P\big((\alpha,\beta)\in R\mid\rho\big)\rangle^{uinv} = P\big((\alpha,\beta)\in R\mid\rho_u\big) = \frac{\mu(R)}{4\pi^2}$$

where $\rho_u(\alpha,\beta)=1/4\pi2$, and $\mu(R)$ denotes the Lebesgue measure of region R.

(2014:12)

When applied to the region corresponding to the longer chord, the probability of longer given by the Universal Average is ⅓.

Similarly (2014:13–14), they treat the midpoint case by applying their method to define the Universal Average on the disk, giving the probability of longer to be ¼.

9.6 A virtue of the Universal Average

Before we move on to a critique, it is worth mentioning a virtue of the Universal Average. Why does the principle of indifference seem true? Consider the case of a die. I am convinced that the principle is correct for this: the counting measure fulfills the normative measure role and correctly gives the probability of a 5 to be 1/6. The Humean objection would be that prior to experience, I have no reason to assign any probabilities to getting a 5, indeed, for all I know it might turn into a dove and fly away. Well there is a point of that kind to be made, so I am making some empirical assumptions, namely that in asking for the probability of a 5 I am asking about the tossing of cubes that do not do that sort of thing. I cannot see any objection to that assumption: it does not seem to be an arbitrary restriction on possibilities relevant to the case and if such possibilities are included then we may have to agree that the probability of any number at all is zero.

Hume may grant me that much and continue, yes, if the material density of the cube is uniform, the counting measure will correctly express my equal ignorance and can therefore be used for the normative measure, but if not, then not. What Aerts and Sassoli de Bianchi give us, I think, is an answer to Hume on this point. The Universal Average, suitably extended for the three-dimensional case, is a method of being meta-indifferent over all possible material densities of the cube. If that meta-indifference results in the normalized counting measure, we have an answer to give Hume.

9.7 Generalizing the Universal Average

As promised, here is a sketch of how Aerts and Sassoli de Bianchi's Universal Average can be generalized. Essentially, their parameterization of the stick-tossing case is a naming, as I have defined a naming. (Indeed, we have a renaming to our treatment of the direction case by r: $(x, \theta) \rightarrow (R(x-1), \theta+\pi/2)$ if $\theta \in [0,\pi/2)$ and r: $(x, \theta) \rightarrow (R(x-1), \theta-\pi/2)$ if $\theta \in [\pi/2, \pi)$.) Since we have a naming, our Corollary 10 shows that any probability measure on the chords corresponds to a probability measure on the naming space that gives the same probabilities to events. We have a measure space on the names $[0, 2] \times [0, \pi]$ using the relevant restrictions of the Lebesgue measure of $\mathbb{R}^2$, σ-algebra $\Sigma = [0, 2] \times [0, \pi] \cap \mathscr{L}$ and measure $\mu = \lambda|_{[0, 2] \times [0, \pi]}$. By the Radon-Nikodym theorem,[127] for any probability measure, P, on the names that is absolutely continuous with respect to μ there is a probability density function $\rho(x, \theta)$ such that

$$P(\sigma) = \int_{0} p d\mu, \quad \sigma \in \Sigma$$

So all probability measures on the chords whose corresponding probability measure on their names under this naming correspond to a member of this family of probability density functions (provided that the measure on the names is absolutely continuous with respect to the Lebesgue measure, which I would expect them to be). We can then define bivalent densities on $[0, 2] \times [0, \pi]$ by the method used for their pebble case two-dimensional bivalent densities and the define the Average Probability and the Universal Average. So in whatever sense their method makes the Universal Average an average over all probability measures for stick tossing or for pebble tossing, in the same sense does their naming of the stick-tossing case make their Universal Average an average over all probability measures on the chords.

Since they don't give their method for treating the two-dimensional space in the paper I will examine further their Universal Average in their stick-tossing case. The problems that arise will arise also for two-dimensional cases because the problems do not depend especially on the dimension but on the various constraints made use of in defining the Average Probability, how that carries forward to the Universal Average and the reason they give for why the Universal Average is a kind of meta-indifference over all the probability measures.

9.8 A danger of triviality for their Crucial Theorem

Aerts and Sassoli de Bianchi's definition of the Average Probability and its limit the Universal Average conceals a certain triviality in their method and their theorem. Each of their step functions on n cells takes only two values.

---

127 Bruckner *et al.* 2008:220, Theorem 5.29.

For $k=1, \ldots, n$, the number of non-zero cells is k and the two values they allow are zero and $n/2k$. I named these step functions *bivalent densities*. Let M be the number of distinct bivalent densities on n cells on $[0,2]$ and let the set $R_n = \{\rho_m : m=1,\ldots, M\}$ be the set of all bivalent densities on n cells. For $i=1,\ldots, n$, let $c_{i,m}$ be the value that $\rho_m \in R_n$ takes on the ith cell, J_i, which interval has characteristic function χ_{J_i} and where $\bigcup_{i=1}^{n} J_i = [0,2]$. Then

$$\rho_m = \sum_{i=1}^{n} c_{i,m} \chi_{J_i}$$

Let $\langle \rho^n \rangle$ be the mean bivalent density on n cells, which is defined by the mean on the cells

$$\langle \rho^n \rangle = \sum_{i=1}^{n} \hat{c}_i \chi_{J_i}, \quad \text{where } \hat{c}_i = \frac{1}{M} \sum_{m=1}^{M} c_{i,m}$$

$P(x \in [a,b]| \langle \rho^n \rangle)$ is the Average Probability! For the proof I use the definition of Lebesgue integration for a simple function on a subset of its domain.[128] The first equality gives their original definition, supplemented with $\rho_n \in R_n$ to make explicit the domain they are summing over and the second restates the sum of that definition in my notation.[129]

$$P(x \in [a,b]| n) = \frac{1}{2^n - 1} \sum_{\rho_n \in R_n} P(x \in [a,b]| \rho_n) = \frac{1}{M} \sum_{m=1}^{M} P(x \in [a,b]| \rho_m)$$

$$= \frac{1}{M} \sum_{m=1}^{M} \int_{[a,b]} \rho_m d\lambda = \frac{1}{M} \sum_{m=1}^{M} \sum_{i=1}^{n} c_{i,m} \lambda \left(J_i \cap [a,b] \right) = \sum_{i=1}^{n} \left(\lambda \left(J_i \cap [a,b] \right) \left(\frac{1}{M} \sum_{m=1}^{M} c_{i,m} \right) \right)$$

$$= \sum_{i=1}^{n} \hat{c}_i \lambda \left(J_i \cap [a,b] \right) = \int_{[a,b]} \langle \rho_n \rangle d\lambda = P(x \in [a,b]| \langle \rho^n \rangle)$$

Consequently the Universal Average is the limit as $n \to \infty$ of $P(x \in [a,b]| \langle \rho^n \rangle)$.

---

128 See, e.g. Capinski and Kopp 2004:76 definition 4.2.
129 I have done this to avoid a potential confusion because of their using the subscript index of ρ differently from me. As they introduce it, the n in ρ_n indexes the number of cells on which ρ_n is a bivalent density whereas I use the index on ρ to keep track of the M individual bivalent densities in R_n. We reconcile the notation by noting that their ρ_n and my ρ_m are variables with the same domain, namely, R_n. $M=2^n-1$ and makes the maths less hairy.

Now we calculate $\langle \rho^n \rangle$. On each J_i we are taking the average, $\hat{c}_i$, of all the $c_{i,m}$. If $c_{i,m} = n/2k$ on J_i there are $n-1$ remaining cells from which to choose $k-1$ cells to be $n/2k$ and hence there are C_{k-1}^{n-1} bivalent densities on which $c_{i,m} = n/2k$ on J_i. Similarly, there are C_k^{n-1} bivalent densities where $c_{i,m} = 0$ on J_i.[130] Consequently, the mean value on J_i of the ρ_m that take the value $n/2k$ on k cells, $\hat{c}_{i,k}$, is

$$\hat{c}_{i,k} = \frac{n}{2k} \frac{C_{k-1}^{n-1}}{C_{k-1}^{n-1} + C_k^{n-1}} = \frac{1}{2}$$

Since this is independent of k, the mean value on J_i of all the ρ_m is the mean of $\hat{c}_{i,k}$ over all $k = 1, \dots n$, so

$$\hat{c}_i = \frac{1}{M} \sum_{m=1}^{M} c_{i,m} = \frac{1}{n} \sum_{k=1}^{n} \hat{c}_{i,k} = \frac{1}{n} \sum_{k=1}^{n} \frac{1}{2} = \frac{1}{2}$$

and hence

$$\langle \rho^n \rangle = \sum_{i=1}^{n} \hat{c}_i \chi_{J_i} = \sum_{i=1}^{n} \frac{1}{2} \chi_{J_i} = \frac{1}{2} \chi_{[0,2]} = \rho_u \quad \text{since} \quad \bigcup_{i=1}^{n} J_i = [0,2].$$

So the mean, $\langle \rho^n \rangle$, of all the bivalent densities on n cells is the uniform probability density function on $[0,2]$, which they call 'ρ_u'. This holds for all n and so the Average Probability, $P(x \in [a,b] | \langle \rho^n \rangle)$, *is the same for all n!*

Consequently, the content of Aerts and Sassoli de Bianchi's Crucial theorem, their line (17), is proved directly:

$$\langle P(x \in [x_1, x_2] | \rho) \rangle^{uinv} = \lim_{n \to \infty} P(x \in [x_1, x_2] | n) = \lim_{n \to \infty} P(x \in [x_1, x_2] | \langle \rho^n \rangle)$$

$$= P(x \in [x_1, x_2] | \langle \rho^n \rangle) = P(x \in [x_1, x_2] | \rho_u) = \frac{x_2 - x_1}{2}.$$

This, however, is not good news. Since the Average Probability is the same for all n, the Universal Average has been defined to be the limit of a constant sequence, each of whose members is the uniform probability measure. Once we realize that this is what is going on, that the Universal Average should turn out to be the uniform probability measure is now no longer a substantial fact but a triviality!

---

130 For a check, the number of ρ_m with k non-zero cells is $C_k^n = C_{k-1}^{n-1} + C_k^{n-1}$ (Pascal's formula for his triangle).

9.9 We need the Meta-indifferent Average

The proven content of the Crucial theorem is *nothing more than* the truth of the equation of line (17) and that makes the theorem trivial, since it is merely the truth that the limit of a particular constant sequence is the constant. Strictly speaking, the clause in the theorem that I emphasized in my quotation of it above 'The universal average, *over all possible probability densities*' is not proven but is an interpolation of what Aerts and Sassoli de Bianchi would *like* to be proven by proving the equation. In other words, that interpolation is an interpretation. It is an interpretation that perhaps had some plausibility before we saw the triviality, but the fact that the equation is trivial in the way we have seen raises some suspicion about the entitlement to that interpretation.

What matters to us is that we want a universal average in the sense of a philosophically significant form of meta-indifference over all the probability measures. To avoid verbal confusion, let us call the philosophically significant universal average the *Meta-indifferent Average*. They are speaking of the probability measure defined by the Universal Average as if it were the Meta-indifferent Average. This is done in the statement of the theorem by the interpolation, but in so doing they face a dilemma between a definition that trivializes the claimed solution and the interpretation they want and think themselves entitled to, an entitlement that we shall see to be flawed in a number of ways.

The Meta-indifferent Average cannot be achieved by mere terminology, a surreptitious definition achieved by the interpolation, in which 'the universal average over all the probability measures' is simply defined to *mean* their Universal Average. Obviously, that is no solution at all since it is really a declaration, rather than a proof, of a very old claim—that the uniform measure is the average of all the probability measures—a declaration made here for no other reason than that it is the average of the bivalent density defined probability measures. Nor can it be achieved by merely *treating* the Universal Average as if it were the Meta-indifferent Average. We need some justification for why it is so. We will now look the bases for their believed entitlement.

9.10 Is the Universal Average justifiable as the Meta-indifferent Average?

For each n, there are uncountably many step functions on the interval [0,2] divided into n intervals. To define their Average Probability Aerts and Sassoli de Bianchi need to restrict, for each n, the number of step functions in its domain to a finite number. They use three restrictions on the domain to define the bivalent densities. First, the intervals are of equal length, second the integral is one, and third, each takes no more than two values, and if two, one is zero.

The middle constraint is justified because bivalent densities are probability densities but the other two are arbitrary. The three constitute an important domain restriction and the entire basis on which they can claim that

> the uniform average (15) [their Average Probability defined by their line (15) as above] being over a finite number of possibilities, it is uniquely defined and does not suffer from possible ambiguities.
>
> (2014:9)

The integral constraint for intervals of equal length, $\lambda(J_i) = 2/n$, gives

$$1 = \int_{[0,2]} \rho_m d\lambda = \sum_{i=1}^{n} c_{i,m} \lambda \left(J_i \cap [0,2] \right) = \frac{2}{n} \sum_{i=1}^{n} c_{i,m}, \quad i.e. \quad \sum_{i=1}^{n} c_{i,m} = \frac{n}{2}$$

So each step function that satisfies the integral and equal interval length constraint is defined by a non-standard partition of $n/2$.[131] Obviously, for each n, there are infinitely many such step functions. So the area and equal interval constraints on their own are insufficient to permit there being an Average Probability derived from these step functions. The restriction to at most two values is also necessary.

Now the stipulation of two values is arbitrary, but if we could prove the same result using any finite number of values up to n, its arbitrariness would be eliminable. However, as soon as we allow the $c_{i,m}$ to take three different values, one zero and two non-zero ones, we no longer have finitely many step functions. This is straightforwardly proved. Let all but the first two $c_{i,m}$ be zero. Then the other two define and are defined by standard partitions of $n/2$ by $0 = x_0 < x_1 = c_{1,m} < x_2 = c_{1,m} + c_{2,m} = n/2$, and there are continuum many partitions of $n/2$ into two pieces.

The stipulation of no more than two values for $c_{i,m}$ is therefore an ineliminable constraint on the step functions used for their result, but its justification is merely heuristic rather than a ground for what is built on it to be the Meta-indifferent Average.

At this point Aerts and Sassoli de Bianchi may say that this doesn't matter. What matters is that

> the $\rho_n(x)$ are dense in the space of probability densities (in the sense specified above).
>
> (2014:9)

---

131 If $0 = x_0 \leq x_1 \leq x_2 \leq \ldots \leq x_n = n/2$, $c_{i,m} = x_i - x_{i-1}$. Non-standard because we use $\leq$ rather than $<$ to allow of some $c_{i,m}$ being zero.

by which they mean that a sequence $\{\rho_n\}$ of bivalent densities can be chosen such that

Let... $[x_1, x_2]\subset[0, 2]$.... $P(x\in[x_1, x_2]|\rho_n)\to P(x\in[x_1, x_2]|\rho)$ as $n\to\infty$.

(2014:7–8)

The density claim is a foundation of their claim that the Universal Average is the Meta-indifferent Average. This foundation is uncertain.

The topological definition of density is that a set X is dense in Y iff $Y=$ the closure of X. Applied to a metric space with metric, d, gives a standard theorem, X is dense in Y iff for all $y\in Y$ there exists a sequence $\{x_n\}$ such that $\lim_{n\to\infty}d(x_n, y)=0$.[132] Applied here gives

Theorem 39

Let K be the set of all probability densities, D be the set of all *bounded* probability densities, and BD the set of all bivalent densities, on [0,2]. Let Π_K be the set of probability measures defined by K, and Π_{BD} the set of probability measures defined by BD. All except K are subsets of the metric space of bounded real valued functions with metric $d(f, g)=\sup_{x\in[0, 2]}|f(x) - g(x)|$. Then

(1) BD is dense in D iff for all $\rho\in D$ there exists a sequence $\{\rho_n\}\in BD$ such that $\lim_{n\to\infty}d(\rho_n, \rho)=0$.

(2) Π_{BD} is dense in Π_K iff for all $P\in K$ there exists a sequence $\{P_n\}\in\Pi_{BD}$ such that $\lim_{n\to\infty}d(P_n, P)=0$.

BD being dense in D does not imply BD is dense in K because D is a proper subset of K, a point that creates a strand of additional difficulties in each problem we show below, but which for brevity we shall leave for the reader to note as we go along.

The point of using a dense subset of a space as Aerts and Sassoli de Bianchi wish to use it is that one can prove certain things for the space by proving it for the dense subset. Aerts and Sassoli de Bianchi prove neither of the right-hand sides of (1) or (2). For example, instead of the convergence they claim to prove (above and 2014:8), they would need to prove that convergence not just for intervals of [0,2] but for all $E\in\mathscr{L}\cap[0,2]$.[133] More importantly, for the density claim they would need to prove that the convergence is not just pointwise but *uniform*, whereas their work attempts only the former. Consequently, their meaning of 'dense in the

---

132 See Bruckner *et al.* 2008:350.

133 If they had proved what they claim to have proved (see below) this can be fixed for the Borel algebra on [0,2], which is the σ-algebra generated by the π-field of intervals (see §14.8), but that algebra is still only a proper subset of $\mathscr{L}\cap[0,2]$. That being said, it might be reasonable to restrict the domain of Π_K to the Borel algebra.

Table 3 The level ω row is the row of limits of what lies below

Level number	Sets of densities	Averages of probability measures
	K	Meta-indifferent Average
???	???	???
ω	$B_\omega = ???$	Aerts and Sassoli de Bianchi Universal Average $\langle P(x \in [a, b]|\rho)\rangle^{\text{univ}} = (b-a)/2$
...	...	...
limit $n \to \infty$	limit $n \to \infty$	limit $n \to \infty$
↑	↑	↑
...	...	...
n	$B_n = \{\rho_{n, m}: m=1, ..., 2^n-1\}$	$P(x \in [a, b]|n) = (b-a)/2$
...	...	...
5	$B_5 = \{\rho_{5, m}: m=1, ..., 31\}$	$P(x \in [a, b]|5) = (b-a)/2$
4	$B_4 = \{\rho_{4, m}: m=1, ..., 15\}$	$P(x \in [a, b]|4) = (b-a)/2$
3	$B_3 = \{\rho_{3, m}: m=1,...,7\}$	$P(x \in [a, b]|3) = (b-a)/2$
2	$B_2 = \{\rho_{2, m}: m=1, 2, 3\}$	$P(x \in [a, b]|2) = (b-a)/2$
1	$B_1 = \{\rho_{1, 1}\}$	$P(x \in [a, b]|1) = (b-a)/2$

space of probability densities' is eccentric and does not licence the assumption that what they prove for the Universal Average suffices to makes it the Meta-indifferent Average.

Now consider the stack of bivalent densities $\rho_{n, m}$ as shown in Table 3. The level number, n, is number of intervals into which [0, 2] is divided. We name the M_n bivalent densities at the nth level by the index numbers m (i.e. the m index used for ρ_m above). For each n, the set $B_n = \{\rho_{n, m}: m=1, ..., M_n\}$. K is the set of all probability densities on [0, 2].[134]

The Average Probability, for each level, is the mean of the probabilities got from the probability densities of that level. The Universal Average is merely the limit of Average Probabilities as the level number tends to infinity. B_ω is therefore the set of probability density functions over whose probability measures the Universal Average is an average. What justifies taking the Universal Average to be that average is the thought that the average of the limit should be the limit of the averages.[135]

What would then justify Aerts and Sassoli de Bianchi in taking their Universal Average to be the Meta-indifferent Average is for B_ω to be K. This may appear true but it is not. Recall the construction of the members of B_n:

134 D above is only a proper subset of K.
135 This kind of thought applied to integrals is the motivation for proving theorems like the monotone convergence theorem and the dominated convergence theorem (Bruckner *et al.* 2008:198 and 203). Aerts and Sassoli de Bianchi have not formulated things in a way that could use those theorems on this point. Nor do they use the monotone convergence theorem where they might have in 2014:8.

They are step functions taking values zero or $n/2k$ on equal sized intervals of length $2/n$. In the limit, therefore, they are functions taking values zero or ∞ on intervals of length 0, whose integral is 1. So B_ω, which is the limit as $n \to \infty$ of B_n, is a set of Dirac delta functions and therefore $B_\omega \neq K$.

The entirety of any argument they give that can be given to support the identity $B_\omega = K$ is that the bivalent densities are dense, in their sense, in K. That doesn't help, because it means only that for all $\rho \in K$ there is a sequence of bivalent densities $\{\rho_n\}$ such that $\lim\limits_{n \to \infty} \int_a^b |\rho - \rho_n|\, dx = 0$, which is irrelevant to the identity.

What is left to justify why the Universal Average is the Meta-indifferent Average is this:

1. The Universal Average is by definition the limit of the horizontal row Average Probabilities.
2. Every probability measure defined by a $\rho \in K$ is the limit of a sequence probability measures defined by a vertical path through the stack of bivalent densities (i.e. a sequence $\{\rho_n\}$, $\rho_n \in B_n$).
3. Therefore the Universal Average is the Meta-indifferent Average.

The argument is invalid. However, so long as we have in the back of our mind that $B_\omega = K$, the argument has considerable cogency. But since $B_\omega \neq K$, the Universal Average is not directly an average over K. In reply, it might be said that the Universal Average is an *indirect* average over K by being an average over approximations, because the family of sets of bivalent densities, $\mathscr{B} = \{B_1, B_2, B_3, ...\}$, is a family of approximations to K. But the explanation is false: $\mathscr{B}$ is not a family of approximations tending to K but tending to B_ω.

Premiss 2 still supplies some support. Yes, because $B_\omega \neq K$, the Universal Average is not meta-indifferent over all probability measures in the straightforward way that it appeared to be, not even indirectly. Nevertheless, there is a sense in which every probability measure defined by a $\rho \in K$ is *represented* by $\mathscr{B}$, in the way described by Premiss 2, and *that* is the way in which the Universal Average is meta-indifferent over all probability measures.

Unfortunately, although Aerts and Sassoli de Bianchi appear to prove Premiss 2, they haven't. What gives the impression that they have is their claimed proof that:

> given a $\rho(x)$, and an interval $A \subset [0, 2]$, we can always find a suitable sequence of cellular $\rho_n(x)$ such that $P(x \in A|\rho_n) \to P(x \in A|\rho)$, as $n \to \infty$, for all A.

(2014:7)

As stated this appears to be the claim that for each $\rho \in K$ there is a *single* sequence $\{\rho_n\}$ doing this job. Indeed, it must be a single sequence since Premiss 2 requires Π_{BD} to be dense in Π_K. The appearance is mistaken.

We need to clarify an ambiguity in the order of quantifiers. As written, up to the final clause gives 'for each $\rho \in K$, for each interval, there exists a sequence $\{\rho_n\}$ such that ...etc'. This won't do for Premiss 2 since it allows that there can be a *different* sequence for each interval and proving only that much does not prove that every probability measure defined by a $\rho \in K$ is the limit of a sequence of probability measures defined by a sequence $\{\rho_n\}$. We might read the final clause as within the scope of the existential quantifier giving us the correct order needed for the truth of Premiss 2 'for each $\rho \in K$, there exists a sequence $\{\rho_n\}$ such that for each interval A... etc'.

The proof they give proves the proposition under the first order of quantifiers, not the second, and so fails to prove Premiss 2. First they put in place the machinery that shows how the integral over the ends of an interval $[a, b]$ that occupy only partial cells of a bivalent density tend to zero. This means we need only worry about proving the convergence on that part of the interval occupying full cells. The critical line to prove $P(x \in [a, b]|\rho) = \lim_{n \to \infty} P(x \in [a, b]|\rho_n)$ is this

we can always choose $\rho_n(x)$ in such a way that $\tilde{n}_i / \tilde{n} \to \int_{S_i} \rho(x)dx$ as

$l \to \infty$, for all $i = 1, ..., m$. This is because for each i, the probability $\int_{S_i} \rho(x)dx$ is a real number with value in the interval $[0, 1]$, and rational numbers of the form $\tilde{n}_i/\tilde{n}$ with $0 \le \tilde{n}_i \le \tilde{n}$, $\tilde{n} > 0$, are dense in $[0, 1]$.

(2014:8)

Each S_i is a $2/m$ long interval of $[0,2]$ so there are m such intervals. What this proves is that for each S_i there exists a sequence $\{\rho_n\}$ for which

$$\int_{S_i} \rho_n(x)dx = \frac{\tilde{n}_i}{\tilde{n}} \to \int_{S_i} \rho(x)dx.$$

So the existential quantification over sequences is within the scope of the universal quantification over the intervals S_i. Therefore the quantifiers are the wrong way round and this allows different sequences for different S_i.

It might be thought that this could be fixed in the following way. For each S_i we have a sequence $\{\rho_{n,i}\}$ where for all n, $\rho_{n,i} \in B_n$. So lets stitch together a single bivalent density sequence that covers all the S_i as follows:

Let χ_{Si} be the characteristic function for S_i and let$\{\pi_n\}$ be the sequence defined by $\pi_n = \sum_{i=1}^{m} \rho_{n,i}\chi_{Si}$, for all n.

But since $\{\rho_{n,i}\}$ can be distinct, there exist π_n that are *not* bivalent densities: although a π_n will be a step function, nothing about this constrains a π_n to be either bivalent, since the non-zero value of $\rho_{n,i}$ need not be the non-zero value of $\rho_{n,j}$, nor need its integral be 1, that is it need not be a probability density function.

Consequently, Premiss 2 is not proved. Instead, all we now have is

1. The Universal Average is by definition the limit of the horizontal row Average Probabilities.
2. The probability that $x \in [a,b]$ given by a probability measure defined by a $\rho \in K$ is the limit of the probability that $x \in [a, b]$ given by a sequence of probability measures defined by a sequence $\{\rho_n\}$ such that $\rho_n \in B_n$
3. Therefore the Universal Average is the Meta-indifferent Average.

Since for a single probability measure in K, this Premiss 2 allows that the probability of different events may require different sequences, the sense in which the Universal Average is supposed to be the Meta-indifferent Average on this basis is now quite obscure.

Now I do not doubt that these objections are of the kind that make scientists think philosophers waste their time arguing about mere conventions. For the Universal Average is, in some etiolated sense, an average over K and there are no other proposals. Indeed, some of the issues we saw along the way, such as that there are continuum many three valued step functions which blocked forming an average over them, could be used to argue that only something like the Universal Average will work for formulating a Meta-indifferent Average.

That last claim is false: in the next chapter I will show how to define a probability measure that is the measure-theoretic expectation of all the probability measures in a set. Furthermore, the danger here is the retreat into mere methodology. No one ever doubted that we could find ad hoc solutions to Bertrand's paradox and so when offering a candidate Meta-indifferent Average, it must be sufficiently justifiable as not ad hoc. For example, although it failed, the justification for using the Maximum Entropy principle as a variety of meta-indifference (not a variety of the Meta-indifferent Average, of course) was much stronger than what we are left with here. So whilst I readily grant the frustration and certainly was myself quite taken with Aerts and Sassoli de Bianchi's proposal, and whilst I also grant that these worries are not absolutely conclusive, nevertheless, they substantially weaken the prospect that the Universal Average, if successfully extricated from van Fraassen's trap and generalized as I showed here, could solve Bertrand's paradox. I now turn to a further and independent reason to think that a successful extrication still will not be a solution.

9.11 The Universal Average cannot avoid the paradox

Recall that we are addressing a generalization of the Universal Average based on the fact that if we have a naming, our Corollary 10 shows that any probability measure on the chords corresponds to a probability measure on the naming space that gives the same probabilities to events. The stick-tossing case gave us a naming by $[0, 2] \times [0, \pi]$.

In the pebble-tossing case Aerts and Sassoli de Bianchi use the square $[0, 2\pi] \times [0, 2\pi]$ for the pebbles giving the angular component in the polar coordinates of the endpoints of the chords. We saw in §3.11 that this is not a naming, contrary to their claim that they have chosen 'some convenient parameters to *univocally* identify two points intersecting the unit circle' (2014:12, my emphasis). We also saw how to fix this to give a naming by the quotient space shown in Figure 10.

Given that naming, we can now generalize the Universal Average for that naming as before. Starting with Corollary 10 we can apply the Radon-Nikodym theorem to probability measures on the naming quotient space to give us the set of all probability densities on the chords, K. We can then define bivalent densities on the quotient space by the method used for their pebble case two-dimensional bivalent densities, even in terms of step functions on right angled triangles instead of squares if that is easier, and then define the Average Probability and the Universal Average.

So we have two distinct ways of generalizing the Universal Average. We already know what each of those ways will give for the probability of longer. In §3.12 we proved that the cross sections were uniform and so the above treatment by the naming will give the same result as their treatment of the stick-tossing case in terms of the cross section: probability of longer = ½. Since the quotient space equivalence classes are defined by the reflection in $x = y$ it will give the same result as their treatment of the pebble case: probability of longer = ⅓. So Bertrand's paradox recurs.

Indeed, we can generalize Aerts and Sassoli de Bianchi's method using *any naming whatsoever* and doing so will deliver whatever the uniform probability measure on that naming space gives for the probability of longer. Consequently, if we generalize Aerts and Sassoli de Bianchi's Universal Average in order to avoid falling into van Fraassen's trap, we drown instead in the Deluge theorem 24.

10 Meta-indifference

10.1 Introduction

In Chapter 5 I discussed the diagnosis of the Distinction strategy: that Bertrand's question is unanswerable as it stands, being a clumsy way of raising whilst confusing distinct well-posed questions with unique answers. I argued that it is not vague or ambiguous: it is simply a general question. The best that a Distinction Strategist can make of the diagnosis is that it is a *generic* question rather than a general question. In response, and without conceding the point, I argued that even if Distinction Strategist *could* reject the general question, the principle of indifference warrants a statistical generalization over the distinct questions which the generic question subsumes. Ignorance of which method of random choice has been used is just more ignorance, and the principle of indifference is supposed to warrant applying equiprobability over equal ignorance, so equiprobability should be assigned to *those* possibilities. I called this *meta-indifference*. Meta-indifference might entail consistent probabilities for the probability of the longer chord, and were it to do so meta-indifference would thereby provide a solution to Bertrand's paradox by the Well-posing strategy.

Left over from that chapter was one option for the Distinction Strategist. I showed that the Distinction strategy fails unless Bertrand's question is a generic question which cannot be a general question *and* meta-indifference fails to entail a probability of a longer chord. In this chapter, among other things, I shall argue that meta-indifference cannot fail to entail a probability of a longer chord. That completes the refutation of the Distinction strategy.

It seems to me there are only two kinds of meta-indifference that *could* do work in addressing Bertrand's paradox and so supply a meta-indifferent Well-Posing strategy. One evaluates all the possible probability measures individually and picks the best on that evaluation and the other takes some kind of average over them all. Before we turn to the arguments, I concede that I cannot prove that no other kind of meta-indifference than those

DOI: 10.4324/9781003456308-10

I address here could be developed. We can, perhaps, see one such glimmering in Jaynes' notion of indifference over problems.

So perhaps there are many ways to be meta-indifferent and perhaps some don't fall under my kinds. Among those there might be some which avoid the arguments I shall make. The question would then be whether they do any useful work in addressing Bertrand's paradox. For these reasons, I am to some extent shadow boxing in this chapter. In the previous two chapters I have addressed the only attempts at meta-indifference I know of. I have shown them to fail and in this chapter I say what can be said in general about the first kind and give a general argument against the second. Consequently, absent any contenders for the first kind, the general argument shows that if the Well-posing strategy can work at all, it can work at the base level. The chapter finishes by addressing Aerts and Sassoli de Bianchi's claim that their Universal Average does not fall to my general argument.

10.2 Meta-indifference by Individual Evaluation

Whether individual evaluation of the probability measures and choosing the best under that evaluation is really *meta-indifference* could be denied. Certainly, it is a way of taking into account all the probability measures rather than picking on one of them for some reason or other. Does that suffice, however, to make it a way of considering second-order equal ignorance? This is unclear.

For example, the evaluation by entropy in Chapter 8 assumed that probability measures capture equal ignorance to varying degrees and defines a measurement of those degrees. Ignorance produces uncertainty and equal ignorance means equal uncertainty and equal surprise whatever the outcome. We measured the uncertainty of each probability measure by its entropy. In broad terms, that is a defensible measure of uncertainty because entropy varies negatively with the concentration of the probability measure. The more the probability measure is predominantly in narrower regions of the possible events, the less uncertain the outcomes are under that probability measure, and also the lower the entropy. Entropy is thereby order correlated with uncertainty. Equal ignorance over a region of events is maximal uncertainty of that region and by the aforementioned correlation, that is correlated with maximum entropy. Hence the best probability measure was defined to be whichever probability measure maximized entropy.

Admittedly, then, entropy isn't exactly second-order ignorance, but it does amount to a second-order evaluation of the extent to which equal ignorance over the events is represented by a probability measure. Furthermore, fussing too much about the terminology of meta-indifference is of little interest. I am happy to grant that any evaluation that can be applied to all the probability measures, that represents in some way a second-order view of the relation of each probability measure to equal ignorance, and for which the evaluation defines an order on the probability measures, is a contender for a Well-posing strategy of a distinct kind. I name that kind *Meta-indifference by Individual Evaluation* because it roughly fits meta-indifference and that will do.

As we saw, the Maximum Entropy principle didn't work for reasons to do both with entropy but also to do with various difficulties that may threaten any such contender. We may find that the proposed method of evaluation doesn't cover all the probability measures and then there may be hard questions over whether what is left suffices for the evaluation to remain a contender. There are dangers of recurrence of the paradox if ways of applying the evaluation vary or if more than one probability measure comes out best. New revenge paradoxes can arise. The moves needed to avoid these dangers may result in a collapse into the original paradox.

Beyond noting the above dangers, I don't have further argument to give. I doubt there is a general argument to be given. As we saw with the Maximum Entropy principle, the devil will be in the detail. For example, we might regard entropy as grading probability measures as being more or less random and there might be alternative ways of defining an order on randomness. With that thought in mind, the middle instance of Well-posing in Chapter 6 might be adapted. Although Wang and Jackson made no attempt to present their distinction between a chord drawn at random and truly random chords as a matter of degree, they might attempt to supplement their attempted solution by defining an ordering of randomness under which the Poisson process, and thereby what they call truly random chords, comes out on top. Absent further contenders for Meta-indifference by Individual Evaluation, we leave this kind of Well-posing strategy as defeated so far as any known contenders go and await future attempts.

10.3 Meta-indifference by Joint Averaging

The second kind of meta-indifference is when, extending equal epistemic status from first-order over events to second-order over the probability measures of events, we hope to get an average over the first-order probability measures of events. For that we need a second-order probability measure over the set of first-order measures. Aerts and Sassoli de Bianchi propose a simple example:

Modified Bertrand's Question (BQ4): We choose at random one of the following three processes, then perform it: (1) we draw at random a chord onto a circle; (2) we draw at random two points onto a circle and draw a chord passing through them; (3) we draw at random a point inside a circle and a chord having such point as its middle point. What is the probability that the obtained chord is longer than the side of the inscribed equilateral triangle?

(2014:15)

Here we are given the count measure for the normative measure in POIS, resulting in a second-order probability measure giving each first-order measure equal probability of ⅓. This then defines the average probability measure for the chords and gives the probability of longer as ⅓(½+⅓+¼)=13/36.

I now show how we can define a Meta-indifferent Average over the probability measures. In Meta-indifference by Joint Averaging, I apply the principle of indifference as developed before, only at the second order. Once again, we need a measure to get a measure and so equal ignorance over the first order measures requires a measure on a σ-algebra of the first order measures giving equal measure to members of that σ-algebra with equal epistemic status. I use that to define a second-order probability measure by which I then define the probability on events by the expectation over the first order probability measures. Treating this all in terms of namings, which is justified by the theorems of §2.6, allows me to insert the following definition into the other apparatus of the FPOIS:

Principle of Meta-Indifference for Sets (POMS): Let Ψ be the set of first order probability measures from all the (Ω, Σ, P), each of which (Ω, Σ) is a naming space of the same event space, $(E, \mathcal{E})$. Let $\mathcal{M}$ be a σ-algebra on Ψ such that for all $\sigma \in \Sigma$ and for all $x \in \mathbb{R}^+$, $\{P : P(\sigma) \in [x, \infty)\} \in \mathcal{M}$.[136] Let ν be a non-zero finite measure on $\mathcal{M}$. Given that we have no reason to discriminate between members of $\mathcal{M}$ with equal measures under ν, the members of $\mathcal{M}$ with equal ν-measure have equal second order probability.

I then define a *Meta-indifferent Probability Measure* for the naming space (Ω, Σ) by the probability measure $P_\Psi : \Sigma \to \mathbb{R}^+$ defined in this theorem:

Theorem 40

Taking $(\Psi, \mathcal{M}, \nu)$ from POMS, for each $\sigma \in \Sigma$ we define $V_\sigma : \Psi \to \mathbb{R}^+$, $V_\sigma(P) = P(\sigma)$. Let $P_\Psi : \Sigma \to \mathbb{R}^+$ be

$$P_\Psi(\sigma) = \int_\Psi V_\sigma \, d\left(\frac{\nu}{\nu(\Psi)}\right)$$

---

136 I need this perhaps apparently odd condition to ensure ν-measurability of a certain function. For $x > 1$ these sets will all be $\varnothing$.

Then V_σ is a random variable on second order probability space $(\Psi, \mathcal{M}, v/v(\Psi))$ and P_Ψ, the expectation of V_σ, is a probability measure for the measurable space (Ω, Σ).[137]

Proof: Corollary 8 gives us the second order probability space $(\Psi, \mathcal{M}, v/v(\Psi))$. Since for all x, $V_\sigma^{-1}([x,\infty)) = \{P:P(\sigma)\in[x,\infty)\}\in\mathcal{M}$, V_σ is a $v/v(\Psi)$–measurable function on the probability space $(\Psi, \mathcal{M}, v/v(\Psi))$. Hence each V_σ is a random variable on $(\Psi, \mathcal{M}, v/v(\Psi))$.[138] We now prove that P_Ψ is a probability measure on (Ω, Σ). Evidently $P_\Psi(\varnothing)=0$. Suppose $S=\{s_1, s_2, s_3, \ldots\}$ is a sequence in Σ. Since for all P, P is σ-additive, $P(S) = \sum_{k=1}^{\infty} P(s_k)$. We define a sequence of functions $W_n:\Psi\to \mathbb{R}^+$ by the first equality in this line:

$$W_n(P) = \sum_{k=1}^{n} V_{s_k}(P) = \sum_{k=1}^{n} P(s_k) \le \sum_{k=1}^{\infty} P(s_k) = P(S) = V_S(P) \le 1$$

So each W_n is dominated by 1 and $W_n\to V_S$. Hence by the dominated convergence theorem (Bruckner *et al.* 2008:203 Theorem 5.14):

$$P_\Psi(S) = \int_\Psi V_S\, d\left(\frac{v}{v(\Psi)}\right) = \lim_{n\to\infty} \int_\Psi W_n\, d\left(\frac{v}{v(\Psi)}\right) = \lim_{n\to\infty} \int_\Psi \sum_{k=1}^{n} V_{s_k}\, d\left(\frac{v}{v(\Psi)}\right)$$

$$= \lim_{n\to\infty} \sum_{k=1}^{n} \int_\Psi V_{s_k} d\left(\frac{v}{v(\Psi)}\right) = \lim_{n\to\infty} \sum_{k=1}^{n} P_\Psi(s_k) = \sum_{k=1}^{\infty} P_\Psi(s_k)$$

So P_Ψ is countably additive. Finally, for all P, $V_\Omega(P)=P(\Omega)=1$, and so

$$P_\Psi(\Omega) = \int_\Psi V_\Omega d\left(\frac{v}{v(\Psi)}\right) = \int_\Psi 1\, d\left(\frac{\mu}{\mu(\Psi)}\right) = \frac{\mu}{\mu(\Psi)}(\Psi) = 1.$$

Hence P_Ψ is a normal measure. QED

Corollary 41

$$P_\Psi(\sigma) = \frac{\int_\Psi V_\sigma dv}{\int_\Psi V_\Omega dv}$$

---

137 $v/v(\Psi)$ is the normal measure derived from v by dividing by $v(\Psi)$.
138 For being a measurable function, see e.g. Bruckner *et al.* 2008:164 Theorem 4.5. For measure theoretic definitions of random variables and their expectation see, e.g., Capinski and Kopp 2004:66 and 114.

Proof: $v/v\,(\Psi)$ is absolutely continuous with respect to v and for all $m \in \mathcal{M}$,

$$\frac{v}{v(\Psi)}(m) = \frac{1}{v(\Psi)}\int_m 1 dv = \int_m \frac{1}{v(\Psi)} dv,$$

hence $1/v(\Psi)$ is the Radon-Nikodym derivative of $v/v(\Psi)$ with respect to v (Bruckner *et al.* 2008:Theorem 5.29). Consequently

$$P_\Psi(\sigma) = \int_\Psi V_\sigma d\left(\frac{v}{v(\Psi)}\right) = \int_\Psi V_\sigma \frac{1}{v(\Psi)} dv = \frac{1}{v(\Psi)}\int_\Psi V_\sigma dv = \frac{\int_\Psi V_\sigma dv}{\int_\Psi V_\Omega dv}$$

where the second equality is by Bruckner *et al.* 2008:224 Theorem 5.31(i) and the last because $\int_\Psi V_\Omega dv = \int_\Psi 1 dv = v(\Psi)$. QED

Remark: The point of this corollary is to prove that Theorem **40** does for POMS the same thing as Theorem **7** and its corollaries did for POIS by exhibiting the parallel between defining the Meta-indifferent Probability Measure from POMS as a quotient of the second-order normative measure v and Corollary 8 defining the Equal Ignorance Probability Measure as a quotient of the first-order normative measure μ.

On these definitions, for each $\sigma \in \Sigma$, $P_\Psi(\sigma)$ is the expected value of the random variable $V_\sigma(P)$, that is to say, it is the expected value of $P(\sigma)$, and hence is the mean of $P(\sigma)$. In this way a Meta-indifferent Probability Measure, P_Ψ, is a mean probability measure over all the probability measures in Ψ.

Although we had to go into some technicalities, the features we used here are unavoidable. They are unavoidable because all our earlier unavoidability theorems apply to the second-order measures just as much as they applied to the first-order measures. They apply just because POMS is just POIS applied to the set of probability measures, Ψ. POMS takes as an input a normative measure, only it is a second-order measure over the probability measures rather than the events, in order to output a second-order probability measure similarly to Corollary 8. Any proposed method of determining an average probability measure over the members of Ψ is committed to a second-order probability measure which, by the POIS Is Unavoidable Corollary **15**, can therefore be begotten from POMS.

Unless there were already a uniform measure on Ψ, or a measure determined by the question as in the earlier example from Aerts and Sassoli de Bianchi, determining equiprobability on Ψ would have to make use of POMS. The normative measure would have to be gotten from the Theorem of Induced Measure, justified by appeal to an equal ignorance function and the sufficiency principle. Clearly, if there is more than one measure on Ψ, and there is no measure which makes sense as *the* uniform measure, Bertrand's paradox recurs because of distinct and contradictory ways of assigning equiprobabilities to members of Ψ.

In the case of the set of chords, $\mathbb{C}$, we have the set of probability measures on $\mathbb{C}$, $\Psi_{\mathbb{C}}$, being treated as a probability space. Because the individual members of $\Psi_{\mathbb{C}}$ are measurable functions, there is no difficulty in constructing probability measures on $\Psi_{\mathbb{C}}$ because there are a variety of standard methods for constructing measures on spaces of measurable functions. For a start, there is a family of norms that can be defined for them. The norms

on the L^n function spaces are defined for $P \in \Psi_{\mathbb{C}}$ by $\|P\|_n = \left(\int |P|^n \, d\lambda \right)^{\frac{1}{n}}$ and $\Psi_{\mathbb{C}} \subset L^n$ since $\|P\|_n < \infty$. There are other norms definable as well (see Rynne and Youngson 2008:chapter 2). From a norm we can define a metric on Ψ by $d(P, P') = \|P - P'\|$. The Borel sets of a metric space are a σ-algebra. There are then methods for using the metric to define what is called a pre-measure and finally a measure (see Bruckner *et al.* 2008:chapter 3) on that Borel algebra. In Chapter 13 we will see those methods deployed so I will not go into them here. The upshot of this is that the Principle of Meta-indifference for Sets and Theorem 40 cannot fail to entail a Meta-indifferent Probability Measure and therefore a Meta-indifferent probability of the longer chord. This point was the point to be proved to place the final nail in the coffin of the Distinction strategy.

Finally, Meta-indifference by Joint Averaging must fail. $\Psi_{\mathbb{C}}$ is itself quite as abstract as $\mathbb{C}$. The method just mentioned gives different measures for different norms and we can also beget measures on $\Psi_{\mathbb{C}}$ by TIM. Each measure on $\Psi_{\mathbb{C}}$ (of which there are infinitely many) will beget a probability measure for $\Psi_{\mathbb{C}}$, and some of those will be contradictory and produce paradoxical probabilities of longer from the Meta-indifferent Probability Measures they define. Therefore Bertrand's paradox recurs at the second-order level. If we try to fix it by a retreat to a third-order Principle of Meta-meta-indifference for Sets, we fall into a vicious regress. So Meta-indifference by Joint Averaging cannot be used to give a solution to Bertrand's paradox by the Well-posing strategy.

10.4 Do Aerts and Sassoli de Bianchi avoid the recurrence?

Aerts and Sassoli de Bianchi explicitly commit their use of the Universal Average to being an application of

what Shackel calls the *principle of meta-indifference*.

(2014:15)

by which they mean the elementary version of POMS that I gave in my earlier paper (Shackel 2007:173). Their Universal Average is certainly at least an attempt at deriving or estimating a Meta-indifferent Probability Measure P_ψ, and therefore an attempt at Meta-indifference by Joint Averaging.

Aerts and Sassoli de Bianchi think

> our analysis precisely avoids the arguments presented by Shackel, as it does not lead to the vicious regress he indicates.
>
> (2014:15)

I showed in the last chapter that their claimed solution, as they present it, is an instance of the Distinction strategy and so falls into van Fraassen's trap. Even if their Universal Average is correct for the questions they ask about stick tossing and pebble tossing, they cannot avoid Bertrand's general question. So they must generalize the application of their Universal Average.

We saw that by formulating the set up for the Universal Average in terms of namings, it can be generalized. Unfortunately, it can be generalized in too many ways. Their stick tossing parameterization gave a naming and so did their pebble tossing parameterization once we took the quotient space. Each way gave a distinct probability of longer. So even if we accept the Universal Average as a successful derivation of a Meta-indifferent Probability Measure P_ψ, it is a perfectly good example of the recurrence of Bertrand's paradox at the second-order level because it cannot avoid coming up with two of them. Therefore, even setting aside the obscurity over whether their Universal Average can count as the Meta-indifferent Average, and contrary to their claim, they do not avoid the vicious regress.

11 Permissivism

11.1 Introduction

The current debate over permissivism versus uniqueness has not led to permissivism being proposed as a solution of the paradox. Nevertheless, according to permissivism a particular pair of the premisses in the Rigorous argument are not both true together, the pair that are denied by the Irrelevance strategy, and permissivism applied to the paradox is an instance of the Irrelevance strategy.

For want of a word, and given the special meaning of uniqueness in the debate over permissivism, all can agree that *singleness* is a requirement of probability. The question at issue is singleness with respect to what. On Feldman's formulation of uniqueness it is singleness with respect to evidence. I, however, formulated the principle of indifference in terms of equal epistemic status defined by equal epistemic reason and explicitly allowed that evidences would be only one possible account of epistemic reasons and equal evidence only one possible account of equal epistemic reason. For example, one way in which it might be thought that reasonable people can disagree despite possessing the same evidence is if false beliefs about rationality can be rational. In that case, equal epistemic reason is not simply a matter of the evidence but also depends on the evaluation of the evidence based on those false beliefs about rationality. So for my purposes I need, and therefore use, a broader account of uniqueness than that in the permissivism debate, namely, unique with respect to epistemic reason.

In this chapter I argue that epistemic objectivity requires uniqueness. The argument I give is novel and places pressure on permissivism in general by showing that permissivism amounts to the denial of the principle of indifference. Discussing the argument in a wider context of the arguments for permissivism would be necessary before I would advance it as a rebuttal of permissivism in general. That discussion, however, is not needed for the question at issue here, namely, whether permissivism solves Bertrand's paradox. What I think the argument shows without need for that wider

DOI: 10.4324/9781003456308-11

discussion is that permissivism doesn't solve Bertrand's paradox and that philosophical theories of epistemic probability that subscribe to both the principle of indifference and epistemic objectivity are committed to uniqueness. In support of this conclusion I shall also discuss Novack's argument that concurs in my conclusion.

11.2 Permissivism for an Irrelevance strategy

The debate over permissivism is between two views:

> Uniqueness: Given one's total evidence, there is a unique rational doxastic attitude that one can take to any proposition.
>
> (Feldman 2007:201)

There are some complications here because we usually focus on salient evidence and ignore background knowledge. There are competing theories of total evidence, although they usually include background knowledge and rational belief.

The opposing view is

> Permissivism: Given one's total evidence, there is sometimes more than one rational doxastic attitude that one can take to any proposition.[139]

I formulated the principle of indifference in terms of equal epistemic status defined by equal epistemic reason and explicitly allowed that evidences would be only one possible account of epistemic reasons and equal evidence only one possible account of equal epistemic reason. If the possessed epistemic reasons include more than the total evidence it looks as if permissivism wins, but wins too easily. For this reason, and whether or not there are other reasons to formulate uniqueness in terms of evidence, the varieties of uniqueness and permissivism that matter for us are

> *Rational uniqueness*: Given one's total possessed epistemic reasons, there is always a unique rational doxastic attitude that one can take to any proposition.

and its opponent as

> *Rational permissivism*: Given one's total possessed epistemic reasons, there is sometimes more than one rational doxastic attitude that one can take to any proposition.

139 For recent general discussion of these see Greco and Hedden 2016 and focused on Bayesian permissivism see Meacham 2014.

It is of these varieties I speak throughout. We should note the implications permissivism has for pairs of propositions. It may be permissible for someone to believe one more than another, or vice versa, or equally, but only one of these three. It may be permissible to believe a proposition or its contrary but not both.

When it comes to uncertain propositions, these positions amount to the assertion or denial of there being a unique rational probability measure given one's total possessed reasons. The distinction is primarily about propositional justification rather than doxastic justification, for which additional procedural factors may need to be in place. We can expand it to include doxastic justification by assuming those factors are in place.

Here we need to distinguish two ways in which uniqueness might be challenged: imprecise probabilities and non-unique precise probabilities. For example, White (2005) discusses whether epistemic normativity may allow rational degrees of belief to be better understood as imprecise probabilities. Imprecise probabilities are appealing to those who think that credences are not sharp but are blurry. For this reason partial belief is held to be better represented by a set of numbers (usually an interval) rather than a number.

This, however, is not really a rejection of uniqueness but a substitution of a different item to represent a unique rational degree of belief.[140] The probability functions for imprecise probabilities are functions with codomain the power set of $[0,1]$ instead of $[0,1]$ itself, so the imprecise probability function itself is not a probability measure, but it may be defined by a set of probability measures.[141] Consequently, imprecise probabilities as such are not a variety of permissivism. Permissivism for imprecise probabilities would be the denial of a unique rational *imprecise* probability function given one's total evidence.

In fact, there is an immediate translation between precise probabilities and imprecise probabilities: wherever we have been speaking of probability measures they speak of imprecise probability functions and the objects that are probabilities follows on. This means we can translate the Rigorous argument against the principle of indifference in terms of imprecise probabilities (assuming that imprecise probabilists want to defend the principle of indifference). The only difference is that we would have to do some extra work on Premiss 10 and the subsequent lines to show that there are two distinct imprecise probability functions for the chords. We would have to supplement our definitions and produce correlate extensions of the

---

140 I say this even if some imprecise probabilists would then wish to say that any degree in the interval is permissible. To my mind that is a way of giving up imprecise probabilities and advancing permissivism about precise probabilities instead.

141 A set of probability measures defining imprecise probabilities is often called the representor. Walley 1991; Cozman 2016.

relevant theorems in Chapters 2 and 3. A sketch of what might underlie the success of such a programme is evident from the Deluge Theorem 24, that the probability of longer can take any value in [0,1]. The point here is that since there are equal ignorance functions giving distinct probabilities, there will be different sets of equal ignorance functions that give us different intervals for the probability of longer, and that may suffice for a version of the Rigorous argument to go through for imprecise probabilities. I am not going to do that work nor am I claiming that it can be done. As I said at the beginning, what we have been doing is complex enough already. I also grant that imprecise probabilists may be able to defend [0,1] as the sole correct probability for any events on the chords. I will return briefly to imprecise probabilities in the final chapter and give there a reference to a different kind of rebuttal to imprecise probabilities as a solution of the paradoxes. In the rest of the chapter I will therefore continue to give my arguments in terms of probability measures.

Premiss 2 of the Rigorous argument (§4.3) is that there is a single epistemically rational probability measure for an event space, upon which premiss depends Line 5, that Bertrand's question has a single answer. The permissivists first argument is that Premiss 2 is really just an expression of uniqueness, or that it only gains cogency on the assumption of uniqueness, so they can simply deny Premiss 2. Denying Premiss 2 suffices for being an instance of an Irrelevance strategy.

But of course, I used the word 'single' with malice aforethought. One and the same event is supposed to have a single probability. To allow otherwise is to allow that, from the angle and direction cases, ½ = the probability of longer = ⅓. Non-single probabilities are therefore contradictions and so to require singleness is only to require consistency.

Non-singleness is therefore not an option: what is an option is whatever it is that singleness is relative to. The early objective Bayesians in part distinguished themselves from subjective Bayesians over this very point. The latter allowed a relativity of singleness to persons that the former found objectionable. For example, the latter allowed and the former doubted that with exactly the same total evidence, you and I might rationally have distinct probabilities for the one and the same event.

Premiss 2, therefore, is not a covert expression of uniqueness: although it is true that it is implied by uniqueness it is not equivalent to it. So the permissivist can only deny Premiss 2 on the interpretation under which 'single' means uniqueness in the special sense of single relative to one's total possessed reasons. They owe us an account of the alternative relativity that they propose.

What if the permissivist simply denies there is any such ground to which singleness is relative? Given total possessed reasons, sometimes it just *is* permissible to have more than one rational doxastic attitude (although not both). Call this *blunt permissivism* and the permissivist who offers an account of the relativity of singleness *grounded permissivism*. Most of what follows will be addressed to the latter, because we will eventually see that these two end in the same place and the argument I give at that point therefore covers them both.

The permissivist can then supplement their first argument as follows: under consistent interpretations, Premisses 2 and 10 of the Rigorous argument are contraries. Premiss 10 is that there are 2 mappings from the chords to 2 measure spaces which are each defined by a property of the chords of which we are equally ignorant, upon which premiss depends Line 13, that each such equal ignorance probability measure gives an answer to Bertrand's question distinct from the other. Essentially, the argument will be that the sense of singleness that suffices for the truth of Premiss 2 undermines Premiss 10 and the sense of equal ignorance needed for the truth of Premiss 10 undermines Premiss 2.

Permissivism holds that there is more than one rational attitude we can take to the probability of longer, but it doesn't say that it is rational to hold all those attitudes at the same time. So the truth of Premiss 2 is only under an interpretation which expresses the latter point. Whereas, the truth of Premiss 10 is only under an interpretation which expresses the former point. For example, there is more than one rational attitude to equal ignorance, and hence 2 equal ignorance mappings each corresponding to one of those rational attitudes, which gives Premiss 10. But relative to that, it is not true that the probability measures are single, so Premiss 2 is false. However, on taking up one of the permissible rational attitudes to equal ignorance, the probability measure is single, which gives Premiss 2, but there is only one equal ignorance mapping on that attitude, so Premiss 10 is false. Hence Premisses 2 and 10 are ambiguous and on each consistent interpretation they are contraries. Thus do we have the permissivist version of the Irrelevance strategy.

11.3 The argument against the permissivist Irrelevance strategy

Recalling that our topic is epistemic probability, whose features include entailments for rational partial and full belief, there is a very short argument against permissivism as a solution. The principle of indifference implies uniqueness, therefore any philosophical theory committed to the principle is committed to uniqueness, so even if the solution works, it is useless to the theories that need it.

This argument is valid and, I shall argue, sound. The critical question is the truth of its premiss. This may appear obviously true. Since the principle gives equal probabilities for events with equal reason, it must entail equal doxastic attitudes towards those events. So far, so good. But of course,

Bertrand's paradox raises a question about the apparent simplicity of this entailment, since the principle has given a pair of contrary probabilities for the longer chord.

The blunt permissivist, in particular, can advance the paradox as a proof that the entailment is only apparent: that it proves instead the existence of a proposition, the chord is longer, for which two different partial beliefs are permissible; alternatively, a pair of contrary probabilistic propositions, either of which is permissible to believe.

The grounded permissivist has a different objection available: that the argument is irrelevant. The principle of indifference is downstream of the evaluation of equal reason. Reasonable people may disagree over that evaluation and different evaluations might be rational. This point can underlie the diagnosis of their Irrelevance strategy. The truth of Premiss 10 is upstream of the evaluation, where the equally rational but distinct evaluations of which events have equal reason are still in play, such as evaluations that promote the tangent angles as having equal reason versus those that promote the chord directions; but upstream, Premiss 2 is false. The truth of Premiss 2 is downstream of this evaluation, when Premiss 10 is false. Hence the premises are contraries under each of their contextually dependent interpretations and the Rigorous argument is unsound.

The grounded permissivist owes us an account of what exactly their singleness is relative to. The blunt permissivist denies any ground to which singleness is relative whilst bluntly holding to permissivism. My argument against the permissivist strategy is this:

1. Singleness is relative to epistemic normativity.
2. Suppose epistemic normativity is subjective.
3. The permissivist strategy succeeds but subjective permissivism is irrelevant to whether Bertrand's paradox refutes the principle of indifference.
4. Suppose epistemic normativity is objective.
5. There is a deontic modal argument against objective permissivism.
6. Objective permissivism offers three grounds to which singleness may be said to be relative, of which only one has any strength.
7. If the strong ground succeeds, at a certain point it also ends up bluntly holding to permissivism.
8. The way bluntly holding to permissivism denies Premiss 2 amounts to abandoning the principle of indifference rather than solving the paradox.
9. Therefore there is no relevant permissivist strategy that succeeds.

The first premiss is simply the kind of account owed by the grounded permissivist. The other premises are the subject of what follows.

11.4 Epistemic normativity

As I noted in the first chapter, epistemic theories of probability are ontically subjective. We are now concerned with the normative commitments
of such ontically subjective theories, and that takes us to the question of
epistemic objectivity versus subjectivity, which is person-independence
versus person-dependence.

For a matter to be epistemically objective is for the rationally correct
state of belief to be independent of persons. It therefore depends only on
the epistemic reasons, where the only such reasons are not themselves person dependent. That is to say, although different persons may possess different reasons, the matter of possession is the only person-dependency
there is. The reasons themselves are not person-dependent: if something is
a reason for one person, it is a reason for anyone else as well. From hereon
I call these objective reasons. Subjective reasons, by contrast, may be a
reason for one person but not for another.[142] For a matter to be epistemically subjective is for the rationally correct state of belief to depend on
persons as well as on objective reasons, where the dependence on persons
may be as such or because there are also subjective reasons.

11.5 Subjective permissivism is irrelevant

Subjective permissivism might also be called interpersonal permissivism.[143]
Premiss 2 is true given the relativity to persons, because the singleness of
rational probability measure is relative to each person. Premiss 10 is true if
we drop the relativity to persons but false if we don't. Given the relativity
of Premiss 2 to a person, either just because of who they are or because of
their subjective reasons, Premiss 10 is false because that relativity gives
them a free choice over which function is equally ignorant for them or
because that relativity picks one of them out. Of course, if this person-rel-
ativity *doesn't* suffice to make Premiss 10 false when taken relative to persons, the strategy has failed before it even got started so we don't need to
consider any possible weakness here. Nevertheless, the evident arbitrariness in this part of the strategy has a consequence.

We have ended up in a position normatively equivalent to subjective
Bayesianism, which places no rational constraint other than that a person's
credence function should be a single probability measure, that is any probability measure is rational. If there are no subjective reasons, who someone

---

142 Objective versus subjective reasons are also used to mark other distinctions, such as that
  between there being a reason and that reason being possessed, or between a state of affairs and a mental state. I am using it to mark the distinction between agent neutral and
  agent relative reasons, without taking a position on whether reasons are state of affairs,
  conditions, mental states or whatever.
143 E.g. see Dogramaci and Horowitz 2016 using this terminology.

is determines somehow or other the equal ignorance function for them and it could be any whatsoever, just however it strikes them. Again, it has to strike them one way rather than another on pain of losing the strategy. Likewise if there are subjective reasons, the subjective reasons that make Premiss 10 false are arbitrary: there is no fact about Bertrand's question that, *in itself*, makes any difference between the various possible equal ignorance functions; just one of them does make a difference, or something else, I don't know what, just does make a difference.

The reason subjective permissivism is irrelevant is that on such an account of epistemic rationality, Bertrand's paradox was never a problem in the first place. Bertrand's paradox is one of the ways the problem of the priors manifests. Subjective Bayesians' response to that problem is to deny it is a problem: rationality does not constrain the priors beyond probabilistic consistency. Any consistent prior is acceptable, so pick any measure on the chords or use any function you like in the TIM to induce the normative measure on the chords and the POIS begotten probability measure is fine, whatever it is. Subjective permissivism has ended up in essentially the same place, at least so far as Bertrand's paradox goes. What is required for their strategy to deny Premiss 10 whilst affirming Premiss 2 means an arbitrary choice among the equal ignorance functions, and hence among the TIM begotten probability measures, is entirely rational. But if an arbitrary choice is entirely rational, then it doesn't matter that Bertrand's question allows of too many answers: any answer, or at least, many answers, are rational, provided each person picks only one.

That no one has ever proposed this as a satisfactory solution to Bertrand's paradox shows that no one who takes the subjective view of epistemic rationality ever thought the paradox worth bothering with. So I concede that such a view can give a successful instance of the Irrelevance strategy. As such, however, it is in turn irrelevant to the philosophical concern we have had with the paradox since it has always been obvious that this kind of epistemic relativism need not trouble itself about inconsistent probabilities provided it could distribute the different probabilities across different rational people. This point is important since it make it clear that Bertrand's paradox has only ever been a concern for or challenge to the epistemic objectivity of probability.

11.6 Objective permissivism fails

Objective permissivism might also be called intrapersonal permissivism. Here is a modal argument against objective permissivism:

1. Suppose permissivism is true.
2. There exist inconsistent propositions p and q and it is permissible to believe p and permissible to believe q (1).

3. It is possible to believe p and believe q. (Premiss)
4. If believing p is permissible and believing q is permissible and it is possible to believe p and believe q then it is permissible to believe p and believe q. (Premiss)
5. It is permissible to believe p and believe q. (2, 3, 4)
6. It is not permissible to believe p and believe q. (Premiss)
7. Therefore permissivism is false. (RAA, 5, 6)

Line 2 follows because permissivism is trivial if all it allows is the permissibility of believing consistent beliefs. Premiss 3 is an empirical fact: we are able to have inconsistent beliefs. Premiss 6 is an instance of the impermissibility of believing inconsistent propositions and permissivism is also committed to its truth. Premiss 4 is an instance of an intuitive deontic modal principle: If A is permissible and B is permissible and it is possible that A and B, then it is permissible that A and B. So the permissivist will have to deny that premiss for doxastic permissibility. I think that may be difficult to defend, but questions of the form of deontic modal principles are notoriously unsettled,[144] so having opened a can of worms, I shall not examine it further.

A number of facts appear to support permissivism. First, reasonable people can disagree despite being possessed of the same reasons. But if they are reasonable, then there is more than one rational opinion. Second, there are competing epistemic values such as narrow and broad coherence, simplicity, explanatory power and being non-presumptuous with respect to what we don't know.[145] There are also competing epistemic duties, such as open-mindedness, appropriate curiosity, being serious and so on. There is no lexical ordering of these and different orderings or weightings may result in different opinions being rational. Third, there are rational false beliefs about rationality and so one may rationally have a different yet rational opinion because of rationally faultless but erroneous evaluation of the reasons.

The first point against objective permissivism is this. The permissivist must tell us what ground is singleness relative to and then explain how that interacts with Bertrand's question to make different probabilities correct. We have three candidate grounds in the last paragraph. Fair enough. But this looks like a very subtle way of falling into van Fraassen's trap, since there is nothing in Bertrand's question for these grounds to get a grip on and make a difference. Taking each in turn, what exactly are two reasonable people supposed to be responding to differently when one goes the angle way and the other the direction? It is very difficult to give any

---

144 For my own view, see Shackel 2018.
145 See Chapter 1.

objective difference that could justify choosing one over the other. This needs help from competing values and duties to make work. The second ground would have to say that applying different ordering or weighting to values or duties will direct us one way or the other. But it is entirely obscure what in Bertrand's paradox this difference in ordering or weighting can be rationally latching onto. The third ground is in the best position, since it allows that both could have a different rational false belief about rationality leading one to go one way and one the other, despite there being no objective difference.

The blunt permissivist simply denies there is any such ground. Given possessed reasons, sometimes it just *is* permissible to have, for example, either of two inconsistent beliefs but impermissible to believe them both, without there being any ground to which the final singleness of rational belief is relative. We will see shortly that the grounded permissivist who offers us a credible ground will end up here as well and the argument I give at that point therefore covers them both.

I shall now argue that permissivism does not succeed on the first and third grounds given in the penultimate paragraph and both the second ground and blunt permissivism abandon the principle of indifference rather than solve the paradox.

The third strikes me as the weakest of the grounds. First of all, it can and has been denied that one can have rational false beliefs about rationality:

Fixed Point Thesis: No situation rationally permits an a priori false belief about which overall states are rationally permitted in which situations.

(Titelbaum 2015:261 see also Shackel n.d.-a)

I think the fixed point thesis, at least in my preferred version—that false beliefs about rationality are irrational—is true. But even if it is not, even if we can have rational false beliefs about rationality, this does not suffice for permissivism. In an absolutely straightforward sense, even if my opinion is in some sense faultless, it is not rational just because it is in part based on a false belief *about rationality*. The tendency to think otherwise is, I think, based on not attending to the distinction between being excused and being justified. That my false belief about rationality is rational may excuse the error that I consequently make, but it does not justify it. In other words, the sense in which it is faultless is not that it is rational but that it is excusable.

The first ground relies on a false assumption, that being reasonable is being perfectly rational. Certainly, being reasonable requires some degree of rational achievement and there may be certain elements that must be present such as not too much dogmatism, a degree of sensitivity to the reasons, simple logical consistency and so on. There are also ethical components required, such as having imaginative sympathy, being moderately tractable rather than pig-headed, being open to hear what others have to say, and so on. But perfection is not required. Consequently reasonable people can disagree even if uniqueness is true and so their disagreement does not support permissivism.

The strongest ground for permissivism is the second one. Competing values and duties without a lexical ordering or general weighting is a familiar problem in ethics. Of course, in ethics there can be a number of permissible actions, but to make a relevant comparison with uniqueness we need to consider not permissible actions but obligatory ones. The reason is that epistemic permissivism says that it is permissible to believe either of two inconsistent beliefs but not both, whereas in ethics if two things are permissible, the only constraint on whether, therefore, both are permissible is whether both are possible. For permissivism, however, the possibility of believing inconsistent but permissible beliefs does not suffice for believing both to be permissible. The relevant analogy for permissivism would be pairs of actions, each of which is obligatory in the sense that it normatively rules out the other despite the pair being jointly possible. Also, to keep the analogy relevant the ethical reasons cannot be subjective.

Objective permissivism in ethics is not thought to be supported by the nature of the competition of competing values and duties. Rather, the view is that the absence of a lexical ordering or general weighting is a fact about the peculiar nature of the values and duties that has no tendency to support the claim that either of two incompatible obligations could be permitted. Rather, for each specific question of what to do, the reasons add up to determine what is obliged and yes, in one case one kind of duty or value may take precedence or outweigh another in evaluating the reasons and in another case a different kind of duty or value may take precedence or outweigh. Indeed, among the arguments for particularism is that reasons can have entirely reversed polarities between cases (Dancy 2004, 2018). What *is* true is that in any particular case it may be hard to evaluate just what the reasons add up to. But the ethical quandary is a quandary of our fallibility, not an indication of the permissibility of both of two incompatible obligations.

I think the same applies in the epistemic case. Although rationality does not include a lexical ordering or weighting of the epistemic values and duties, in any particular case the combination of the possessed reasons and their evaluation in the light of applicable values and duties entails there is a unique rational doxastic attitude to be taken. So the mere fact of competing values and duties without a general ordering or weighting is neutral between permissivism and uniqueness.

What exactly, then, is it about the nature of the competition that is supposed to allow more than one rational doxastic attitude? First of all, the permissivist doesn't have to say that in all cases there is more than one rational doxastic attitude, just that in some cases this is true. Suppose we have such a case. The permissivist doesn't have to say that any way whatsoever of ordering or weighing the values and duties in evaluating the reasons applies, just that more than one way does. So consider the allowable ways. If there is an ordering between them for the case at hand then one may come out on top and we lose permissivism. So there must be at least two between which there is no difference in rational status, they each make permissible a doxastic attitude but the pair of doxastic attitudes is inconsistent and therefore impermissible for both to be held. In short, to give us permissivism the allowable ways must result in making rational at least two inconsistent doxastic attitudes.

At this point, any choice between the allowable ways is arbitrary. This arbitrariness cannot be a matter of our fallibility since that would not suffice for the rationality of more than one. The arbitrariness itself must be rational. Any allowable way of ordering or weighing the values and duties in evaluating the reasons in this particular case is as rational as any other. Every way is as good as every other and the possessed reasons don't settle the matter in any particular direction.

It is at this point that we have rejoined blunt permissivism since now, given possessed reasons, it just *is* permissible to have either of two inconsistent doxastic attitudes but impermissible to have them both. Although particular allowable ways makes each permissible, there is nothing further, no ground, to which the adoption of a single one of them is relative. In other words, *we have no reason to discriminate between the two possibilities.*

The difference I have introduced here is that on occasion this is equal reason *at the second order*. There may be distinct evaluations of equal reason at the first order, such as in Bertrand's paradox the possessed reasons being taken to give equal reason to tangent angles or to directions. But there is no reason to discriminate between those evaluations, which is equal reason at the second order.

Consequently, when we have arrived at the place where there is no ground to which singleness is relative, that place is the very ground for the application of the principle of indifference to the case. What the objective permissivist does with this ground is to say, in such a case, not that the possibilities have equal probability but that it is permissible to believe in any of the possibilities. This applies at the first-order and the higher levels that take us to meta-indifference. So the permissivist is committed to allowing us either to believe the chord will be longer, or not longer, committed to allowing the probability to be either ⅓ or ½, and so on.

This strikes me as a reductio of objective permissivism as it stands. Even if not, and more importantly for us, if this is the basis on which the nature of the competition of values and duties justifies permissivism, and I don't see any other, or if the permissivist denies a singleness ground altogether, then yes, the objective permissivist denies Premiss 2 of the Rigorous argument. But their way of doing so amounts to *giving up the principle of indifference* at some order of application. I suppose you might say this is still an instance of the Irrelevance strategy, since it gives up the principle of indifference for a reason other than Bertrand's paradox. Nevertheless, giving up the principle is not solving the paradox! Consequently, rational permissivism implies giving up the principle of indifference and hence the principle of indifference implies rational uniqueness.

A power of this argument is that it is generalizable for any grounded permissivism. If a permissivism offers a singleness ground, at some point it must end up with some range of permissible but incompatible doxastic attitudes about possibilities we have no objective epistemic reason to discriminate between. So in the face of an attempted rebuttal to this argument, whatever singleness ground is used to give that rebuttal, the version of permissivism will eventually end up in the same place as blunt permissivism that denies any such ground. Admittedly, the demonstration that it ends up in the same place may be hard to construct, but that it will end up in the same place is at least cogent from these general considerations.

11.7 Novack's argument

Part of Novack's defence of the principle of indifference shows that we have come to the same conclusion. Novack concerns himself with 'an intuitive notion of a belief state that's tilted'(2010:656), that is, of believing one proposition more or less than another, which allows also of believing them equally. However, he is

> not assuming that degrees of belief ought rationally to conform to the probability calculus ("belief-probabilism"), that degrees of belief are the only interesting kinds of belief, or even that degrees of belief exist at all.

> (2010:657)

Consequently his

> principle of indifference... says that if one has no more reason to believe A than B (and vice versa), then one ought not to believe A more than B (nor vice versa).

> (2010:655)

Novack defines

> Uniqueness.... there's always exactly one mandatory directional attitude
> regarding contraries.... The negation of Uniqueness I call 'Permissivism".
>
> (2010:672)

On Novack's analysis

> the territory contested by Uniqueness and Permissivism... lies entirely
> within the realm of evidential symmetry.
>
> (2010:673)

By evidential symmetry he means equal reason. Within that territory, since
'the 'ought' here is the all-things-considered 'ought' of epistemic rationality' (2010:657), his principle of indifference rules out any attitude other
than equal belief for evidentially symmetric contraries, and therefore implies Uniqueness. Hence, since Permissivism is the negation of Uniqueness,
Permissivism implies the falsity of his principle of indifference.

Where I think Novack has been a bit quick, and why I took a far more
extensive route through permissivism, is because his territorial claim is
false. Novack's argument for his territorial claim is

> If we turn our attention to settings of evidential asymmetry, then it's
> clear that Uniqueness rules here. Since '$P >_e Q$' just means that the sum
> total of everything epistemically relevant points to P, one is epistemi-
> cally required to have a belief-state that's tilted in favour of P. Any
> consideration that would serve as an escape clause, allowing one not to
> lean towards P, is by definition not epistemically relevant to the ques-
> tion of P and Q.
>
> (2010:672-3)

It is not clear that the permissivist must grant this because it may beg the
question for reasons I discussed above. In defining his principle of indifference, Novack defines

> 'reason' is meant to include all epistemically relevant factors, so it cov-
> ers more than just, e.g., empirical evidence. It covers everything that's an
> epistemically rational reason for belief, and nothing else.
>
> (2010:656)

The 'nothing else' in the quotation is important. The sum total of everything epistemically relevant just means all the epistemic reasons. The blunt

permissivist might simply say that sometimes, although the reasons point one way, either tilt in belief is permissible. The grounded permissivist has a stronger argument. Which way the reasons point depends on their evaluation. That evaluation is subject to the competition of values and duties which lack a general ordering or weighting. There may be one way of applying the values and duties making believing P more than Q permissible and another way making believing Q more than P permissible, where neither way is more rational than the other. Novack is therefore ignoring blunt permissivism and assuming the falsity of what I called the strongest ground for permissivism. I do not assume that falsity.

11.8 Epistemic objectivity requires uniqueness

I have now proved the premises of my argument against the permissivist instance of the Irrelevance strategy and have therefore shown the strategy fails. The argument is not conclusive but is strong. Subjectivist permissivism can succeed but once we are clear what its success amounts to, it is clear that it is irrelevant because the challenge posed by Bertrand's paradox has only ever been to philosophical theories of probability that claim to be epistemically objective. The deontic modal argument against objective permissivism depends on a deontic modal principle and such principles are much controverted. Here is not the place to attempt to settle the controversy for my modal principle so I leave it in place as a cogent but not conclusive threat to objective permissivism. I do so because the second line of argument addressing both grounded and blunt permissivism is close to being conclusive, if not against objective permissivism as such at least against objective permissivism as a solution of Bertrand's paradox. Yes, perhaps one can be an objective permissivist, but the price is to give up the principle of indifference altogether. I do not see a way round this since the line of argument strikes me as entirely generalizable in the way I explained.

So the upshot of this chapter is that the principle of indifference entails rational uniqueness, Bertrand's paradox has only ever been a challenge to philosophical theories of probability that are epistemically objective and thus any such theory must either abandon the principle of indifference or satisfy uniqueness. Consequently, if the principle of indifference entails two distinct rational probabilities for a proposition, an epistemically objective theory of probability committed to the principle of indifference entails that rational probabilities are unique and non-unique. Hence for such theories, non-uniqueness is merely a polite word for a double inconsistency, from the distinct rational probabilities for the same event and from the wider theoretical commitments. Non-uniqueness is therefore contradictory for epistemically objective theories of probability that are committed to the principle of indifference.

12 Uniqueness a Criterion of Identity

12.1 Introduction

In the literature there are two criticisms of the earlier version of my Principle of Indifference for Continua (Shackel 2007) that fault it *qua principle of indifference* for producing paradoxical probabilities. The claim made is, in essence, that it cannot be the correct principle of indifference because it does not confine the domain of normative measures sufficiently, and its failure in this respect then manifests in inconsistent probabilities. Consequently, this critique redefines the principle of indifference as a something that by its true definition, whatever that is, gives unique probabilities for possibilities of which we are equally ignorant.

Philosophically, this critique amounts to the claim that uniqueness is a criterion of identity for the principle of indifference and is an instance of the Irrelevance strategy. Premiss 2 of the Rigorous argument is true because the true definition of the principle of indifference satisfies uniqueness. Under that definition Premiss 10 might be true but the step from Line 12–13 is invalid because the definition entails a constraint on normative measures such that if more than one satisfies the constraint, all produce the same probability measure begotten of TIM and POIS. If the definition can for some reason evade using TIM for the chords and produce the normative or probability measure directly this still applies due to the unavoidability theorems. We can determine a domain of equal ignorance functions from the unique probability measure. That unique probability measure provides the antecedent of the left to right direction of the Equal Ignorance Function Is Unavoidable theorem 28 whose consequent is not that there is a unique equal ignorance function but only that at least one exists. However, even if there is more than one, since the probability measure is unique any equal ignorance functions in that domain will only produce one and the same probability measure from their corresponding measure space when used in TIM. Hence in either case, Premiss 10 might be true whilst Line 13 is false.

DOI: 10.4324/9781003456308-12

12.2 Uniqueness a criterion of identity

In 2007 I gave this definition of a subordinate principle of indifference:

> *Principle of Indifference for Continuum Sized Sets*: For a continuum sized set X, given a σ-algebra, Σ, on X and a measure, μ, on Σ, and given that we have no reason to discriminate between members of Σ with equal measures, then we assign equiprobability to members of Σ with equal measures: for all x, y in Σ, if $\mu(x)=\mu(y)$ then $P(x)=P(y)$.
>
> (Shackel 2007:159)

The reader can see by inspection that this is a slightly clumsier version of the *Principle of Indifference for Continua* given in Chapter 1. November objects:

> The problem with Shackel's formalization of IP [Indifference Principle] lies in his formalization of IP's presumption [of event comparability— see next section for his definition] ... [which] relies on a strong assumption that is not part of Kolmogorov's theory.... that each σ-algebra is given with a "regular" measure defined over it.... Due to this assumption, Shackel's formalization faces a serious problem: mathematically there can be infinitely many measures defined over any given σ-algebra, so it is unclear which one of these measures should be taken as the "regular" measure in Shackel's mathematical formalization of IP. Without some way of choosing one of these "regular" measures, Shackel's mathematical formalization of IP's presumption is incomplete. This means that Shackel's formalization of IP remains ill-defined and hence, despite its appeal, cannot serve as a mathematical formalization of IP.
>
> (2019a:22)

I neither claimed nor assumed that any such regular measure exists, of course. The only constraint on the domain of normative measures is that 'we have no reason to discriminate between members of Σ with equal measures'. In Chapter 3 I proved that November is correct in saying 'there can be infinitely many measures' when I proved the Deluge Theorem 24. Nor is there anything wrong with how comparability is formalized (see next section). I can only imagine that November thinks I made this assumption because of what *he* assumes about the nature of the principle. The grounds for this objection are given earlier

> Another important implicit assumption concerning IP is that... given a set of events, an application of IP results in a unique assignment of

probabilities to these events... IP asserts that equivalent events have specific equal probabilities and no other equal probability values.

(November 2019a:8)

Since there can be infinitely many probability spaces whose σ-algebra component is the same, a mathematical formalization of IP has to be some mathematical way of uniquely selecting exactly one probability space from a given set of same-events spaces.

(2019a:18)

So I am being faulted for the definition of POIS being ill defined because it fails to restrict the domain of normative measures that are the inputs to the principle sufficiently to restrict the output probability measure to a single measure. What is wrong with this is that the principle of indifference assumes otherwise. November does not explain the basis of that assumption, but he explicitly states its necessity:

a mathematical formalization of IP has to be some mathematical way of uniquely selecting exactly one probability space from a given set of same-events spaces.

(2019a:18)

A remark on the paradoxes shows he thinks it is a matter of the principle's identity:

it would be more accurate to say that the problem with IP-related paradoxes is that the mathematical formalization of IP is still an open question, rather than claim that they are not well-defined.

(2019a:13)

In other words, the correct mathematical formalization of the principle will not produce the paradoxes because it will satisfy uniqueness. Therefore any proposed formalization of the principle that fails to satisfy uniqueness is failing to represent the principle in that respect. Hence uniqueness is a criterion of identity of the principle of indifference.

Gyenis and Rédei similarly fault my earlier principle for failing to restrict the domain of normative measures:

What is... the principle of indifference in connection with such infinite probability spaces?.... Shackel 2007, who aims at an analysis of Bertrand's paradox in abstract measure theoretic terms, realizes the importance of this question but does not offer a convincing specification of the principle of indifference: Shackel just assumes a measure, μ, on S and stipulates that the probabilities $p(A)$ be given by $p(A) = \mu(A)/\mu(X)$....

254 Uniqueness a Criterion of Identity

But there are infinitely many measures μ, on S that could in principle be taken as ones that define a probability p. Which one should be singled out that yields a p that could in principle be interpreted as expressing epistemic indifference about elements in X?

(2015a:354–5)

The need for a prior measure originates in the Need-a-Measure-to-get-a-Measure principle. We saw in §1.9 the reason the POIS has this structure is that the principle of indifference takes events and equal ignorance as inputs. The Principle of Indifference Representation thesis 5 shows that these inputs have their representation in POIS as the σ-algebra of events and the normative measure on that σ-algebra that gives equal measure to events of which we are equally ignorant. Gyenis and Rédei are quite right that POIS fails to single out a normative measure. They continue

it is impossible to formulate an indifference principle on such a measurable space.

(2015a:354–5)

In saying this, they too take it that the true principle of indifference *does* entail a unique probability measure. Hence they too take uniqueness to be a criterion of identity for the principle of indifference. November also acknowledges Gyenis and Rédei's prior exposition with which he agrees (2019a:22).

As explained in the introduction, uniqueness being a criterion of identity is an instance of the Irrelevance strategy. We shall now look at what further these authors have to say about supplying a principle that satisfies uniqueness, examine any such principle for its success in satisfying uniqueness and then I shall give my argument against this Irrelevance strategy.

12.3 November's position

November gives two instances of the Irrelevance strategy. When he says

the problem with IP-related paradoxes is that the mathematical formalization of IP is still an open question.

(2019a:13)

and later

a mathematical formalization of IP has to be some mathematical way of uniquely selecting exactly one probability space from a given set of same-events spaces.

(2019a:18)

he is implying that uniqueness is a criterion of identity, although as we shall see, perhaps he does not intend to. To that extent, anyway, he subsequently articulates the application of the strategy explicitly. Recall from §2.9 that for November, a mathematical formalization is 'a set of constraints (C)' and his same-event spaces are probability spaces for a given set of events rather than what we mean by event spaces.

> The major question concerning IP's mathematical formalization as a set of constraints on same events spaces is whether... there [is] a set of constraints (C) which manages to constrain every set of same-events spaces in such a way that only one space in each of these sets satisfies C... [If yes] then.... all IP-related paradoxes are not genuine paradoxes and the aforementioned debate is settled.
>
> (2019a:24)

So that is the Irrelevance strategy: Premiss 2 is true but the step from Premiss 10 is invalid because Line 13 is false: the true formalization of the POI produces only one probability of longer.

November continues:

> [If no]... [then] there is at least one genuine IP-related paradox.... [But this] can also be seen as a solution to IP-related paradoxes in the sense that it refutes one of their premises. Recall that IP-related paradoxes rely on the assumption that IP is always sufficient for a unique assignment of probabilities to events. However, a negative answer to the existence question would be a mathematical proof that this assumption is plain wrong. This means that a negative answer would refute one of the premises of IP-related paradoxes and thus solve them.
>
> (2019a:24)

It is hard to make sense of exactly what is going on here. First, if he is now conditionally conceding to there being genuine paradoxes he is giving up uniqueness as a criterion of identity for the principle and hence it is not a failure of the mathematical representation of POI that the set of constraints (C) fails to determine unique probabilities. But then, that they don't is supposed also to be a solution because the paradoxes have uniqueness as a premiss which, he is now saying, would be shown to be false. But in that case, and contrary to what he just said, there *aren't* any genuine paradoxes. Anyway, recalling that Premiss 2 of the Rigorous argument is about singleness, this is equivalent to denying that premiss. So in calling this a solution he is now articulating a permissivist version of the Irrelevance strategy.

That being said, he then appears immediately to give it up:

> A negative answer to the existence question [i.e., if no] would also imply
> that IP is not a valid principle in the sense that applying it does not al-
> ways result with a unique assignment of probabilities to events. As a
> result, the discussion surrounding IP should change: Instead of focusing
> on whether or not IP is a valid principle (since a negative answer to the
> existence question shows that it is not), the discussion should be focused
> on describing the cases where it is safe to use IP, if there are any.
>
> (2019a:25)

I do not propose to attempt to resolve these apparent contradictions and
perhaps he is merely noting the various approaches that could be taken. He
concludes that 'currently the existence question is still unanswered'
(2019a:26) and does not offer us a set of constraints and a formal principle
of indifference to examine further.

12.4 GR-principle of indifference needs to be wide

We now turn to considering whether Gyenis and Rédei have provided a
principle of indifference that satisfies uniqueness. Recall the principle we
derived on behalf of Gyenis and Rédei in Chapter 7:

> *GR-principle of indifference*: If X is a compact topological group and if
> the group action expresses epistemological indifference about the ele-
> mentary random events in X, then the probabilities of the events are
> given by the Haar measure on X.

The virtue of their proposal is that the normalized Haar measure for a
compact topological group is unique and so this principle satisfies uniqueness for compact topological groups. That, however, is not sufficient.

Call a principle of indifference that suffices to define probability for continua in general a wide principle. POIS is wide. If the GR-principle is confined to compact topological group event spaces, it is not wide. In Chapter
7 we saw some reasons why it might not be. As a narrow principle, any
satisfaction of uniqueness is done only by restricting the domain of application, so is done at the cost of radically diminishing the significance of
their proposal.

We are concerned with probability for continua in general and so are
concerned with wide principles. Gyenis and Rédei are claiming to have
given us a 'General classical interpretation' with a 'General principle of
indifference' (2015a:356). To do that the GR-principle of indifference
needs to be wide. We saw in §7.14 the basis on which they advance it as a

wide principle, namely, the method by which they extend their proposal to event spaces that are not compact topological groups. I showed that it is really just another way of using TIM with a Haar measure space being used as the inducing space. We shall now consider whether, as a wide principle, the GR-principle of indifference satisfies uniqueness.

12.5 Understanding how the group action expresses epistemic indifference

Gyenis and Rédei do not explain *how* the group action expresses epistemic indifference. In Chapter 7 I explained this and stated the GR-Representation lemma 30: essentially, that we are epistemically indifferent between the elements of X (which are, or represent, the elementary random events) iff the group action maps any element to any element.

Now it might be thought that the group action expressing epistemic indifference can't mean only this much, that is only mean that we are equally indifferent between each individual elementary event as each other. We saw in Chapter 2, since we are talking about continuum sized sets of possibilities, the equal probability of elementary events does *nothing* to fix the probabilities of compound events. For example, the compound events [2,3] and [2,4] in [2,6] are comprised of the same number of elementary events, each having probability zero, but under the probability measure begotten of the Lebesgue measure they have probabilities of ¼ and ½ respectively. Consequently the most natural interpretation of expressing epistemic indifference must instead be about indifference among *sets* of elementary events that are mapped to each other by the group action.[146] At this point, however, we meet a difficulty. Understood in this way, the group action can both express indifference where there is none and under generate equal probability.

The set of Borel sets of X, S is the σ-algebra of which the Haar measure is a measure. In mapping any element uniquely to any element, the group action also maps Borel sets of X to Borel sets of X. For any x in X and B in S, the group acting on B on the left gives $xB = \{y \in X : \exists b \in B \& y = xb\}$ and on the right gives Bx and it is possible that $xB \neq Bx$.[147] Suppose we have such a case. Suppose first that we are epistemically indifferent between B and xB but not between B and Bx: nevertheless, the group action can map B to both xB and Bx and so it expresses indifference where there is none.

146 Indeed, as I showed in Chapter 2, it must go this way for continuum sized event spaces and for it to do this we needed what I called the generality of the way I formulated the principle of indifference.

147 Only for non-Abelian groups. The group action mapping Borel sets to Borel sets is granted only for the sake of argument. If it does not do so then the group action cannot express epistemic indifference.

Suppose instead that we are epistemically indifferent between all three so they should have the same probability. The Haar measure is either left invariant[148] or right invariant[149] but it need not be both. Consequently, assuming without loss of generality a left invariant measure, it is possible that $p_H(B) = p_H(xB) \neq p_H(Bx)$. So in this case, although the group action expresses indifference in the sense of mapping them to each other, it can fail to result in equal probability. Sometimes this is avoidable by taking the group action only on the left or the right, but it is possible for there to be two sets such that avoiding this for one requires taking the group action on the left and for the other on the right.

The upshot is that what may appear to be the most natural interpretation of how the group action expresses epistemic indifference is not available. That in itself would be at least an obscurity in GR-classical probabilism. There needs to be some way of expressing our epistemic indifference between Borel sets between which we are indifferent. So it would appear that the proposal fails to do what the POIS does explicitly: state how we represent indifference between members of the σ-algebra.

It would take us off track to look at technical fixes. Instead I am going to explain why Gyenis and Rédei could, and perhaps intend to, set aside both the most natural interpretation and the difficulties we have just seen and confine themselves to the group action mapping any element uniquely to any element as the entirety of how the group action expresses epistemic indifference.

The justification would be that our indifference over the elementary events is all we need to express because the rest of the work is done by the fact that if we induce a compact topological group on standard cases that are not, in fact, compact topological groups, such as the interval of the real line, the Haar measure induced is the normalized Lebesgue measure of the standard cases, thereby giving us what we expect from a principle of indifference. The *expression* of epistemic indifference over compound events is therefore *not* done by the group action and not done at all. Instead, the Haar measure *defines* (rather than expresses) epistemic indifference over compound events by the Borel sets that represent them having equal Haar measure. In doing this the Haar measure gets it right (as this approach would have it) because it is uniform in standard cases and so analogously uniform in non-standard cases. On this defence, then, the Haar measure does *not generally* play the full prior normative measure role,[150] since it

---

148 $p_H(B) = p_H(xB)$

149 $p_H(B) = p_H(Bx)$

150 Not generally because we will see in the next section special cases in which it can play the full role.

expresses epistemic indifference only over the elementary events. Call playing only so much of the role the *half-normative-measure role*.

It might be argued that this defence shows a great virtue of Gyenis and Rédei's proposal: the Haar measure is the natural measure for compact topological groups and it results in fixing the probabilities which then fixes epistemic indifference. So one might say that, because compact topological groups have a natural measure, which measure need only play the half-normative-measure role, their principle does for compound events implicitly what the POIS does explicitly.

Certainly, this defence requires rather more articulation, and may yet face difficulties similar to those recently mooted. A problem here is that it is open to getting the epistemic indifference over compound events wrong,[151] which the POIS won't. Worse, one might say that defining epistemic indifference between compound events by their having equal Haar measure is just getting the order of definition the wrong way round. For any measure of a continuum, it is trivial that the elementary events have the same measure and their measure does not determine the measure of compound events; so where our equal epistemic status does its work for continua is precisely in the indifference between compound events determining their equal probability. Consequently it cannot be right to reverse that. These problems are Gyenis and Rédei's burdens to meet but are left unaddressed by them.

12.6 Bertrand's paradox recurs for the GR-principle of indifference

The POIS's 'no reason to discriminate between sets of equal normative measure' and Gyenis and Rédei's group action 'expressing epistemic indifference' are both instances of representing equal ignorance by a specified mathematical property. Not only do Gyenis and Rédei not explain *how* the group action expresses epistemic indifference, neither do they explain what is *meant* by the group action expressing epistemic indifference. We can give an argument by which the group action can be taken to represent equal ignorance based on what we saw in Chapter 7 of how it does so. Equal ignorance over X is having no reason to discriminate between elementary random events in X. Consequently *what it is* for 'the group action [to] express... epistem[ic] indifference'(2015a:356) is for the group action to map one elementary random event in X to another such event if and only if we are equally ignorant of both (GR-Representation lemma 30). The principle of indifference requires that events over which we are equally ignorant have the same probability. The Haar measure is the unique measure up to

---

151 I.e. determining indifference between a pair of sets that we know are not, or vice versa.

a constant that is 'invariant with respect to the group action'(2015a:355) and hence satisfies what the principle of indifference requires for the elementary events. We saw in the last section why we cannot extend this to representing equal ignorance of compound events as well.

This is both rather trivial and creates some threats. Starting with the former, the entirety of the group action expressing epistemic indifference is its permuting the elementary events. It expresses epistemic indifference because given a set of elementary events between which we are equally ignorant, then we are equally ignorant between them however the set is permuted. Likewise the Haar measure results in each element having the same probability however the set is permuted. All that has been achieved by this is to add an idle cog to a trivial entailment of the principle of indifference. Instead of the trivial 'the elementary events of equal ignorance have equal probability' we now have 'the elementary events of equal ignorance under any permutation by the group action have equal probability under any permutation by the group action'.

The cog is idle because when the principle of indifference is applied to continua, all the elementary events under any standard measure of a continuum have the same measure of zero,[152] so although it is certainly true that if you permute them they have the same measure, nothing of substance is added by this fact. But this triviality is what the GR-principle of indifference amounts to once we have removed its mathematical finery. And so the question arises, why even bother with the idle cog? Since any standard measure gives equal measure to the elementary events of continua, so too does the Haar measure. Any standard measure therefore represents equal ignorance over the elementary events equally as well as the Haar measure does. So the *group action* expressing epistemic indifference adds nothing and is essentially idle. As we shall now see, the threat originates in this triviality.

Although Gyenis and Rédei wish to reject POIS, the TIM is Unavoidable theorem 23 means that they cannot, since for any probability measure their principle defines on the events, there is a way of begetting it by TIM and POIS. In this sense, we can consider the way in which the GR-principle of indifference is an instance of POIS: it is an instance in which the normalized Haar measure plays both the prior normative measure role and the posterior probability measure role. However, in playing its prior role in general, it represents equal ignorance only over the elementary events and it fails to represent equal ignorance over compound events (so plays only

---

152 Any standard measure, because you can gerrymander a measure for some inequality among the elementary events, but most ways of doing so end up with a measure that is trivially infinite for most sets in the σ-algebra and therefore not useable as a normative measure.

the half-normative-measure role) and instead defines their equal ignorance in its posterior role.

The problem here is that because any standard measure for continua will give equal measure of zero to the elementary events, *any* standard measure can play the half-normative-measure role. Consequently, any epistemic status achieved by the Haar measure on this basis is achieved by any such measure. So if the principle of indifference allows the normative measure to achieve only this much of its normative role, that is to fulfil only the half-normative-measure role, and for that normative measure then to *define* rather than *represent* equal ignorance over compound events, *any arbitrary measure* suffices *just as well* as any other. So now, the basis on which we saved the Haar measure from the apparent fault of being unable to *represent* (rather than define) equal ignorance over compound events, if justifiable, is equally justifiably applied to allow in any standard measure whatsoever. In this way we end up losing uniqueness because even if the GR-principle itself defines a unique probability measure, the sufficiency of fulfilling the half-normative-measure role drowns us in the Deluge Theorem 24.

Furthermore, supposing we are considering a particular empirical situation, it becomes very difficult, and perhaps impossible, to explain why a prior normative measure that grants equality to compound events in line with our understanding of the situation is the only justifiable application of the Principle of Indifference. Consequently, the threat tends to undermine the most intuitively correct applications of the GR-principle of indifference, namely, those cases in which the empirical situation has a symmetry of compound events which is well represented by the group action. In such a case, and for that reason, the Haar measure could represent equal ignorance of compound events and therefore fulfil the *full* normative measure role. But now that any prior measure is justifiable provided only that it plays the half-normative-measure role, any other measure whatsoever is equally justifiable, even in these cases for which the claim that only the Haar measure gets it right is most plausible. In these ways, the threat is a subtle recurrence of Bertrand's paradox.

12.7 The mistake about the Principle of Indifference

Even if there is some way of avoiding the just articulated failure of uniqueness, there remains a further problem arising from where we were driven to in §12.5. To avoid the problem of misrepresenting equal ignorance among compound events we had to take the Haar measure to define both what the compound events are (events with names in the Borel algebra) and what equal ignorance between them is (events whose names have equal Haar measure). In doing this Gyenis and Rédei are committed to the mistake that I warned against in Chapter 1.

It is no job of a principle of indifference to define what the possibilities are or to define what it is to lack reason to discriminate between them. Those two factors are logically prior to the job of the principle, which is to define probabilities after these are given. But if a principle of indifference for continua allows the normative measure to *define* the compound events and equal ignorance between them rather than *represent* them, then it has defined exactly what is supposed to be logically prior to it. What are supposed to be inputs to the principle end up being outputs of the principle.

Principles that fail to respect these logical priorities make this mistake iff they are wide. If they are narrow, they can avoid them. As we saw at the end of §12.6 on recurrence, the Haar measure can fulfil the full prior normative measure role provided the domain of its application is restricted to compact topological groups where the group action also maps compound events of equal ignorance to one another. So for narrow principles the logical priority is respected by the philosophical context in which the principle is presented by narrowing its domain, rather than in the principle itself, which may appear to contravene that priority. However, the GR-principle of indifference needs to be wide if it is to be philosophically significant.

12.8 Falling into van Fraassen's trap

Having faulted my earlier principle for continua, Gyenis and Rédei remark

> It is clear that *without some further structure* on an infinite X, it is not possible to single out any probability measure on S,[153] and hence it is *impossible to formulate an indifference principle* on such a measurable space.[154]

> (2015a:354–5, my emphasis)

So in their view, since the latter is impossible, then any indifference principle for continua must *itself* impose sufficient structure on X, which is the base space not a σ-algebra, to determine a unique probability measure. The *real* principle of indifference for continuum sized sets gives us the conditions for a unique probability measure. Therefore we have to find a way to identify such conditions. Only when we have done so, and then put them in the antecedent of the principle whilst offering some analogy for the claim that one of them 'expresses epistemological indifference', will we

---

153 Recall that S is their σ-algebra on a base space X.
154 That it is impossible to single out any probability measure without something extra is proved by the Deluge theorem 24.

have formulated the true principle. In so doing, they are falling into van Fraassen's trap. Recall

> Most writers... [say] we must be told... *which* events are equiprobable... that response asserts that in the absence of further information we have no way to determine the initial probabilities.
>
> (van Fraassen 1989:305)

Admittedly what Gyenis and Rédei want to do is more mathematically sophisticated than the examples van Fraassen is responding to. Nevertheless, their remark that 'It is clear that without some further structure on an infinite X, it is not possible to single out any probability measure on S' (2015a:355) is just their version of 'in the absence of further information we have no way to determine the initial probabilities'. On this, van Fraassen's trap closes:

> In other words, this response rejects the Principle of Indifference altogether. After all, if we were told *as part of the problem* which parameter should receive a uniform distribution [*pari passu* for Gyenis and Rédei, which structure to place on X to make it a compact topological group, thereby picking out the Haar measure of that group], no such Principle would be needed.
>
> (van Fraassen 1989:305, my emphasis)

What Gyenis and Rédei do by defining the principle of indifference in terms of the Haar measure amounts to telling us which compound events are equiprobable. That is falling into Van Fraassen's trap: they merely appear to have constrained or reformulated the principle of indifference in a way that solves its difficulties, when in fact they have covertly abandoned it and in its place substituted a covert and ad hoc stipulation of equal probability.

12.9 Confusing a criterion of success for a criterion of identity

I now offer my argument against the uniqueness Irrelevance strategy. The proposal that uniqueness is a criterion of identity is fundamentally mistaken about the nature of the principle of indifference. As we saw in Chapter 1, the identity of the principle is grounded in epistemology and defines probabilities in terms of the possibilities having a certain feature, namely, lacking epistemic reason to distinguish between them. It takes the possibilities and equal ignorance as inputs and probability is its output. By contrast, in examining Gyenis and Rédei's theory we saw that taking uniqueness as a criterion of identity made the compound events and equal ignorance outputs of their principle.

In the previous chapter we saw that uniqueness arises independently of the principle, being a requirement that epistemic objectivity places on a philosophical theory of probability that uses the principle of indifference. So uniqueness is a criterion of epistemic objectivity that applies to the principle of indifference but that is independent of and logically prior to the principle. What this means is that determining a unique probability measure is not a matter of the principle of indifference's *identity* but of its *success*, and of the success of a philosophical theory that uses it at being an epistemically objective theory. So in demanding that of its nature the principle picks out a unique probability measure November (whether by intention or not) and Gyenis and Rédei mistake a criterion of success for a criterion of identity.

13 Symmetry

The Forlorn Hope

13.1 Introduction

The Distinction strategy has been shown to fall into van Fraassen's trap. The Irrelevance strategy has little plausibility on its face and no one has fulfilled enough of a burden of proof to even keep it as a contender. We are left with the Well-posing strategy and have proved that, since meta-indifference by joint averaging produces a recurrence of the paradox, if it can work at all, it can work at the base level. This means it can never work unless we can find something additional we can appeal to that is not just another covert arbitrary restriction of Bertrand's question but is appropriately internal to the chords. At this point, the only thing that remains available is to appeal once more to symmetry, only we must do so in a way of greater generality than the appeal from Jaynes.

The most general ground remaining for a solution is that a relation between symmetry and the equal epistemic status of the chords can rule out inconsistency. This is the hope from symmetry. I show that for it to succeed we must be able to represent the symmetry of events mathematically and there must be a mathematical constraint that is justified by representing the normative ground of the principle, which constraint must then guarantee that the principle of indifference will not produce paradoxical probabilities, in other words, and recalling that we defined consistent probabilities to be non-paradoxical probabilities, it must *guarantee consistency*, for short. A constraint that does this I call a Justified Guarantee.

The set of symmetries of an object satisfies the mathematical axioms for being a group. There is one attempt to use groups to define the principle of indifference in the literature, the classical interpretation of probability from Gyenis and Rédei. I show that their earlier defusal of Bertrand's paradox, even if it were successful, is incomplete without a Justified Guarantee and, indeed, Gyenis and Rédei offer one, although not named as such. I show that their Justified Guarantee fails.

DOI: 10.4324/9781003456308-13

I then turn to considering the hope from symmetry in full generality. I formulate this possibility by use of the completely general mathematical group theoretic representation of symmetry and prove a number of representation theorems. I define the Principle of Symmetric Indifference, a principle implied by the Principle of Indifference for Sets and the representation theorems. I show that solving the paradox in these terms is equivalent to defining a mathematical condition on renamings to be the Justified Guarantee. I give a mathematical proof that the natural condition to try, that a renaming should be a group isomorphism, fails to guarantee consistency. I then prove that any other condition on renaming can either be justified or guarantee consistency, but not both, and that therefore the hope from symmetry is forlorn.

13.2 The hope from symmetry

One explanation of the equal epistemic status appealed to in the principle of indifference has been an underlying symmetry among the possibilities. For example, the equal possibility of each side of a coin or each face of a die seems to be constituted by their rotational symmetries, that of each card in a deck by its shuffling symmetries (i.e. permutations), of dominoes or balls in a jar by sliding symmetries (i.e. translations), and so on.[155] There are also symmetries in how we refer to the possibilities that should leave their epistemic status, and hence probability, unchanged. For example, whether we refer to Heads and Tails by 'Heads' and 'Tails' or vice versa, or by any other pairs of symbols, or by any interchanges among any such sets of symbols, should make no difference. These symmetries are all permutations of symbols so permutations of naming symbols must neither affect epistemic status nor probabilities. Since such permutations are renamings, as we defined them in §2.4, that they should not affect probabilities is a restatement of the constraint on renamings (that they be probability isomorphisms Theorem 4) that we derived from Paris' Renaming principle.

Plainly, the symmetries of possibilities and symbols are closely interlinked but the distinction between them is rarely drawn explicitly. It is evident that any symmetry among the possibilities can be represented by a symmetry among the naming symbols. For example, the rotational symmetries of the vertices of an equilateral triangle can be represented by the permutations defined by addition modulo 3 on using 0, 1 and 2 as names for the vertices. For this reason a symmetry grounding equal epistemic status can always be treated in terms of a naming symmetry (which is just a renaming in which the domain and codomain are the same) and this is

---

155 See also Carnap 1950:ch.8 and his remarks about symmetrical invariance pp. 488–9.

how I shall treat it here. Likewise, I will show how the paradoxes can be treated in terms of renamings.

Consequently, a hope has been that if we could formulate the symmetry constraints adequately, we would solve Bertrand's paradoxes. So, for example, Rosenkrantz (Rosenkrantz 1977:65) proposed that Jaynes' invariance symmetries (Jaynes 1968) impose constraints sufficient to solve von Kries' water/wine paradox. As we saw in an earlier chapter, Jaynes himself proposed that the rotational symmetries of the circle and the translational symmetries of the plane impose constraints sufficient to solve Bertrand's chord paradox (Jaynes 1973, especially the quotation from page 488 given here in Chapter 5).[156]

I shall call this the hope for a *Symmetry-grounded Principle of Indifference* and a philosophical theory of probability that hopes to defend its use of the POI on this basis a *Symmetry-grounded probabilism*. We saw earlier van Fraassen articulate a broad ground for the necessity of respecting symmetry in his

> *Symmetry requirement*: Problems which are essentially the same must receive essentially the same solution.
>
> (van Fraassen 1989:236)

This is a vaguer notion of symmetry than that I have been speaking of, but plainly symmetries in my sense constitute ways in which problems can be essentially the same. The hope for a Symmetry-grounded Principle of Indifference is the hope that

> *Symmetry is sufficient*: If we formulate adequate symmetry constraints then the principle of indifference will produce consistent probabilities rather than paradoxical probabilities, including for Bertrand's paradoxes.

13.3 The Justified Guarantee

To succeed in this hope the symmetry of events must be represented mathematically, thereby representing the epistemic status of events, and so doing must result in our being able to define a mathematical condition that

---

156 Jaynes and Rosenkrantz are examples of authors appealing to symmetry, but they are themselves Bayesians hoping to solve the problem of the priors rather than classical or logical probabilists. For views on symmetry and probability see Strevens 1998; Bartha and Johns 2001; Zabell 2005.

avoids the paradoxes by blocking paradoxical probabilities and guaranteeing consistency instead.

The condition itself must be defined in some manner that is not philosophically arbitrary (on pain of a mere ad hoc avoidance of paradox) but to be a mathematical representation of the normative ground in the principle of indifference, namely, the equality of epistemic status. To be its representation, the condition must be both necessary and sufficient for the normative ground. I shall call a condition that has these two features, of being a guarantee of consistency and justified by representing the normative ground in the principle of indifference, a *Justified Guarantee*.

13.4 The hope is forlorn

It is the purpose of this chapter to show that the hope from symmetry is forlorn.

For the sake of an instance of the principle of indifference in the literature formulated in group theoretic terms, I turn once again to the classical interpretation of probability proposed by Gyenis and Rédei. Their use of compact topological groups and the Haar measure is the only mathematically sophisticated development of the appeal to symmetry I know of. It makes fully explicit one way in which classical probabilism can use groups if we hope to use symmetry to save the principle of indifference from the paradoxes. Gyenis and Rédei's putative Justified Guarantee is articulated by what they call their 'non-trivial strategy'. I shall show why their strategy fails and also why its defeat does not remove all hope.

I will then turn to treating the hope from symmetry in complete generality. To do so I address symmetry within mathematical group theory. I prove some representation theorems in order to formulate a subordinate principle of indifference, the Principle of Symmetric Indifference, in group theoretic terms. Being a group isomorphism is the natural condition for a Symmetry-grounded probabilism to propose for the Justified Guarantee. I show that this fails. I then show no other condition can be a Justified Guarantee. Consequently, no Symmetry-grounded probabilism can provide the required Justified Guarantee, and consequently the hope for a Symmetry-grounded Principle of Indifference is forlorn.

13.5 Gyenis and Rédei's defusal is insufficient without a Justified Guarantee

Here I show that Gyenis and Rédei's putative Justified Guarantee is not an optional extra for their classical probabilism but is needed if it is to be a consistent. First we need to see briefly how Gyenis and Rédei's classical interpretation could be a Symmetry-grounded probabilism. Plainly they wish it to be, for why else use compact topological groups? They do not give an argument for why the group action of a compact topological group has epistemic significance but include in their definition of their principle of indifference that it expresses epistemic indifference. They do not explain what that would be.

The role of Gyenis and Rédei's particular kind of group, compact topological, is that such groups have unique natural measures, that is measures of their own, internally rather than externally related, which are their Haar measures. It is this uniqueness based in symmetry, rather than anything else about the Haar measure, that may do the work when we are concerned to avoid the inconsistent probabilities that arise when incompatible measures of a group are equally philosophically defensible.

Gyenis and Rédei do not distinguish events from their names. If the elements of topological group X are used to represent events, Gyenis and Rédei routinely identify the events with those elements. So they treat the problem of paradoxical probabilities entirely in terms of how they arise from what they call relabellings.

It might be thought that Gyenis and Rédei's defusal of Bertrand's paradox obviates any need for a Justified Guarantee for their classical interpretation to succeed as a Symmetry-grounded probabilism. Of course, I believe I have shown in Chapter 7 that their defusal doesn't work. Nevertheless, I want to show that even if I am wrong and it worked in its own terms, their classical probabilism couldn't fulfil the hope for a Symmetry-grounded probabilism on the basis of that defusal alone, but it requires a Justified Guarantee as well.

Recall that the essence of their defusal is that labelling invariance is violated but that doesn't matter because labelling irrelevance is what matters and labelling irrelevance does not entail labelling invariance. The reason that this defusal, if successful, won't defend a Symmetry-grounded probabilism from inconsistency is that on its own it does not go far enough. At one and the same time it admits that relabellings producing inconsistent probabilities is a theorem in measure theory but claims we can reject the consequent paradoxes provided we respect labelling irrelevance. That means we must introduce a restriction on relabellings to those that are acceptable. Now acceptability cannot simply be defined by respecting labelling *irrelevance*. After all, if we could do that we could just as well have defined it by respecting labelling *invariance*. Since these are both criteria of consistency, in either case, to define acceptable relabellings by the criterion (i.e. to those that don't violate the criterion) is just to restrict relabellings to consistent relabellings. And if we can do that, we might just as well have started there rather than bother with this whole literature on the paradoxes. We didn't start there because it is a merely ad hoc evasion of the paradoxes.

Worse, it is not even clear that it would work. After all, going back to Bertrand's paradox, we have three representations which are inconsistent. Certainly, if we now have defined acceptable relabellings as those which are consistent, that is fine so far as it goes. But which of the three representations do we *start* with? Which is the right one from which any further constrained and therefore consistent relabelling is to start?

So even if Gyenis and Rédei's defusal works in its own terms, it is incomplete and there is a job left over, the job of providing a non–ad hoc restriction on relabellings which is both logically prior to and results in relabellings satisfying whichever criterion applies. Evidently that just is the very job of providing a Justified Guarantee. Although Gyenis and Rédei don't discuss these points explicitly, since they do offer what they call the non-trivial strategy, which is supposed to fulfil the roles of a Justified Guarantee, presumably they would agree. So whether or not Gyenis and Rédei's defusal of Bertrand's paradox works, their classical probabilism still needs its Justified Guarantee, which is therefore not an optional extra if it is to be a consistent Symmetry-grounded probabilism

13.6 Gyenis and Rédei's Justified Guarantee fails

Gyenis and Rédei propose what they call a non-trivial strategy to

> block the emergence of the general Bertrand's paradox [i.e. guarantee consistency].... by impos[ing] some extra condition on relabellings

> it is unreasonable to expect a relabelling to preserve probabilities unless the relabelling also preserves our epistemic status with respect to the elementary events; after all, the principle of indifference states that p_H is the empirically correct probability only if the group structure of X expresses epistemic neutrality. So the following stipulation is in the spirit of the principle of indifference:

> **Definition:** The relabelling h between probability spaces (X, S, p_H) and (X', S', p'_H) preserves the epistemic status if [and only if] it is a group isomorphism between X and X'.

(2015a:361)

It is common for mathematicians to give definitions using a conditional when a biconditional is meant. We will see shortly that if we took this as a conditional it would not provide the strategy intended, so we read it as a biconditional as indicated and in the following speak of its left-to-right and right-to-left directions.

Preserving probabilities is guaranteeing consistency, that is fulfilling the first role of a Justified Guarantee. Gyenis and Rédei's motivating philosophical idea is this: if a relabelling is to preserve the output of the principle of indifference, that is to preserve probabilities, then it must first preserve the critical input to the principle of indifference, our 'epistemic status with respect to the elementary events', namely, our epistemological indifference, which is represented by 'the group action express[ing] epistemological indifference' (2015a:356), which clause comes from their definition of their principle of indifference.[157] So an acceptable relabelling preserves the epistemic status expressed by the group action and therefore the condition we need (the Justified Guarantee, in our terms) must be a constraint on relabellings that does that. Then the application of the philosophical idea is that preserving epistemic status is to be *defined as* being a group isomorphism.

The preservation of epistemic status is a philosophical requirement and since 'preserving epistemic status' does not mean being a group isomorphism, the last step is rather close to Humpty-Dumptying[158] (covertly changing the subject by a stipulative redefinition), producing an ad hoc rather than a justified guarantee of consistency. Instead, and no doubt this is what was intended, the proposal should be that preservation of epistemic status is to be mathematically *represented* by the relabelling being a group isomorphism. Understanding the condition that a relabelling be a group isomorphism as a constraint that mathematically represents the preservation of our epistemic status makes this part of their proposal an attempt to fulfil the second feature of a Justified Guarantee. Hence, recalling our earlier remark, Gyenis and Rédei's non-trivial strategy is a putative Justified Guarantee and we will call their definition the GR-Justified Guarantee.

By the definition of a group isomorphism, a relabelling that is a group isomorphism preserves the group structure and therefore the group action. The group action expresses the epistemic status of elementary events. Therefore such a relabelling preserves the epistemic status of elementary events. This gives us the right-to-left direction of the GR-Justified Guarantee.

Gyenis and Rédei then say

As the probability measures p_H and p'_H are completely determined by the respective group actions, re-labellings that preserve the epistemic status [i.e. that are group isomorphisms by the GR-Justified Guarantee] are isomorphisms between the measure spaces.'

(2015a:362)

---

157 We must understand 'the group structure of X expresses epistemic neutrality' as meaning what is stated in the original definition.

158 See Carroll 1962; Shackel 2005.

We'll call this *Gyenis and Rédei's Lemma*: if a relabelling is a group isomorphism and the measures are Haar measures defined by the groups then it is a probability isomorphism. Without the lemma their Justified Guarantee does no guaranteeing.

So right-to-left of the definition both preserves epistemic status and preserves probabilities, hence gives us consistent probabilities.[159] But it is not enough.

What is needed for the non-trivial strategy to succeed in blocking 'the emergence of the general Bertrand's paradox' (2015a:361), thereby guaranteeing consistency, is that the philosophically motivated restriction of relabellings, which restriction defines the class of acceptable relabellings, results in all acceptable relabellings preserving probabilities. Given that the philosophical restriction is to relabellings that preserve the epistemic status of elementary events, for the strategy to succeed requires the left-to-right direction of the GR-Justified Guarantee. Gyenis and Rédei do not give an explicit argument, but here is how it must go:

1. Acceptable relabellings preserve the epistemic status of elementary events.
2. Relabellings that preserve the epistemic status of elementary events are group isomorphisms.
3. Group isomorphisms between probability spaces with Haar measures are probability isomorphisms.
4. Probability isomorphisms preserve probabilities.
5. So acceptable relabellings preserve probabilities.

The first premiss is the motivating philosophical idea, the second premiss is the left-to-right direction of GR-Justified Guarantee, the third is Gyenis and Rédei's lemma and the fourth follows from the definition of a probability isomorphism.

So for their non-trivial strategy to work, what they need is *not* the right-to-left direction of the GR-Justified Guarantee but the left-to-right direction. I shall now show that Gyenis and Rédei's lemma entails that the second premiss is false and hence that the GR-Justified Guarantee is false and the non-trivial strategy fails.

We need to recall that a relabelling is not defined between just any old pair of probability spaces but between 'two probability spaces *describing*

---

159 This is what is meant by 'a non-trivial strategy to preserve labelling invariance' (2015a:349), which is a misleading phrase. Labelling invariance is defined by them to mean *any* relabelling is a probability isomorphism and this they have proved false. What their non-trivial strategy amounts to is only that *some* relabellings, namely group isomorphisms, are probability isomorphisms.

the same phenomenon' (2015a:357 my emphasis). There is a very good reason that this is part of the definition. Without it the General Bertrand's paradox would have no connection to any paradox, since a relabelling would be allowed from a probability space that describes the phenomena to one that doesn't. In that case, the relabelling need not represent an arbitrary but epistemically equally good renaming of the events of interest, but can be instead a function between the labelling of the events of interest and the labelling of some other, completely different, events. As we noted before, a representation of poker hands could be relabelled to a representation of bridge hands. No one has ever thought the variance of probabilities in such a relabelling could be paradoxical because no one ever thought it should preserve probabilities in the first place!

Consequently, for two spaces even to be eligible for relabellings between them, they must each describe the same phenomenon, and do so in the way that GR-classical probabilism prescribes. The elementary random events of the phenomenon are represented by a compact topological group, X. Our epistemic status is represented by the 'group action express[ing] epistemological indifference about the elementary random events' (2015a:356). From this we derive a probability space describing the phenomena, (X, S, p_H), using the Haar measure, p_H, and the Borel algebra, S, of the compact topological group, X.

So when we have two spaces, (X, S, p_H) and (X', S', p'_H), that describe the same phenomenon, *each* compact topological group represents the *same* elementary events as the other, *each* 'group action expresses epistemological indifference about the elementary random events' (2015a:356)[160] and each space does all this logically prior to the relabelling. Each space is therefore *already* an expression of 'our epistemic status with respect to the elementary events' (2015a:361) before the relabelling. Since our epistemic status is expressed by each space before any relabelling, a relabelling of one by the other cannot change our epistemic status with respect to the elementary events but must preserve it.

The reason it must preserve it is this: if we apply any relabelling from X to X', we end up with a new description of the phenomenon by (X', S', p'_H). By the definition of a relabelling, the original description by (X', S', p'_H), represented the same phenomenon in the same way as (X, S, p_H) did: by its group action expressing 'epistemological indifference about the elementary random events'. The relabelling will preserve the epistemological indifference between any pair of elementary random events because, by the GR-Representation Lemma 30, if the group action of X maps their labels to one another under their original labelling by X then the group action of X'

---

160 Obviously a self relabelling must preserve epistemic status, since it is one and the same group action expressing it, but I am wanting to cover this in full generality.

maps their labels to one another under the new labelling by X'. So *any* re-labelling 'preserves our epistemic status with respect to the elementary events' (2015a:361). Here is a proof:

Theorem 42

Any relabelling in GR-classical probabilism preserves epistemological indifference between the elementary events.

Proof: Let E be the set of elementary events, let $l:E \to X$ and $l':E \to X'$ be label-lings of elementary events by compact topological groups whose group actions express epistemological indifference between elementary random events, and let $h:X \to X'$ be a relabelling. Suppose $d, e \in E$. If we are epistemo-logically indifferent between them then $l(d)$ is mapped to $l(e)$ by the group action (GR-Representation lemma applied to X). So to show that relabelling preserves epistemological indifference we need to show that if $l(d)$ is mapped to $l(e)$ by the group action on X then $h \circ l(d)$ is mapped to $h \circ l(e)$ by the group action on X'.

Suppose $l(d)$ is mapped to $l(e)$ by the group action on X. Since $l(d)$ is mapped to $l(e)$ then we are epistemologically indifferent between d and e (GR-Representation lemma applied to X). The composition of the labelling l and the relabelling h is just another labelling $h \circ l:E \to X'$. Consequently $h \circ l(d)$ in X' labels d and $h \circ l(e)$ in X' labels e.

Since we are epistemologically indifferent between elementary events d and e and since the group action of X' already expresses epistemological indif-ference between the elementary random events, the group action of X' maps X'-labels of d to X'-labels of e (GR-Representation lemma applied to X'). Hence the group action on X' maps $h \circ l(d)$ to $h \circ l(e)$. Hence h preserves epis-temological indifference between the elementary events d and e. QED

Note that this theorem does *not* depend on us being epistemologically in-different between *all* the elementary events. Epistemological indifference might partition E, where we are indifferent between the members of the same cell but not between members in different cells. It is this possibility that makes it seem that a relabelling might fail to preserve indifference.

We know from Gyenis and Rédei's *Proposition 1* above (§7.7) that there are relabellings that are *not* probability isomorphisms, and I have just proved that they, too, preserve epistemological indifference. We have Gyenis and Rédei's lemma, that relabellings of compact topological groups that are group isomorphisms are probability isomorphisms of Haar mea-sures. Consequently, those *Proposition 1* relabellings are not group iso-morphisms. Hence there are relabellings that 'preserve our epistemic status with respect to the elementary events' (2015a:361) that are not group iso-morphisms. Whence Premiss 2 and the aforementioned GR-Justified Guarantee are both false.

The philosophical motivation for the non-trivial strategy was that acceptable relabellings preserve epistemic status with respect to the elementary events. The proposed strategy is that being a group isomorphism is a necessary and sufficient condition for preserving epistemic status. Their condition is sufficient for a relabelling to preserve epistemic status, and to preserve probabilities. However, to *represent* preserving epistemic status, and to *guarantee consistency* by blocking relabellings that don't preserve probabilities, their condition must also be necessary, and it is *not* necessary. Consequently Gyenis and Rédei's non-trivial strategy, their putative Justified Guarantee, is neither justified nor a guarantee, and so the strategy fails.

Showing that GR-classical probabilism fails in this way does not suffice, however, to rule out all hope. We do not have to confine symmetry to compact topological groups and we can make equal ignorance a stronger constraint within the principle of indifference than epistemological indifference is in the GR-principle of indifference. So I now turn to treating the hope from symmetry in complete generality.

13.7 Symmetry in group theory

A symmetry is a function that maps a symmetric object to itself perfectly. A symmetry is therefore a bijection of the constituents of that object to itself, which is to say that it is a permutation. In mathematics, symmetries are studied in terms of group theory, because the complete set of symmetries of anything with symmetries satisfies the group axioms.

Definition

A *group* $(G, \bullet)$ *of symmetries of* X is a set of permutations $g:X \rightarrow X$ closed under an associative operation $\bullet: G \times G \rightarrow G$, where G contains an identity for $\bullet$ and every element in G has an inverse in G under $\bullet$.

Permutations are also called transformations. From hereon by 'group' I mean symmetry group because the symmetry of events is our hope.[161] Consequently, since the elements of G are permutations of X, the group operation is defined by function composition, that is $g \bullet h$ is defined by $(g \bullet h)(x) = g \circ h(x) = g(h(x))$, which is always associative. For that reason I will use the composition symbol '$\circ$' for both function composition and group operations. Every group on X is a subgroup of the set of all the permutations of X.

---

161 Cayley's theorem says every group is equivalent to a symmetry group (Jacobson 2009:38).

The action of those symmetries on whatever they are symmetries of is called the group action. For example, the 0 and π rotational symmetries of a coin are a group and the group action is what those rotations do to the coin, that is leave it alone or turn it over. For the die, the group action of the group of rotations is what those rotations do to the die; for example, the quarter-turn symmetries each rotate the die by one quarter.

Formally, the group action is a function from the cross product of the set of symmetries and the symmetric object to the symmetric object that respects the group operation. Since our group elements are permutations:

Definition

$k:G \times X \rightarrow X$, $k(g, x) = g(x)$ is *the group action*.
Under this definition, k fulfills the group action axioms thus:

Identity axiom: if j is the identity permutation $k(j, x) = j(x) = x$

Operation axiom: $k(g \bigcirc g', x) = (g \bigcirc g')(x) = g((g'(x)) = k(g, g'(x)) = k(g, k(g', x))$.

It is possible for $G = X$. For example, the integers are a group under addition, where each integer defines a symmetry, such as $1: Z \rightarrow Z$, $z \rightarrow z + 1$ and the group action as a whole is the addition operation $(z, z') \rightarrow z + z'$.

Of interest for our purposes is that the group action partitions X and maps members of a single cell of that partition to each other.

Orbits Lemma 43

Given a group, G, on X, let the relation $\approx$ be the relation defined by the group action. That is, $x \approx y$ iff for some $g \in G$, $x = k(g, y) = g(y)$
  (a) $x \in$ Orbit (y) iff $x \approx y$, and $\approx$ is an equivalence relation.
  (b) The orbits of the group partition X.
  (c) Any pair in X are in the same orbit iff they are mapped to one another other by the group action.

Proof: (a) By definition Orbit $(y) = \{x \in X: x = g(y), g \in G\}$ hence $x \in$ Orbit (y) iff $x \approx y$. The identity is in G so $\approx$ is reflexive. Every element in G has an inverse so $\approx$ is symmetric. G is closed under the group operation of function composition so $\approx$ is transitive.[162] (b) Since $\approx$ is an equivalence relation it partitions X and from (a) the partitions are the orbits. (c) LTR If $x, y \in$ Orbit (z) then $x \approx z$ and $y \approx z$ hence $x \approx y$ whence $x = g(y) = k(g, y)$ for some $g \in G$. RTL: If $x = k(g, y)$ then $x = g(y)$ for some $g \in G$, hence $x \approx y$, whence $x \in$ Orbit (y). QED

---

162 Shafarevitch points out the elegant relation 'that... [$\approx$] is reflective, symmetric and transitive... is just a rephrasing of the three properties in the definition of a transformation group' (Shafarevich 1990:100).

Notation: For each group, $\approx$ denotes the equivalence relation defined by the group action. If we need to keep track of distinct group equivalence relations for groups G and H we will use the notation $G{:}\_\approx\_$ and $H{:}\_\approx\_$.

13.8 The Representation of Event Symmetries

Any symmetry among events can be represented by a symmetry of names, because with our definition of a naming any symmetry among events defines a permutation of names of those events.

Theorem 44

Let S be a group of symmetries on $\mathscr{E}$ of event space $(E, \mathscr{E})$, let (X, Σ) be a measurable space and $n{:}\mathscr{E}{\to}\Sigma$ a naming. Let G on Σ be such that for each $s{\in}S$, there is a $g{\in}G$, $g{:}\Sigma{\to}\Sigma$ defined by

$$\forall e \in \mathscr{E} \; g{\circ}n(e) = n{\circ}s(e) \tag{1}$$

and we define the operation $\bullet{:}G{\times}G{\to}G$ by

$$(g \bullet g'){\circ}n(e) = n{\circ}s{\circ}s'(e). \tag{2}$$

Then each g is a permutation of Σ, there is a unique $g{\in}G$ for each $s{\in}S$, $\bullet$ is function composition and G is a group.

Proof: Checking that each g is well defined on Σ, since n is a bijection, for all x in Σ there is an e in $\mathscr{E}$ for which $x{=}n(e)$ and hence $g(x){=}g(n(e)){=}n(s(e)$. Since s is a permutation, each g is a composition of bijections and is therefore a bijection from Σ to itself and is therefore a permutation. To prove uniqueness of g for each $s{\in}S$, suppose, for a contradiction, both g and g' are defined by (1) but $g'{\neq}g$. Then there exists y: $g'(y){\neq}g(y)$ and there exists $e{:}y{=}n(e)$. So $g'(y){=}g'{\circ}n(e){=}n{\circ}s(e){=}g{\circ}n(e){=}g(y){\neq}g'(y)$. Hence each g is unique. $\bullet$ is function composition because members of G are permutations: $s{\circ}s'{\in}S$, so by (1) there is a $g''{\in}G$: $g''{=}n{\circ}s{\circ}s'{=}g{\bullet}g'$ so G is closed under $\bullet$ and

$$\begin{aligned} g{\circ}g'{\circ}n(e) &= g{\circ}n{\circ}s'(e) && \text{from (1)}\\ &= n{\circ}s{\circ}s'(e) && \text{from (1)}\\ &= g \bullet g'{\circ}n(e) && \text{from (2).} \end{aligned}$$

Since S is a group and we defined the operation $\bullet$ on G in the obvious way to define the operation on G by the group operation on S, G is a group. If j is the S identity then the g defined by j is the G identity because $g{\circ}n(e){=}n{\circ}j(e){=}n(e)$. Let g be defined by s and h be defined by s^{-1}. Then

$$\begin{aligned} g{\circ}h{\circ}n(e) &= n{\circ}s{\circ}s^{-1}(e) && \text{from (2)}\\ &= n(e) = n{\circ}s^{-1}{\circ}s(e) = h{\circ}g{\circ}(e) && \text{from (2)} \end{aligned}$$

so h is the inverse of g. The group operation is function composition so is associative. QED

Remark: What I have done here is for each symmetry, s, of the events $\mathcal{E}$ I have defined a permutation, g, on their names Σ, to be the map that maps the name of each event e to the name of the event $s(e)$ that the symmetry s, maps event e to.

Definition

A *Group Representation of Event Symmetry* is $(G, E, \mathcal{E}, S, n, X, \Sigma)$ as defined in Theorem **44**.

I can now prove a representation theorem.

Event Symmetry Representation Theorem 45

Given a Group Representation of Event Symmetry $(G, E, \mathcal{E}, S, n, X, \Sigma)$:
  (a) S and G are isomorphic groups.
  (b) The orbits of S and G are in a one-to-one correspondence.
  (c) Any pair of events are symmetric iff their names are mapped to one another by the group action of G.
  (d) For all d, $e \in \mathcal{E}$, $S{:}d \approx e$ iff $G{:}n(d) \approx n(e)$
  (e) Furthermore, (c) is equivalent to (d).

Proof: We define $h{:}S \to G$ using line (1) definition of members of G:

$$h(s) = g, \text{ the } g \in G \text{ defined by } \forall e \in \mathcal{E} \ g \circ n(e) = n \circ s(e) \tag{4}$$

(a) From the last theorem, for each $s \in S$, there is $g \in G$ and each such g is unique so h is an bijection. Then from the definition of the group operation in (2)

$$\begin{aligned} h(s \circ s') \circ (e) &= n \circ s \circ s'(e) \quad \text{from (4)} \\ &= g \circ g' \circ n(e) \quad \text{from (2)} \\ &= h(s) \circ h(s') \circ n(e) \quad \text{from (4)} \end{aligned}$$

So $h(s \circ s') = h(s) \circ h(s')$ and therefore h is a group isomorphism.
(b) Suppose $e, d \in \mathcal{E}$. Then $S{:}e \approx d$ iff $e = s(d)$ for some $s \in S$ iff

$$n(e) = n \circ s(d) \text{ iff } n(e) = h(s) \circ n(d) \text{ for an } h(s) \in G \quad \text{from (2)}$$
$$\text{iff } G{:}n(e) \approx n(d).$$

Consequently n defines a bijection between the orbits of S and the orbits of G by the mapping Orbit$(e) \to$ Orbit$(n(e))$.
(c) Pairs of symmetric events and their names are in corresponding orbits of S and G and hence this follows by the Orbits Lemma **43c**.
(d) This was proved during the proof of (b).

(e) The equivalence of (c) and (d) is because (c) is merely a restatement of (d), in terms that will be useful to us later. QED

Lemma 46

Any renaming $r{:}(X, \Sigma) \to (Y, \Psi)$ of a Group Representation of Event Symmetry $(G, E, \mathcal{E}, S, n, X, \Sigma)$ gives a Group Representation of Event Symmetry $(H, E, \mathcal{E}, S, r \circ n, Y, \Psi)$ of the same event space and H is isomorphic to G.

Proof: Since a renaming is a bijection, $r \circ n{:}\ \mathcal{E} \to \Psi$ is a naming. Consequently we can apply Theorem 44 to $(E, \mathcal{E}, S, r \circ n, Y, \Psi)$ to give the Group Representation of Event Symmetry $(H, E, \mathcal{E}, S, r \circ n, Y, \Psi)$. Theorem 45 gives isomorphism. QED

13.9 The Group Representation of Equal Epistemic Status

Recall that events of equal epistemic status is our expression for events between which we have no reason to discriminate, understood to generalize over any account anyone wishes to use in the principle of indifference. The upshot of the last couple of sections is that in attempting to use symmetry to solve the paradoxes, we can represent any symmetry of events by group theory, and so the paradoxes can be solved by symmetry iff they can be solved by a suitable use of groups in constructing a probabilism's use of the principle of indifference. So if symmetry explains the equality of epistemic status of events then the defining principle of a Symmetry-grounded probabilism will make use of a group representation of that symmetry and attribute equal probabilities to events that are equivalent under the symmetry as represented by a mathematical group.

The equality of epistemic status of events can be a complex matter, in the sense that there may be many different pluralities of events having equal epistemic status without events in one plurality being comparable with events of others (prior to the determination of probabilities by the principle of indifference, which may subsequently provide comparability, of course). So when we speak of the equal epistemic status of events, we are by no means limited to simple cases such as a die, in which all the atomic events have equal epistemic status. The state of the equal epistemic status of events of an event space $(E, \mathcal{E})$ may involve many distinct pluralities in both E and $\mathcal{E}$. (The reader may recall that it is because of this I formulated our principle of indifference to have the feature I called generality.)

Fortunately, as will become evident from the representation theorem to follow, once the group representation of equal epistemic status is in place, the Orbits lemma, via the theorems of the last section, essentially does all the work of keeping track of the complexity of that state. Consequently, in speaking of the equal epistemic status of events I will be speaking of states of any complexity without having to articulate or define any such complexities.

Lemma 47

Let $\equiv$ be the *relation of equal epistemic status explained by symmetry*. Then $\equiv$ is an equivalence relation.

Proof: Any event has equal epistemic status with itself and if an event has equal epistemic status with another then the other has equal epistemic status with it, so equal epistemic status is reflexive and symmetric, *a forteriori*, $\equiv$ is. Suppose events c and d have equal epistemic status and d and e have equal epistemic status, in each case explained by symmetry. Symmetries are transitive, so c has equal epistemic status to e. QED

The Equal Epistemic Status Group Theorem 48

Given event space $(E, \mathcal{E})$, if symmetry explains the equalities of epistemic status of events in $\mathcal{E}$ then
- (a) there is a group, S, of symmetries on $\mathcal{E}$, defined by the equalities of epistemic status;
- (b) the cells in the $\equiv$-partition are the orbits of the group and $\equiv$ is the equivalence relation $S:\_\approx\_$defined by the group action of S, that is for all $d, e \in \mathcal{E}, d \equiv e$ iff $S:d \approx e$;
- (c) events of equal epistemic status are in the same orbit of the group action and vice versa.
- (d) Furthermore, (b) is equivalent to (c).

Proof:

(a) $\equiv$ partitions $\mathcal{E}$ into sets of events having equal epistemic status. Let $D \subset \mathcal{E}$ be a cell of the partition. Let SYM_D be the set of all permutations of D. SYM_D is a group. Then let S_D be the set of permutations of $\mathcal{E}$ defined by SYM_D as follows: For each $sym \in SYM_D$, there is $w \in S_D$, $w:\mathcal{E} \rightarrow \mathcal{E}$

$$w(e) = \begin{cases} sym(e) & \text{if } e \in D \\ e & \text{if } e \in \mathcal{E} - D \end{cases}$$

Call the $w \in S_D$ the *permutations based on D*. Let W be the union of all the S_D.

Since the identity on D belongs to each SYM_D, the identity on $\mathcal{E}$ is in W. If $w \in W$ then for some D, w is defined by some $sym \in SYM_D$. Since SYM_D is a group, $sym^{-1} \in SYM_D$ which defines $v \in S_D$ as above. Then

$$w \circ v(e) = \begin{cases} sym \circ sym^{-1}(e) = e & \text{if } e \in D \\ e & \text{if } e \in \mathcal{E} - D \end{cases}$$

and likewise $v \circ w(e) = e$, so v is the inverse of w.

To get the group, we define S to be the closure of W under function composition. Then S is associative since composition is. The identity is in W so is in S. We need to check S for inverses. If $s \in S$ then there exist $\{w_1, w_2, w_3, ..., w_n\} \in W$ such that $s = w_1 \circ w_2 \circ w_3 \circ ... \circ w_n$. The inverses of the w_k are in W so let

$r = w_n^{-1}\mathrm{O}...\mathrm{O}w_3^{-1}\mathrm{O}w_2^{-1}\mathrm{O}w_1^{-1}$. $r \in S$ and $r\mathrm{O}s = s\mathrm{O}r$ is the identity so r is the inverse of s.

   (b) $s \in S$ are composed of some number of $w \in W$, and each w is based on a cell of the $\equiv$-partition. If two constituents of s are based on different cells then, since the cells are disjoint and each is the identity on the other's cell, they commute. We can therefore define a minimal constitution of s by commuting constituents based on different cells until they meet any constituents based on the same cell, when their composition produces another $w \in W$ based on the same cell to replace them. Hence without loss of generality we can assume that for each s, and each cell K of the $\equiv$-partition, there is at most a unique constituent of s, $w^s_k \in W$, based on K.

Whence for each $s \in S$, for each K-cell–based constituent of s, $w^s_k \in S_K$, and for all $d \in K$, $s(d) = w^s_k(d)$ and $\{w^s_k : s \in S\} = S_K$. So for all $d \in \mathcal{E}$ and $s \in S$, if d is in cell K

$$\mathrm{Orbit}(d) = \left\{ e \in \mathcal{E} : e = s(d), s \in S \right\} = \left\{ e \in \mathcal{E} : e = w^s_k(d), s \in S \right\}$$

$$= \left\{ e \in \mathcal{E} : e = w(d),\ w \in S_K \right\} = K$$

The final equality holds because S_K is defined by SYM_K, the set of *all* permutations of K. Hence if $e \in K$, there is a $sym \in SYM_K$ such that $e = sym(d)$ and so there is a $w \in S_K$ for which $e = w(d)$, so $e \in \{e \in \mathcal{E}: e = w(d),\ w \in S_K\}$. If $e \in \mathcal{E} - K$, then, since for all $w \in S_K$, $w(d) \in K$, there is no $w \in S_K$ for which $e = w(d)$ so $e \notin \{e \in \mathcal{E}: e = w(d),\ w \in S_K\}$. So the orbits of S are the cells of the $\equiv$-partition and by the Orbits Lemma 43(a) $\equiv$ is therefore the equivalence relation $S{:}\underline{\approx}\underline{\ }$ defined by the group action of S.

   (c) Since $\equiv$ is the equivalence relation $S{:}\underline{\approx}\underline{\ }$, events between which we have no reason to discriminate are in the same orbit. By the Orbits Lemma 43(c), events are in the same orbit iff they are mapped to one another by the group action. So events between which we have no reason to discriminate are mapped to one another by the group action.

   (d) The equivalence of (b) and (c) is because (c) is merely a restatement of (b), in terms that will be useful to us later. QED

Definition

Given event space $(E, \mathcal{E})$, hereafter we call a group, S, as defined in Theorem 48 an *Equal Epistemic Status Group* of $(E, \mathcal{E})$.

The Equal Epistemic Status of Events Representation Theorem 49

Let $(G, E, \mathcal{E}, S, n, X, \Sigma)$ be a Group Representation of Event Symmetry, where S is an Equal Epistemic Status Group. Then

   (a) G is isomorphic to the Equal Epistemic Status Group;

   (b) the orbits of G represent the sets of events of equal epistemic status;

(c) G represents the equal epistemic status of events by the group action mapping names of events to one another iff the events have equal epistemic status;

(d) For all $d, e \in \mathscr{E}$, $d \equiv e$ iff $G: n(d) \approx n(e)$.

(e) Furthermore, (c) is equivalent to (d).

Proof: Follows from the corresponding parts of Theorem 45 when S is the Equal Epistemic Status Group, making use of the identity of $\equiv$ and $S:\_{\approx}\_$ from Theorem 48. QED

Definition

An *Equal Epistemic Status Representation Group* $(G, E, \mathscr{E}, S, n, X, \Sigma)$ is a Group Representation of Event Symmetry where the symmetry group of events, S, is an Equal Epistemic Status Group of $(E, \mathscr{E})$.

13.10 The Principle of Symmetric Indifference

We want to define a normative measure on the basis of the symmetry of events and now that we have our group representation of that we can do so.

Definition

Given a measurable space (X, Σ), a *Measure Determined by a Group* of permutations of X is a non-zero finite measure, μ, on X, such that members of Σ mapped to one another by the group action have equal μ-measure.

Symmetry Measure Theorem 50

Given an Equal Epistemic Status Representation Group and a measure determined by its group, the measure fulfills the normative measure role.

Proof: By Theorem 49(c) the names of events of equal epistemic status are mapped to one another by the group action and since the measure, μ, is determined by the group, the names of events of equal epistemic status have equal μ-measure. QED

I can now state:

The Principle of Symmetric Indifference: (POSI) Given an Equal Epistemic Status Representation Group $(G, E, \mathscr{E}, S, n, X, \Sigma)$ and a measure determined by its group, μ, members of Σ with equal μ-measure have equal probability.

This follows from the POIS and the Symmetry Measure Theorem 50. As before, the probability measure for the events is then got from the measure determined by the group using Theorem 7.

For the rest of this chapter, all probability measures are assumed to be got from POSI. Since the normative measure in POSI is a measure determined by the group and the probability measure is determined by the normative measure, any POSI begotten probability measure is also a measure determined by a group.

13.11 The Classical Probabilism Proposition

It is worth drawing out an implication of what I have proved about group representation of equal epistemic status explained by symmetry. The following proposition gets its name from the historical tendency to regard symmetry itself as the ground of classical interpretations of probability. What I have done above explains why that is natural if equal epistemic status is itself an equivalence relation. In Chapter 1 I acknowledged the controversy on that point, but also noted that if the converse principle is true, that equal probability entails equal epistemic status, then it must be so. The converse principle is defensible for epistemic probability and, indeed, is the principle defended by those who advance what I called the reversed principle of indifference (i.e. who define epistemic equality by equal probability). I also noted that despite his scepticism, Keynes endorsed the converse principle.

The Classical Probabilism Proposition 51

If equal epistemic status is an equivalence relation then the representation theorems above apply to all events, the Principle of Symmetric Indifference is of equally general application as POIS and all probabilisms using the principle of indifference are, at least so far as they do, Symmetry-grounded probabilisms. Consequently, the forthcoming results of this chapter also apply completely generally.

Explanation: The use made of the relation of equal-epistemic-status-explained-by-symmetry, ≡, in proving the representation theorems was always what was proved by Lemma 47, namely that it is an equivalence relation. So if equal epistemic status is quite generally an equivalence relation, everything in this chapter carries over to equal epistemic status in general.

13.12 Justified Guarantees for a Symmetry-grounded Probabilism

We now turn to the question of the hope from symmetry in the full generality defined by the Principle of Symmetric Indifference. By the Event Symmetry Representation Theorem 45 we can treat symmetries of the events by symmetries of names and consequently by renamings. Since the probabilities of events are begotten of POSI, inconsistent probabilities for events are inconsistent probabilities for renamings (by Theorem 12).

Consequently, we can guarantee consistent probabilities for events iff we can guarantee consistent probabilities for renamings and so we have the

> *Justified Guarantee Equivalence*: there is a Justified Guarantee for events iff there is a Justified Guarantee for renamings.

So I shall do the work on a Justified Guarantee for renamings and use the equivalence to get the general result.

For a Symmetry-grounded probabilism to guarantee consistency it must place a mathematical condition on renamings so that renamings satisfying the condition preserve probabilities, whereby they will satisfy a suitably restricted version of Paris' Renaming principle. Such a condition cannot be philosophically arbitrary—since if ad hoc stipulations were acceptable we need never have been troubled by the paradoxes in the first place—but must represent a philosophical condition grounded in relevant epistemic principles. In the case of a Symmetry-grounded Probabilism, the Principle of Indifference, via its subordinate principles POIS and POSI, provides the epistemic ground of the equal epistemic status of events. So the condition must represent for renamings the equal epistemic status of events.

By the Equal Epistemic Status of Events Representation Theorem 49(c), the equal epistemic status of events is represented by the group action mapping names of events to one another iff the events have equal epistemic status. So to represent the equal epistemic status of events the condition must be necessary and sufficient for the renaming to preserve that representation of equal epistemic status.

What is it for the renaming to preserve that representation of equal epistemic status? Simply for the new group under the renaming to map new names to one another whenever the old group mapped the old names to one another and vice versa. Recalling that 49(d), for all d, $e \in \mathscr{E}$, $d \equiv e$ iff G: $n(d) \approx n(e)$ is equivalent to 49(c), this give us

Definition

Given an Equal Epistemic Status Representation Group $(G, E, \mathscr{E}, S, n, X, \Sigma)$, a renaming $r: (X, \Sigma) \rightarrow (Y, \Psi)$ *preserves the equal epistemic status of events* iff for all d, $e \in \mathscr{E}$, $G:n(d) \approx n(e)$ iff $H:r \circ n(d) \approx r \circ n(e)$, where H is the group defined in Lemma 46.

So we want a mathematical condition on renamings, K, such that

> *Justified Feature*: Renamings preserve equal epistemic status of events iff they are K.

and to exclude paradoxical probabilities and guarantee consistency we want

Consistency Guarantee Feature: Renamings preserve probabilities, if they are K.

Justified Feature makes K a mathematical condition both necessary and sufficient for a renaming to preserve equal epistemic status of events and hence K is a mathematical representation for renamings of the equal epistemic status of events. Then the condition does its job by the (contrapositive of the) left-to-right direction ruling out renamings for the philosophical reason that they do not preserve the equal epistemic status of events. The right-to-left ensures that a renaming that is K is doing the philosophical job of preserving the epistemic status of events. The Consistency Guarantee Feature makes K sufficient to guarantee consistency. So a condition satisfying the Justified Feature and the Consistency Guarantee Feature together is a Justified Guarantee.

Consider

The *Group Isomorphism Condition*: a renaming satisfies the Group Isomorphism Condition just in case it both defines and is defined by a group isomorphism.

Given the representation of symmetry by groups, the Group Isomorphism Condition is the obvious and natural proposal for a Justified Guarantee for a Symmetry-grounded probabilism, just because it is so straightforwardly based on the motivation for a such a probabilism. The symmetries explain the equal epistemic status of events, and groups explain symmetries, giving us the Equal Epistemic Status Group. Groups on names represent event symmetries, the Principle of Symmetric Indifference uses a measure determined by the group of an Equal Epistemic Status Representation Group to define the probabilities. So a relation between the two groups in a renaming that preserves the group structure should preserve both equal epistemic status and preserve probabilities, so is exactly what we want—and that is exactly what a group isomorphism does. The preservation of equal epistemic status makes being a group isomorphism justified. And provided a renaming that is a group isomorphism is a probability isomorphism, it preserves probabilities and hence is a guarantee of consistency. So this gives us the Justified Guarantee we were looking for.

Given that motivation it is difficult to see what other condition could be given as good a philosophical ground. Nevertheless, having analysed it further I shall then discuss unknown conditions of unknown motivation in terms of their bare fulfilment of the defining roles of a Justified Guarantee.

13.13 The Group Isomorphism Condition is not a Justified Guarantee

I now show the Group Isomorphism Condition fails.

Lemma 52

Any renaming of an Equal Epistemic Status Representation Group preserves the epistemic status of events.

Proof: Let $r: (X, \Sigma) \to (Y, \Psi)$ be a renaming of Equal Epistemic Status Representation Group, $(G, E, \mathscr{E}, S, n, X, \Sigma)$. By Lemma **46** and Theorem **49** the renaming gives us an Equal Epistemic Status Representation Group $(H, E, \mathscr{E}, S, r \circ n, Y, \Psi)$ of the same event space. A renaming $r:(X, \Sigma) \to (Y, \Psi)$ preserves epistemic status just in case $G{:}n(d) \approx n(e)$ iff $H{:}r \circ n(d) \approx r \circ n(e)$. By theorem 49(d) $G{:}n(d) \approx n(e)$ iff $d \equiv e$ iff $H{:}r \circ n(d) \approx r \circ n(e)$. QED

Corollary 53

Any renaming of an Equal Epistemic Status Representation Group that satisfies the Group Isomorphism Condition preserves the equal epistemic status of events.

Proof: Follows immediately *a forteriori*. QED

Lemma 54

Let G be a group of permutations of Σ
    (a) any renaming $r:(X, \Sigma) \to (Y, \Psi)$ defines an isomorphic group H on Ψ by $j{:}G \to H$, $j(g) = r \circ g \circ r^{-1}$.
    (b) Given namings $n{:}\mathscr{E} \to \Sigma$, $m{:}\mathscr{E} \to \Psi$, for $g \in G$ any group isomorphism $k{:}G \to H$ defines a renaming $r{:}\Sigma \to \Psi$, $r = k(g) \circ m \circ n^{-1} \circ g^{-1}$.

Proof: (a) Let $H = \{h = j(g){:} g \in G\}$. Each $h{:}\Psi \to \Psi$ is a composition of bijections and is therefore a permutation of Ψ. Because

$$j\left(g \circ g'\right) = r \circ g \circ g' \circ r^{-1} = r \circ g \circ r^{-1} \circ r \circ g' \circ r^{-1} = j(g) \circ j(g')$$

H is closed under function composition, if $e \in G$ is the identity of G then $j(e)$ is the identity of H and the inverse of $h = j(g)$ is $j(g^{-1})$. Therefore H is a group and j is a group homomorphism. j is an injection since $j(g) = j(g')$ only if

$$r \circ g \circ r^{-1} = r \circ g' \circ r^{-1} \text{ only if}$$
$$r^{-1} \circ r \circ g \circ r^{-1} \circ r = r^{-1} \circ r \circ g' \circ r^{-1} \circ r \text{ only if.}$$
$$g = g'$$

Let $i{:}H{\rightarrow}G$, $i(h)=r^{-1}\circ h\circ r$. Then likewise i is an injection.

$$i\circ j(g)=i\left(r\circ g\circ r^{-1}\right)=r^{-1}\circ r\circ g\circ r^{-1}\circ r=g$$

and likewise $j\circ i(h)=h$ so i is the inverse of j and j is a bijection.
(b) r is a composition of bijections and so is itself a bijection, and is therefore a renaming. QED

Lemma 55

Any renaming defines a group isomorphism that it is defined by.

Proof: Let $r: \Sigma{\rightarrow}\Psi$ be a renaming of a naming $n{:}$ $\mathscr{E}{\rightarrow}\Sigma$ and G be a group on Σ. From Lemma 54(a) there exists a group H on Ψ for which $j{:}G{\rightarrow}H$, $j(g)=r\circ g\circ r^{-1}$ is a group isomorphism defined by r. $r\circ n{:}$ $\mathscr{E}{\rightarrow}\Psi$ is a naming. For $g{\in}G$ let $q{:}\Sigma{\rightarrow}\Psi$, $q=j(g)\circ r\circ n\circ n^{-1}\circ g^{-1}$, which we know from Lemma 54(b) is the renaming defined by j. Then

$$q = j\left(g\right)\circ r\circ n\circ n^{-1}\circ g^{-1} = r\circ g\circ r^{-1}\circ r\circ n\circ n^{-1}\circ g^{-1}$$
$$= r\circ g\circ r^{-1}\circ r\circ g^{-1} = r. \quad \text{QED}$$

Lemma 56

Any renaming of an Equal Epistemic Status Representation Group satisfies the Group Isomorphism Condition.

Proof: Follows immediately from the definition of the Group Isomorphism Condition and Lemma 55. QED

The Group Isomorphism Condition Theorem 57

A renaming of an Equal Epistemic Status Representation Group preserves the equal epistemic status of events if and only if it satisfies the Group Isomorphism Condition.

Proof: LTR Lemma 56 RTL Corollary 53 QED

Theorem 58

There exists a renaming of an Equal Epistemic Status Representation Group that is not a probability isomorphism.

Proof: Let $(G, E, \mathscr{E}, S, n, X, \Sigma)$ and $(H, E, \mathscr{E}, S', m, Y, \Psi)$ be a pair of Equal Epistemic Status Representation Groups of the same event space and let (X, Σ, P_X) and (Y, Ψ, P_Y) be POSI begotten. Let $e{\in}\mathscr{E}$ and let $P_X(n(e)){\neq}P_Y(m(e))$. We know from Bertrand's paradox that the last is possible using event symmetries of the direction case and the angle case.[163]

---

163 Since I showed how the direction case is named in the angle case name space I could have used a renaming of $(G, E, \mathscr{E}, S, n, X, \Sigma)$ to itself, but it is easier to follow the proof by using distinct representation groups.

Let $q: \Psi \to \Psi$ be a probability isomorphism and let $r: \Sigma \to \Psi$, $r = q \circ m \circ n^{-1}$. Since n and m are namings and q is a bijection, r is a renaming. So by Lemma 1(e) r preserves measurability, that is $\sigma \in \Sigma$ iff $r(\sigma) \in \Psi$ and $\varphi \in \Psi$ iff $r^{-1}(\varphi) \in \Sigma$. But

$$P_X\big(n(e)\big) \neq P_Y\big(m(e)\big) = P_Y\big(q \circ m(e)\big) = P_Y\big(r \circ n(e)\big).$$

So $P_X(n(e)) \neq P_Y(r \circ n(e))$ and hence r does not preserve probabilities and so is not a probability isomorphism. QED

Corollary 59

There exists a renaming of an Equal Epistemic Status Representation Group that satisfies the Group Isomorphism Condition that does not preserve probabilities.

Proof: Follows immediately from Lemma 56 and Theorem 58. QED

Proposition 60

The Group Isomorphism Condition is not a Justified Guarantee

Proof: By Theorem 57 the Group Isomorphism Condition is both necessary and sufficient for a renaming to preserve equal epistemic status representation. Hence it has the Justified Feature of a Justified Guarantee and is thus a mathematical representation for renamings of the equal epistemic status of events. However, by Corollary 59 satisfying the Group Isomorphism Condition is insufficient to preserve probabilities, and hence it is not a guarantee of consistency and so does not have the Consistency Guarantee Feature of a Justified Guarantee. Hence the Group Isomorphism Condition is not a Justified Guarantee. QED

13.14 Unknown Justified Guarantees

As promised, I now consider unknown conditions in terms of their bare fulfilment of the defining roles of being justified by representing equal epistemic status and being a constraint on renamings that guarantees consistent probabilities. Let K be the condition we seek to constrain renamings.

1. Renamings preserve the equal epistemic status of events iff they are K. (Justified Feature)
2. Renamings preserve probabilities, if they are K. (Consistency Guarantee Feature)
3. Renamings preserve the equal epistemic status of events iff they satisfy the Group Isomorphism Condition. (Theorem 57)
4. Renamings are K iff they satisfy the Group Isomorphism Condition. (1, 3)

5. There exists a renaming that satisfies the Group Isomorphism Condition that does not preserve probabilities. (Corollary **59**)
6. There exists a renaming that is K and that does not preserve probabilities. (4, 5)
7. A premiss is false. (RAA, 2, 6)

Premisses 3 and 5 are theorems so the contradiction between lines (2) and (6) shows that either (1) is false or (2) is false. Consequently no condition on renamings can have both the Justified Feature and the Consistency Guarantee Feature and therefore there is no condition on renamings that is a Justified Guarantee. By the Justified Guarantee Equivalence, there is no condition on the symmetries of events that is a Justified Guarantee.

13.15 Symmetry cannot save the principle of indifference from Bertrand's paradox

An explanation for the equality of epistemic status appealed to in a principle of indifference can be a symmetry among the random events. Consequently, an abiding hope has been that a Symmetry-grounded probabilism might get around the paradoxes if it could formulate a constraint in terms of symmetry, a constraint which guarantees consistency whilst not being ad hoc but justified by its representation of the equality of epistemic status.

Since in mathematics symmetry is represented and analysed by group theory, the hope can be addressed in terms of the Principle of Symmetric Indifference. Since a symmetry among events can be represented by a symmetry among the symbols naming the events, and since the paradoxes can thereby be represented in terms of renamings, the needed Justified Guarantee is equivalent to a Justified Guarantee on renamings.

The Group Isomorphism Condition is the natural proposal for a Justified Guarantee for a Symmetry-grounded probabilism, and hence the natural proposal for vindicating the hope offered by symmetry. Yet I proved that the Group Isomorphism Condition does not guarantee probabilistic consistency.

I then turned to considering presently unknown conditions. I showed that any such condition was equivalent to the Group Isomorphism Condition iff it was justified by representing the normative ground in the principle of indifference. Consequently, it followed that no condition could be both justified and a guarantee of consistency.

This result is conclusive, but I do not claim it is *absolutely* conclusive because I cannot prove absolutely conclusively that there is no other way of justifying a constraint in the context of the principle of indifference than by its representing the equality of epistemic status when we are using symmetry to explain that equality. The burden here, however, would be on the Symmetry-grounded probabilist to advance a justification based on something other than the equality of epistemic status, to explain exactly how that justification is derived from the defining Principle of Symmetric Indifference, or to offer a different principle that is yet based on symmetry, and to advance it in such a way that I cannot modify my arguments to cover the new justification. I find it very doubtful that this could be done, most especially because equality of epistemic status is the only normative ground in the principle of indifference itself and we are concerning ourselves with Symmetry-grounded probabilism that uses the principle where symmetry explains the equality of epistemic status. But I cannot prove that it cannot be done and so I point out this dubious escape route, which I now set aside.

Consequently, there is no Justified Guarantee and so no Symmetry-grounded probabilism can guarantee the consistency of probabilities in a philosophically justifiable way. The hope for symmetry to solve the problem posed by Bertrand's paradox is therefore forlorn. Finally, if equal epistemic status is an equivalence relation then by The Classical Probabilism Proposition 51 this result is entirely general: any hope to solve Bertrand's paradox is forlorn.

14 Unearthing the Root

14.1 Introduction

Returning once again to measure theory allows us to unearth the root of the failed attempts at solutions. The way that informative measures are developed argues for a distinction between state spaces that have a natural measure and state spaces that don't. We then see that Bertrand's original four paradoxes fall into two groups on either side of the naturalness distinction. The plenitude paradoxes, as I call them, have too many natural measures and the paucity paradoxes have none. Prior discussions of Bertrand's paradox have failed to see this distinction. This distinction allows me to explain why the chord paradox is rightly, renownedly, known as Bertrand's paradox, even if not so known for this reason.

14.2 Measure theory

Measure theory is the general theory of assigning numbers to subsets of a set in such a way as to respect and to extend our pretheoretical ideas of measurement, ideas such as that measurements are positive quantities and the measurement of an item that is the mereological sum of other items is the sum of their measurements. Here I shall lay out a skeletal account of what matters for us, a foundational part of abstract measure theory, adding some flesh to the part of that foundation of particular interest for our purposes.

The foundations are the theorems that constitute the modern extension of a method developed by Carathéodory (Carathéodory 1914, 1963). The importance of Carathéodory's work, and consequently these theorems, is in part that they capture what underlay the work of Jordan, Borel, Lebesgue and others during the attempt to overcome the limitations of the Riemann integral.[164] There grew an understanding that the theory of the integral

---

164 See Hawkins 1970 for a history and Bruckner *et al.* 2008:20–52 for a summary of the mathematics of this history.

DOI: 10.4324/9781003456308-14

rested on the measurement of subsets of the real numbers ($\mathbb{R}^n$, in general) and on how that measurement related to 'measuring' functions, from which the scare quotes may now be removed in the light of measure theory's definition of what such measuring of a function amounts to. Speaking somewhat loosely, the limitations of the Riemann integral were in part due an inability to measure certain subsets of the real numbers, an inability now fixed by measure theory.

14.3 Defining particular measures

I gave the definitions of measurable spaces and measures in Chapter 1. They are all very well, so far as they go, but given a set Ω how is a particular measure defined on it? In general, this is done by a three-stage method, the premeasure stage in which a non-negative function is defined on a family of subsets of Ω, the outer measure stage in which the non-negative function of the family is used to define an outer measure on the power set of Ω and the measure stage in which a measure is defined on a σ-algebra subset of the power set.

With this method we can get infinitely many distinct measures on a set, most of them arbitrary and without mathematical significance.[165] In general, the reason we want to define a measure for a particular space is that we already have a family of subsets of the space for which subsets there is a natural measurement of their size, but the family does not include all the subsets we need to measure. I shall have more to say later about what naturalness means here. In the meantime, we want a measure that at least satisfies these criteria: (1) it agrees with the natural measurement on the original family, (2) it does so uniquely and (3) its σ-algebra includes all the subsets we need to measure. It is not hard to extend the family to a σ-algebra (we can take any σ-algebra containing it) nor to define a measure in terms of the natural measurement of size. But the result is not in general guaranteed to satisfy the criteria. There are, however, conditions which can ensure satisfaction of at least the first two. We shall need these definitions:

> **Definitions:** A family of sets, $\mathcal{K}$, is a *cover* of a set S iff $S \subset \bigcup \mathcal{K}$ and a *finite subcover* is a finite subset of a cover.
>
> Throughout, $\mathcal{S}$, $\mathcal{A}$ and $\mathcal{F}$ are *families* of sets that cover Ω.
> $\mathcal{S}$ is a *semialgebra* on Ω iff $\emptyset$, $\Omega \in \mathcal{S}$, $\mathcal{S}$ is closed under finite intersections and complements are finite disjoint unions of members of $\mathcal{S}$.

---

165 We can, of course, arbitrarily define any σ-algebra on Ω we like by taking a set of subsets and using the σ-algebra generated by it (see below) and manufacture a measure for that σ-algebra, but this takes us even further astray from the point of a measure, hence the point of the method, as is about to be explained.

$\mathscr{A}$ is an *algebra on* Ω iff $\varnothing \in \mathscr{A}$ and $\mathscr{A}$ is closed under finite unions and complements.

The *algebra generated by a semialgebra* is the smallest algebra containing the semialgebra, which is the intersection of all algebras containing the semialgebra.

$\mathscr{M}$ is the *σ-algebra generated by a family of sets* $\mathscr{F}$ iff $\mathscr{M}$ is the smallest σ-algebra containing $\mathscr{F}$.

A function, $g{:}E{\to}C$, *extends* a function, $h{:}D{\to}C$ iff $D \subseteq E$ and g and h agree on D, that is for all $x \in D$, $g(x)=h(x)$.

A function $g{:}\mathscr{F}{\to}[0, \infty]$ is *finitely additive* iff for any finite number of disjoint elements in $\mathscr{F}$, g of their union=the sum of their individual values. g is *countably additive* iff for any sequence of disjoints elements in $\mathscr{F}$, g of their union=the sum of their individual values.

A function $g{:}\mathscr{F}{\to}[0, \infty]$ is a *premeasure* on $\mathscr{F}$ iff $g(\varnothing)=0$ and g is countably additive on any sequence of disjoint sets in $\mathscr{F}$ whose union is in $\mathscr{F}$.

A function $g{:}\mathscr{F}{\to}[0, \infty]$ is a *σ-finite* function iff there exists a countable set of elements in $\mathscr{F}$ whose union is Ω and for each element in that countable set g takes a finite value.

If the original family is, or is extended to, a semialgebra, and we can use the natural measurement of size to define a premeasure on that semialgebra that is then extended uniquely to a premeasure on the algebra generated by the semialgebra, then the second and third stages of the method will give us a measure (on the σ-algebra generated by that algebra) that will satisfy the first criterion. If the premeasure or measure are σ-finite we can satisfy the second and usually we can satisfy the third.

For our purposes it is the premeasure stage that is crucial since it does the work of capturing the natural measurement of size. For this reason I will demonstrate how it goes for a geometric example. I will not demonstrate the other stages for our example since at that point we can simply apply the general theorems to derive the measure for our example. Thus, once I have developed the first step of the first stage for our example I will then exhibit the theorems for the other stages, whose application to our example will then be shown.

14.4 Geometric example

As an example, we are going to consider defining a measure on the whole line. Here I am going to keep this geometric rather than use the 'identification' with the real numbers, since I am concerned to draw attention to the way in which the natural measurement of line segments, namely their lengths, is captured. So we start with

Natural measurement facts:
A: The length of a line segment is the distance between its ends.

B: Line segments, whether they contain their endpoints or not, have the same length if they have the same endpoints.

C: Lengths of line segments are finitely additive, i.e. the length of the mereological sum of a finite collection of non-overlapping line segments is the sum of their individual lengths.

D: The sum of the lengths of a pair of overlapping line segments is greater than the length of the segment that is their mereological sum.

E: The length of a subsegment of a line segment is less than or equal to the length of the line segment.

F: Any line segment has contiguous line segments of any length.

G: Half-lines and the whole line, being longer than any line segment, are infinitely long.

This much suffices, eventually, to give us a measure on the whole line. I will now demonstrate the first step of the first stage.

> **Definitions:** A *segment* is the empty segment or a line segment or a half-line or the whole line. A segment is *bounded* iff it is the empty segment or a line segment. The *boundary* of a segment is its endpoints. A segment is *closed* iff its boundary belongs to it and *open* iff its boundary does not belong to it. Segments *overlap* iff they have more than an endpoint in common. Segments are *disjoint* iff they don't overlap. A segment is a *subset* of another if it is a subsegment of the other.
>
> A *union* of segments is the mereological sum of them and their endpoints. The *intersection* of segments is the subsegment common to them all. The *complement* of a segment, s, is the whole line, L, with the segment removed, $L - s$. A union of segments is a *subset* of another iff every segment in the first is a subsegment of a segment in the union of the second.
>
> If $\mathscr{F}$ is a family of unions of segments, an element in $\mathscr{F}$ is *bounded* iff it is a subset of a bounded segment. The *difference* of two elements, x, y, in $\mathscr{F}$, is $x - y = x \cap (L-y)$.
>
> $\mathscr{S}$ is the set of segments.

Lemma 61

Let $f: \mathscr{S} \to \mathbb{R}^+$ be a function that extends the natural measurement of line segments, that is if s is a line segment, $f(s)=$ the length of s and for which f maps half and whole lines to ∞. For all s, k in $\mathscr{S}$, (a) If s is a subsegment of k then $f(s) \leq f(k)$. (b) Let a segment, s, be a finite union of segments, $s = \bigcup_{i=1}^{m} s_i$. Then $f(s) \leq \sum_{i=1}^{m} f(s_i)$, equal when a disjoint union, and f captures natural measurement fact C for $\mathscr{S}$.

Proof: (a) If k is unbounded we are done, by natural measurement fact G. If s is unbounded both are unbounded and $f(s)=f(k)=\infty$. If both are bounded then by natural measurement fact E, $f(s) \leq f(k)$. (b) If s is unbounded both equal ∞ since at least one segment, s_i, of the union must be unbounded. If s is bounded and the segments disjoint we have equality, otherwise less than equality, by the definition of f. So f captures natural measurement fact C. QED

Lemma 62

Let f be as in Lemma **61** and $T=\{s_1, s_2, s_3, ...\}$ be a sequence of disjoint segments whose union is a segment, S, in $\mathscr{S}$. Then $\sum_{i=1}^{\infty} f(s_i) = f(S)$ (i.e. f is countably additive on $\mathscr{S}$).

Proof: If necessary, reorder T so that for every n, the union of $\{s_1, s_2, s_3, ... s_n\}$ is a segment (i.e. choose s_1 with an endpoint in common with S and place them in the sequence by matching endpoints). The union of $\{s_1, s_2, s_3, ... s_n\}$ is a subsegment of S so by Lemma **61** for all n, $\sum_{i=1}^{n} f(s_i) \leq f(S)$ and so $\sum_{i=1}^{\infty} f(s_i) \leq f(S)$.

Now we prove the other direction. Suppose first that S is bounded. Let c be a closed subsegment of S such that $f(c)=f(S) - \varepsilon$, $0<\varepsilon<f(S)$. Let K be a sequence of open segments such that for all i, $s_i \subseteq k_i$ and $f(k_i) = f(s_i)+\dfrac{\varepsilon}{2^i}$.

Then K is an open cover of c and c, being closed, is compact, and hence there is a finite subcover $\{k_i\}_m$, the union of which is a segment since c is a segment.[166] Applying Lemma **61** we get

$$f(S)-\varepsilon = f(c) \leq f\left(\bigcup_{i=1}^{m} k_i\right) \leq \sum_{i=1}^{m} f(k_i) = \sum_{i=1}^{m}\left(f(s_i)+\frac{\varepsilon}{2^i}\right)$$

$$\leq \sum_{i=1}^{\infty}\left(f(s_i)+\frac{\varepsilon}{2^i}\right) = \sum_{i=1}^{\infty} f(s_i)+\varepsilon$$

Taking the limit as ε tends to zero, $f(S) \leq \sum_{i=1}^{\infty} f(s_i)$.

If S is unbounded then take any increasing sequence of bounded segments $B=\{b_1, b_2, b_3, ...\}$ such that they all share the same centre point and for all n, $f(b_{n+1})=f(b_n)+2$. For each n, each $s_i \cap b_n$ is disjoint and so $S \cap b_n = \bigcup_{i=1}^{\infty}(s_i \cap b_n)$ is a bounded disjoint union. Using what we have just shown for bounded disjoint unions:

$$f(S \cap b_n) \leq \sum_{i=1}^{\infty} f(s_i \cap b_n).$$

---

166 c is compact since $[0,1]$ is a compact space and there exists a continuous function $g:[0,1] \to c$. See Willard 2004:theorem 17.7.

Since each $s_i \cap b_n$ is a subsegment of s_i, by Lemma 61 $f(s_i \cap b_n) \le f(s_i)$

$$\sum_{i=1}^{\infty} f(s_i \cap b_n) \le \sum_{i=1}^{\infty} f(s_i).$$

$S \cap b_n$ is an increasing sequence of segments such that there exists an N, for all $n \ge N, f(S \cap b_{n+1}) \ge f(S \cap b_n) + 1$. Consequently $f(S \cap b_n) \to \infty$ and we have shown that for all n,

$$f(S \cap b_n) \le \sum_{i=1}^{\infty} f(s_i), \text{ So } \sum_{i=1}^{\infty} f(s_i) = \infty. \quad \text{QED}$$

Proposition 63

Let $f: \mathscr{S} \to \mathbb{R}^+$ be a function that extends the natural measurement of line segments. Then

(a) The set of segments, $\mathscr{S}$, is a semialgebra.
(b) f is a premeasure on $\mathscr{S}$;
(c) any finitely additive extension of f captures natural measurement facts A–G;
(d) any extension of f is σ-finite.
(e) the natural measurement of line segments is σ-finite on line segments.

Proof:
(a) The empty segment, $\varnothing$, and the whole line, L, are in $\mathscr{S}$. The intersection of any pair of segments is a segment so by induction any finite intersection of segments is a segment. The empty segment and the whole line are mutual complements. The complement of a half line is a half line. The complement of a line segment is a pair of disjoint half lines. So any complement of a segment is finite disjoint union of segments.
(b) f is defined for all line segments and captures natural measurement facts A and B. Line segments exist of all lengths and lengths are positive. Since the empty segment is shorter than any line segment, for all $\varepsilon > 0, f(\varnothing) < \varepsilon$. Therefore $f(\varnothing) = 0$. By Lemma 62, f is countably additive on $\mathscr{S}$. So f is a premeasure on $\mathscr{S}$.
(c) f captures facts A, B, E, F and G by being defined by segment length. By Lemma 61 f captures fact C for $\mathscr{S}$ so any finitely additive extension of f captures fact C. f captures D because for segments s, t

$$f(s \cup t) = f(s-t) + f(s \cap t) + f(t-s) \le f(s-t) + 2f(s \cap t) + f(t-s)$$
$$= f(s) + f(t)$$

(d) Take a line segment s and a segment b_1. Applying natural measurement fact F, there exists an increasing sequence of bounded segments $B = \{b_1, b_2, b_3, \ldots\}$ such that they all share the same centre point and for all $n, f(b_{n+1}) = f(b_n) + 2f(s)$. Then for all $n, f(b_n) < \infty$ and $L = \bigcup B$, so f is σ-finite and hence so is any extension of f.

(e) Since f is σ-finite its restriction to line segments is σ-finite, but its restriction to line segments is the natural measurement of line segments. QED

We have now completed the first step of the first stage for our example. We can proceed to our measure on the whole line using abstract measure theory and so will return to the example after setting that theory out.

14.5 Abstract theory: extend premeasure from semialgebra to algebra

Lemma 64

Let $\mathscr{S}$ be a semialgebra on Ω. Any finite union of members in $\mathscr{S}$ is a finite union of disjoint members in $\mathscr{S}$. Any finite intersection of finite unions of members in $\mathscr{S}$ is a finite union of disjoint members in $\mathscr{S}$.

Proof: Suppose $A, B \in \mathscr{S}$. Their union is

$$A \cup B = (A - B) \cup (A \cap B) \cup (B - A)$$

where

$$A - B = A \cap (\Omega - B), \quad B - A = B \cap (\Omega - A)$$

$A - B$, $A \cap B$ and $B - A$ are disjoint. Closure under finite intersection means $A \cap B \in \mathscr{S}$. Complements of members of $\mathscr{S}$ are finite unions of disjoint members in $\mathscr{S}$. Therefore the distributivity of intersection over union and closure under finite intersection means $A - B$ and $B - A$ are each a finite union of disjoint members of $\mathscr{S}$. By induction, the result follows.

Suppose $A_1, A_2, \ldots A_n \in \mathscr{S}$ and $B_1, B_2, \ldots B_m \in \mathscr{S}$. Due to the last paragraph, without loss of generality we can assume $A_1, A_2, \ldots A_n$ are disjoint and $B_1, B_2, \ldots B_m$ are disjoint. The intersection of their unions is

$$\bigcup_{i=1}^{n} A_i \cap \bigcup_{j=1}^{m} B_j = \bigcup_{i=1}^{n} \bigcup_{j=1}^{m} A_i \cap B_j$$

Since the As and Bs are disjoint and $\mathscr{S}$ is closed under finite intersection, all the $A_i \cap B_j$ are in $\mathscr{S}$ and are disjoint. So the intersection of their union is a finite union of disjoint members of $\mathscr{S}$. By induction, the result follows. QED

Lemma 65

Let $\mathscr{A}$ be the set of all finite unions of disjoint members in semialgebra, $\mathscr{S}$. Then $\mathscr{A}$ is the algebra generated by semialgebra, $\mathscr{S}$.

Proof: ∅ is in $\mathscr{A}$. We need to show $\mathscr{A}$ to be closed under finite unions and complements. Let A_1, A_2,... $A_n \in \mathscr{A}$ and A be their union. A is a finite union of finite unions of disjoint members of $\mathscr{S}$, so it is a union of finitely many members of $\mathscr{S}$, so by Lemma **64** A is a finite union of disjoint members in $\mathscr{S}$ and so $A \in \mathscr{A}$.

Let $A \in \mathscr{A}$ be the union of disjoint $s_1, s_2,... s_n \in \mathscr{S}$. Complement

$$\Omega - A = \Omega - \bigcup_{i=1}^{n} s_i = \bigcap_{i=1}^{n}\left(\Omega - s_i\right)$$

Since each $\Omega - s_i$ is a finite union of members of $\mathscr{S}$ (by definition of semi-algebra), $\Omega - A$ is a finite intersection of finite unions of members of $\mathscr{S}$ and so by Lemma 64 it is a finite union of disjoint members in $\mathscr{S}$ and so is in $\mathscr{A}$.

Any algebra containing $\mathscr{S}$, being closed under finite unions, must contain all the finite unions of disjoint members of $\mathscr{S}$ and may contain more, hence $\mathscr{A}$ is the smallest algebra containing $\mathscr{S}$. QED

Theorem 66

A premeasure, f, on a semialgebra, $\mathscr{S}$, extends to a unique premeasure, π, on the algebra, $\mathscr{A}$, generated by the semialgebra.

Proof: By Lemma 65 the algebra generated by $\mathscr{S}$ is the algebra that is the set of all finite unions of disjoint elements in $\mathscr{S}$. Since we want π to extend f, for all $s \in \mathscr{S}$ we define $\pi(s) = f(s)$. In defining π on $\mathscr{S}$ we have defined it for all $A \in \mathscr{A}$ thus: π is finitely additive because any $A \in \mathscr{A}$ is a finite disjoint union of elements, say $s_1, s_2,... s_n \in \mathscr{S}$, so

$$\pi\left(A\right) = \pi\left(\bigcup_{i=1}^{n} s_i\right) = \sum_{i=1}^{n}\pi\left(s_i\right) = \sum_{i=1}^{n}f\left(s_i\right)$$

We show that this is well defined, that is show that $\pi(A)$ doesn't depend on which set of disjoint members of $\mathscr{S}$ when united is A. Suppose $A = \bigcup_{i=1}^{n} s_i = \bigcup_{j=1}^{m} t_j$.

Each $s_i \cap t_j$ is a disjoint member of $\mathscr{S}$ and $s_i = \bigcup_{j=1}^{m} s_i \cap t_j$, $t_j = \bigcup_{i=1}^{n} s_i \cap t_j$. Consequently

$$\pi\left(\bigcup_{i=1}^{n} s_i\right) = \sum_{i=1}^{n}\pi\left(s_i\right), = \sum_{i=1}^{n}\pi\left(\bigcup_{j=1}^{m} s_i \cap t_j\right) = \sum_{i=1}^{n}\sum_{j=1}^{m}\pi\left(s_i \cap t_j\right)$$

$$= \sum_{j=1}^{m}\sum_{i=1}^{n}\pi\left(s_i \cap t_j\right) = \sum_{j=1}^{m}\pi\left(\bigcup_{i=1}^{n} s_i \cap t_j\right) = \sum_{j=1}^{m}\pi\left(t_j\right) = \pi\left(\bigcup_{j=1}^{m} t_j\right)$$

We show that π is a countably additive on $\mathscr{A}$. Suppose $A \in \mathscr{A}$ is the countable union of the disjoint sequence $\{A_1, A_2, A_3,...\} \subseteq \mathscr{A}$. Suppose first that

$A \in \mathscr{S}$. Then, since each A_i is a finite disjoint union of $s_1, s_2, \ldots s_n \in \mathscr{S}$ and f is countably additive on $\mathscr{S}$, and since π extends f

$$\pi\left(\bigcup_{i=1}^{\infty} A_i\right) = \pi(A) = f(A) = f\left(\bigcup_{i=1}^{\infty} A_i\right) = f\left(\bigcup_{i=1}^{\infty}\bigcup_{j=1}^{n} s_{ij}\right) = \sum_{i=1}^{\infty}\sum_{j=1}^{n} f(s_{ij})$$

$$= \sum_{i=1}^{\infty}\sum_{j=1}^{n} \pi(s_{ij}) = \sum_{i=1}^{\infty} \pi(A_i)$$

where we used f being countably additive on $\mathscr{S}$ for the fifth equality. So π is countably additive on $\mathscr{S}$.

If $A \notin \mathscr{S}$ then there are disjoint $t_1, t_2, \ldots t_n \in \mathscr{S}$ such that A is their union.

Each $A_i \cap t_j$ is a disjoint member of $\mathscr{S}$, $t_j = \bigcup_{i=1}^{\infty} A_i \cap t_j$, $A_i = \bigcup_{j=1}^{n} A_i \cap t_j$ and

using π being countably additive on $\mathscr{S}$ for the fifth equality:

$$\pi\left(\bigcup_{i=1}^{\infty} A_i\right) = \pi(A) = \pi\left(\bigcup_{j=1}^{n} t_j\right) = \sum_{j=1}^{n} \pi(t_j) = \sum_{j=1}^{n} \pi\left(\bigcup_{i=1}^{\infty}(A_i \cap t_j)\right)$$

$$= \sum_{j=1}^{n}\sum_{i=1}^{\infty} \pi(A_i \cap t_j) = \sum_{i=1}^{\infty}\sum_{j=1}^{n} \pi(A_i \cap t_j)$$

$$= \sum_{i=1}^{\infty} \pi\left(\bigcup_{j=1}^{n}(A_i \cap t_j)\right) = \sum_{i=1}^{\infty} \pi(A_i)$$

We finish by showing the uniqueness of π: Any function, g, that is a pre-measure extending f to $\mathscr{A}$ agrees with f on $\mathscr{S}$ and is countably additive and therefore finitely additive. So for any $A \in \mathscr{A}$, A is a finite union of disjoint $s_i \in \mathscr{S}$ so

$$g(A) = g\left(\bigcup_{i=1}^{n} s_i\right) = \sum_{i=1}^{n} g(s_i) = \sum_{i=1}^{n} f(s_i)$$

$$= \sum_{i=1}^{n} \pi(s_i) = \pi\left(\bigcup_{i=1}^{n} s_i\right) = \pi(A).$$

14.6 Abstract theory: outer measure on power set

Definition: μ^* is an *outer measure* on Ω iff
1. μ^* is a function $\mu^*: P(\Omega) \to R^+$ with $\mu^*(\varnothing) = 0$.
2. μ^* is monotone: for $X, Y \in P(\Omega)$, $X \subset Y$, $\mu^*(X) \leq \mu^*(Y)$
3. μ^* is countably subadditive, that is if X is a sequence of subsets in Ω then

$$\mu^*(\bigcup X) \leq \sum_{n=1}^{\infty} \mu^*(X_n)$$

Theorem 67

Let $\mathscr{F}$ cover Ω, let $g:\mathscr{F}\rightarrow[0, \infty]$ be a function for which $g(\varnothing)=0$, and for all $X\subset\Omega$ let

$$\mu^*(X)=\inf\left\{\sum_{n=1}^{\infty}g(F_n): F_n\in F, X\subseteq\cup F\right\}$$

where the infimum is taken over all sequences F in $\mathscr{F}$. Then μ^* is an outer measure on Ω. (Halmos 1974:42; Bruckner *et al.* 2008:87).

Corollary 68

Given an algebra of sets, $\mathscr{A}$, the σ-algebra, $\mathscr{M}$, generated by $\mathscr{A}$ and a function $\pi:\mathscr{A}\rightarrow[0, \infty]$ for which $\pi(\varnothing)=0$, there exists an outer measure μ^* on the σ-algebra generated by the algebra, $\mathscr{A}$.

Proof: Since $\mathscr{M}$ is the smallest σ-algebra including $\mathscr{A}$ and $\mathscr{A}\subset P(\Omega)$ therefore $\mathscr{M}\subset P(\Omega)$, from which the result follows immediately from the theorem. QED

14.7 Abstract theory: extension theorems

Modern treatments call a variety of related theorems Carathéodory's extension theorem. I have followed the naming used by some of the literature here. We start with *Carathéodory's criterion*:

A set, $M\subset\Omega$, is *Carathéodory measurable* under μ^* iff for all $X\in P(\Omega)$

$$\mu^*(X)=\mu^*(X\cap M)+\mu^*(X-M)$$

Since $X=(X\cap M)\cup(X-M)$, what this definition does is select all those $M\subset\Omega$ which are additive rather than subadditive with all the other subsets of Ω. As we shall now see, confining the outer measure to all such M gives a measure.

Theorem 69

Carathéodory's extension theorem Given an outer measure, μ^*, on Ω, let $\mathscr{M}$ be the family of Carathéodory measurable sets under μ^*. Then $\mathscr{M}$ is a σ-algebra and the restriction of μ^* to $\mathscr{M}$ is a measure, μ. (Folland 2009:42; Bruckner *et al.* 2008: 84; Tao 2011:theorem 1.7.3)

Applying this to Corollary **68** gives:

Theorem 70

Hahn–Kolmogorov extension theorem Given a premeasure, π, on an algebra of sets, $\mathscr{A}$, then there exists a measure, μ, on the σ-algebra generated by the algebra, $\mathscr{M}$, that extends π. (cf. Folland 2009:43; Tao 2011:theorem 1.7.8)

Corollary 71

If π is σ-finite then μ is unique.

Proof: Suppose v is a measure on $\mathscr{M}$ that also extends π. Let $M \in \mathscr{M}$. For any sequence $A_\omega \subset \mathscr{A}$ covering M.

$$v(M) \le v\left(\bigcup_{n=1}^{\infty} A_n\right) \le \sum_{n=1}^{\infty} v(A_n) = \sum_{n=1}^{\infty} \pi(A_n)$$

Recalling that μ is the restriction of the outer measure μ^* defined by the infimum taken over all sequences in $\mathscr{A}$

$$\mu^*(M) = \inf\left\{ \sum_{n=1}^{\infty} \pi(A_n) \colon A_n \in A_\omega, \ M \subseteq \cup A_\omega \right\}$$

we have that $v(M) \le \mu(M)$.

Being extensions of π, both μ and v are σ-finite. Hence there exists a countable sequence, $S_\omega \subset \mathscr{M}$ whose members have finite μ-measure, whose union is Ω and for all $M \in \mathscr{M}$, by countable additivity,

$$\mu(M) = \mu\left(\bigcup_{n=1}^{\infty} M \cap S_n\right) = \sum_{n=1}^{\infty} \mu(M \cap S_n)$$

$$v(M) = v\left(\bigcup_{n=1}^{\infty} M \cap S_n\right) = \sum_{n=1}^{\infty} v(M \cap S_n)$$

where for all n, $v(M \cap S_n) \le \mu(M \cap S_n) < \infty$. So to show $v = \mu$ we need only show that for all $M \in \mathscr{M}$ of finite μ-measure, $\mu(M) = v(M)$.

If $\mu(M) < \infty$, there exists a disjoint sequence $A_\omega \subset \mathscr{A}$ covering M for which

$$\mu(\cup A_\omega) = \sum_{n=1}^{\infty} \mu(A_n) = \sum_{n=1}^{\infty} \pi(A_n) = \sum_{n=1}^{\infty} v(A_n) = v(\cup A_\omega), \quad \text{and so}$$

$$\mu(\cup A_\omega) = \mu\big((\cup A_\omega) - M\big) + \mu(M) = v(\cup A_\omega) = v\big((\cup A_\omega) - M\big) + v(M)$$

If $v(M) < \mu(M)$ then $v((\cup A_\omega) - M) > \mu((\cup A_\omega) - M)$, contradicting the earlier result,[167] hence $v(M) = \mu(M)$. QED

---

167 Use $X = (\cup A_\omega) - M$ in the earlier result, that for all $X \in \mathscr{M}$, $v(X) \le \mu(X)$.

14.8 Natural Measurement Begotten Measures

We finish our measure theory results with a theorem about measures begotten from natural measurements via the methods. The corollary to the theorem completes our geometric example.

The Natural Measurement Begotten Measure Theorem 72

Let $\mathcal{S}$ be a semialgebra of the space X for we have a natural measurement function and let $f\colon \mathcal{S}\to\mathbb{R}^+$ be a σ-finite premeasure that extends the natural measurement function. Then there exists a unique measure on X that extends f.

Proof: This follows from Theorem **66**, the Hahn-Kolmogorov extension Theorem **70** and Corollary **71**. QED

Corollary 73

Let $f\colon \mathcal{S}\to\mathbb{R}^+$ be a function that extends the natural measurement of line segments, that is for any line segment s, $f(s)=$ the length of s. Then there exists a unique measure on the whole line that extends f and captures natural measurement facts $A - G$.

Proof: Proposition **63** shows that f satisfies the hypothesis of Theorem **72** and captures the natural measurement facts $A - G$. QED

Remark: Unsurprisingly, perhaps, when the whole line is identified with the real numbers, this unique measure is the Lebesgue measure on the real numbers.

Definitions: A measure space $(X, \mathcal{M}, \mu)$ is a *natural measure* iff it is a Natural Measurement Begotten Measure

A measure space $(X, \mathcal{M}, \mu)$ is *complete* iff whenever $\mu(M)=0$ and $A\subset M$, then $A\in\mathcal{M}$.

We need this as a lemma:

Corollary 2.46

Every complete σ-finite measure space is its own... Carathéodory extension. (Bruckner *et al.* 2008:101)

Fixed Point Theorem 74

A complete natural measure is a fixed point for the Natural Measurement Begotten Measure theorem.

Proof: Let $(X, \mathcal{M}, \mu)$ be a complete natural measure. $\mathcal{M}$ is both a semialgebra and an algebra and μ is a premeasure so we can use them as such in the

Natural Measurement Begotten Measure theorem. Then by Thomson *et al.*'s Corollary 2.46, the unique measure on X that extends μ is $(X, \mathcal{M}, \mu)$. QED

Remark: Whenever we get a complete measure from the Natural Measurement Begotten Measure theorem, and we usually do get a complete measure, that measure is a fixed point in the sense that further applications of the theorem don't extend the measure to a larger σ-algebra but keep us with the first Natural Measurement Begotten Measure that we got.

Before we turn to the reason for our excursion into this much theory, I want to mention an independent matter, the significance of this theory for satisfying Salmon's Ascertainability criterion. It is indicated by Bingham's comment:

> When doing probability as applied mathematics, we have in mind some real-world situation generating randomness, and seek to use this apparatus to analyze this situation. To do this, we must assign probabilities to enough events to generate the relevant σ-fields, and thence to all relevant events – i.e., all relevant measurable sets – by the standard Carathéodory extension procedure of measure theory.
>
> (Bingham 2010:21)

Being able to apply Carathéodory's method and Theorems 72 and 74 means that in order to provide a normative measure for POIS it suffices to have a premeasure on a semialgebra of events (or of their names). The premeasure would have to give equal premeasure to events in the semialgebra that have equal epistemic status and satisfy the hypothesis of Theorem 72. Starting in that way, one would need to prove that the resulting measure also gave equal measure to events of equal epistemic status in the σ-algebra.

Furthermore, the weakest family on which we can obtain this kind of uniqueness result (named a π-field by Dynkin 1965:201) is one which has only the first feature of a semialgebra, namely, being closed under finite intersection. A finite measure on the σ-algebra generated by such a family is unique (see Williams 1991:19). Consequently, to provide a normative measure for POIS it may suffice to have a finite premeasure on such a family.

The upshot of this is that we can obtain the stronger version of FPOIS I mooted in §2.8 by weakening its input. Instead of the input of an equal reason partition of the σ-algebra of events we need only an equal reason partition of a semialgebra that generates that σ-algebra. Then a natural measurement function of that semialgebra, or of the corresponding semialgebra in the name space, that gives equal measurements to the members of equal reason cells and the Natural Measurement Begotten Measure theorem give us the normative measure. The strongest FPOIS will be when a suitable natural measurement of a π-field is available.

14.9 Natural measures

The arguments I wish to make in the rest of the chapter have a significant theoretical component: that measure theory is the best sense we have been able to make of measurement and so, even if what I have now to say goes beyond our common sense conceptions, it is legitimate to do so in this way because the way it goes beyond is a matter of making use of our best theory of what the common sense conception of measurement comes to. For example, we may common sensically fail to distinguish two measures because they have related, but distinct, domains. That is a blunder in measure theory, and since measure theory is our best theory, it is a blunder, *tout court*. In general, then, I shall take measure theory as strongly dispositive on questions that arise. With this, as I shall call it, *Best Theory* premiss in place, the method we have just looked at tells us many things of importance about Bertrand's paradox and will eventually allow me to make and defend a fundamental distinction in probability paradoxes.

The justification for what I shall say about the specialness of a natural measurement—and therefore subsequently about the Natural Measurement Begotten Measure—comes from two directions. The first direction is what makes it special from the point of view of common sense when analysed metaphysically.

In general, taking the measurement of things is assigning numbers to them.[168] We can, of course, assign arbitrary numbers to things, but such arbitrary assignments are without significance. Many useful measurements tell us of the thing's external relations. When, however, we want numbers that tell us something about the nature of the thing measured, we want more than that. A measurement that is necessarily related to what it is the measurement of is not necessarily a measurement of its nature—see Fine, K. 1994 against equating necessity with essence—even if it may be said to have some weak degree of naturalness. These I shall call *merely necessary measurements*. Event spaces or representations have too many such measurements to be of any use when what we are looking for is a measurement that can be properly called a natural measurement. A natural measurement needs to be a measurement of the nature of the thing, and if there are several then a measurement of what is in some sense fundamentally constitutive is the most natural measurement.

It is in this sense that I earlier defined the natural measurement of line segments to be their length. The length isn't merely *a* property of the segment: the segment just is that length of line extracted from the whole line by its particular endpoints. It has many necessary relations to other geometric objects and other properties about its embedding within a space

---

168 See Suppes and Zinnes 1963 on fundamental measurements.

and although measurements of them may be natural, they are not constitutive in the way of the length, which is therefore the most natural measurement.

The second direction is the special role that being a natural measurement plays in defining a measure as we saw in the last section.

Proposition 63 showed how the finitely additive natural measurement of line segments is σ-finite, defines a premeasure on the semialgebra of segments and that any extension of that premeasure both captured the facts about the natural measurement of line segments and is σ-finite. The theorems from abstract measure theory extend that premeasure first to a premeasure on the algebra generated by the semialgebra of segments, then to an outer measure on the whole line and finally to a measure on the whole line. Since the measure is an extension of the premeasure on the semialgebra of segments, it both captures the natural measurement facts and is σ-finite, and because of the latter it is the unique measure that captures them.

I treated the final part of our geometric example as a corollary to the Natural Measurement Begotten Measure Theorem 72 to make clear that where we ended up was where we would end up with any natural measurement of a space. Measure theory allows us to extend any natural measurement of a (semialgebra or algebra of a) space of objects of interest to a unique measure. In measure theory as understood by a mathematician, the unique measure begotten by the most natural measurement is recognised as *the* measure of that space and which now gets it right in measuring the space quite generally. Getting it right means that the measure answers exactly what we intuited prior to the theory but which was not answered by the limited extent of the natural measurement nor by the prior theory of measure—whose limitation caused the problems that arose for the Riemann integral, as mentioned earlier. The recognition is based in the understanding of the wider mathematical context and noting consilience and reflective equilibrium. Had this recognition that it got it right not happened, the natural measurement used would cease to have been accepted as the most natural and a different starting place sought.

One may impose other measures for various special purposes, such as measures of properties other than the most natural measurement, but none are recognised as having the same metaphysical status. The contrast in the light of the recognition is routinely made by mathematicians in their reference to a measure begotten of the most natural measurement begotten as *the* natural measure of the space at hand. For example, it is possible to define a measure on the Euclidean plane by starting from the premeasure of squares given by their diameters (Bruckner *et al.* 2008:section 3.2). But this is not the natural measure of the Euclidean plane, it has the wrong dimensionality, fails to extend the diameter function properly and 'the class of measurable sets is incompatible with the topology on [the Euclidean plane]:

open sets need not be measurable' (Bruckner *et al.* 2008:111). The natural measure of the Euclidean plane is the measure got from the premeasure of squares given by their areas. Their areas are, of course, their most natural measurement, and the natural measure of the Euclidean plane is the area measurement begotten measure.

So the most natural measurement is theoretically special because it plays the special role it does in defining a unique measure, which unique measure is recognised by mathematicians as measuring the space in terms of *what it is*. Furthermore, since the natural measurement agrees with the natural measure on anything in the space with a natural measurement, and since it was not rejected when the natural measurement begotten measure was seen for what it was, the natural measurement then receives additional support as being special just because it agrees with the measure that is now recognised to be the natural one. In other words, the natural measurement had a prior naturalness which led to its being used to beget its unique measure, but which might have been rejected had not the begotten measure been recognised as being the correct measure that was sought. With that recognition the natural measurement has also an additional posterior naturalness, or alternatively has its prior naturalness affirmed, for agreeing with the now recognised natural measure. Thus do we have a reflective equilibrium.

So the most natural measurement has its common sense metaphysical naturalness that is supported by our a Best Theory premiss because it receives measure-theoretic endorsement grounded in a wider mathematical understanding. Consequently, although there may be natural measures begotten of other natural measurements of a space, the measure begotten of the most natural measurement is accepted as being *the* natural measure for the space.

14.10 Naturalness and Bertrand's paradoxes

We now return to examining the Bertrand's original four paradoxes in the light of the existence of natural measures. Some of the discussion attends more to natural measurements than to natural measures, but the reader should bear in mind that whenever there is the former there is the latter due to Theorem 72.

The big distinction that naturalness draws between the paradoxes is that in the square paradox there are too many natural measures and in the chord paradox there are none. The rest of the chapter will be spent in justifying this difference, showing how it manifests in a variety of ways and explaining why we are deceived into missing it altogether. The contrast will be drawn mainly between the square paradox and chord paradox.

Before we start, I shall now categorize Bertrand's four paradoxes in terms of naturalness.

The plenitude paradoxes have too many natural measures to play the normative measure role in the POIS:

The Square paradox (as modified by me below): side length and area measurements.

The Celestial Sphere paradox: angle and solid angle measurements.

The paucity paradoxes have no natural measures to play the normative measure role in the POIS:

The Horizon paradox (neither angle with horizon nor area where perpendicular pierces the sphere are natural measures of the planes)

The Chord paradox (neither angle with tangent nor distance of midpoint from centre are natural measures of the chords)

In what follows I make regular use of our Best Theory premiss without citing it on all occasions.

14.11 Natural measures and the chords

Our Best Theory premiss has an important role in explaining why there is no natural measure of the chords. As we have seen, what does the work in producing a natural measure is a natural measurement on a semialgebra or algebra. With only that much, the Natural Measurement Begotten Measure theorem assures us that the σ-algebra generated by the semialgebra or algebra has a unique measure extending the natural measurement, giving us a natural measure. So the critical element in having a natural measure is the natural measurement on a semialgebra or algebra—which disjunct I shall now omit for brevity.

I am going to start by contrasting our geometric example with the chords. The base space of our geometric example is the whole line, L. The semialgebra of segments that we start with, $\mathscr{S}_L$, consists of subsets of the line, namely, line segments, the line and all the half lines. So the semialgebra is a particular subset of the power set of the line, $\mathscr{S}_L \subset \mathbb{P}(L)$. The elements in $\mathscr{S}_L$ have length, which is why we chose this semialgebra in the first place, and so length is the natural measurement on the semialgebra $\mathscr{S}_L$. The natural measure, μ_L, is a function with its domain the σ-algebra, $\mathscr{M}_L$, generated by the semialgebra, $\mathscr{S}_L$, so its domain is not the line, L, but a subset of the power set of the line, $\mathscr{M} \subset \mathbb{P}(L)$.

The base space for the chords is the set of chords, C. Each individual chord is a subset of $\mathbb{R}^2$ and so the base space is not a subset of $\mathbb{R}^2$ but of the power set, $P(\mathbb{R}^2)$. We want to end up with a natural measure, μ_C, that is a function with its domain the σ-algebra, $\mathcal{M}_C$, which is a subset of the power set of the chords, which means it is a subset of the power set of the power set of $\mathbb{R}^2$: $\mathcal{M}_C \subset P(C) \subset P(P(\mathbb{R}^2))$.

What we are looking for, then, is a semialgebra, $\mathcal{S}_C$, which generates $\mathcal{M}_C$ and which therefore is also a subset $\mathcal{S}_C \subset P(C) \subset P(P(\mathbb{R}^2))$, and whose elements are subsets of C which elements have a natural measurement.

First attempt: individual chords have length so that is the natural measurement. But chords are not subsets of the set of chords, so can't be elements in $\mathcal{S}_C$ and therefore cannot generate the σ-algebra $\mathcal{M}_C$ for the measure on C.

Second attempt: subsets of the set of chords that have areas, for example, a subset, D, of parallel chords. But subsets of the chords do not have areas. For a set to have an area it must be a subset of $\mathbb{R}^2$, but D is not a subset of $\mathbb{R}^2$ but of $P(\mathbb{R}^2)$ so it doesn't have an area.

Third attempt: Take the union of a subset of parallel chords, $\bigcup D$, which union does have an area. Yes, it has an area, but $\bigcup D$ is not a subset of the set of chords so can't be an element in $\mathcal{S}_C$ and therefore such sets cannot generate the σ-algebra $\mathcal{M}_C$ for the measure on C.

Fourth attempt: the singleton sets of chords can be elements in $\mathcal{S}_C$ so let's call the length of their members the natural measurement of the singleton. Obviously this is a kind of cheat, even if an understandable cheat. It can be rejected on the ground that even if a natural measurement of one thing is some kind of a measurement of something containing it, prima facie it is only externally related and so cannot be a natural measurement. At best it is a merely necessary relation, in the sense that, necessarily, if wooden box A is inside wooden box B and A is has a volume of 1 cubic foot then B has a volume of at least 1 cubic foot. *A forteriori*, when the kind of containment is the purely abstract one of set membership, there is no reason why the natural measurement of the element is any kind of measurement of the set, let alone a natural one. So this is a kind of blunder, equivalent to saying that my singleton set is naturally six foot tall because I am.

It is also a useless cheat since C is not a singleton and even if we added it in, we still don't have a semialgebra, since the complement of a singleton in C is not a finite disjoint unions of members of C. OK, suppose we add all those complements in. Now we have a semialgebra but what is the nature of the σ-algebra and its measure that we end up with? First, the set of longer chords is not in the σ-algebra since it can't be got from countable unions and complements of elements in the semialgebra. Worse, anything but a finite or a carefully selected countable union of singletons will have

infinite measure and this means we couldn't use the measure as the normative measure in the POIS.

Fifth attempt: Sets of chords can be elements in $\mathscr{S}_C$ so lets call the area of
their union the natural measurement. Again, this is a kind of cheat and in
this case the measurement is not even of something that is contained in the
set, since (apart from singleton sets) the region that is the union of chords
is not an element of the set. What we have is a whole, the region, and a set
whose elements are parts of the whole, which whole is the mereological
sum of those parts. It can be rejected on the ground that a natural measurement of a whole may well be a natural measurement of the mereological
sum of its parts (since that is what it is), but it is not a measurement of the
set of its parts, which need have no measurement at all. To extend our
earlier example, the mereological sum of my cellular parts might be six
foot tall, but the set of my cellular parts has no height of any kind.

Once again, it is also a useless cheat, but for a different reason. What we
are doing here is using a function $f:\mathbb{P}(C) \to \mathbb{P}(D)$ to the measure space
$\{D, \mathbb{P}(D) \cap \mathscr{L}_2, \lambda_2|_D\}$, mapping sets of chords to their union in $\mathbb{R}^2$. This is the
case we used in demonstrating that Bertrand's procedure was internally
inconsistent (Chapter 3) and the subsequent analysis showed that its use
wasn't an application of the Theorem of Induced Measure. On top of that,
there is a problem that I didn't point out earlier: the measure space can't be
used as the normative measure in the POIS because it doesn't give equal
measure to events of which we are equally ignorant. It doesn't do so because the function from the chords to this measure space is not an equal
ignorance function. The function maps the event of getting a longer chord
to the whole disk and the event of getting any chord at all to the whole disk
and so these events receive equal measure. But we are not equally ignorant
between the event of getting a longer chord and the event of getting any
chord so the function is not an equal ignorance function.

I don't have other attempts to propose. Essentially what is going wrong
here is that we need a semialgebra of subsets of C, which subsets have a
natural measurement. But the subsets of C don't have a natural measurement and so we have no way of deriving a natural measure for the set of
chords. Consequently we do not have a natural measure to fulfil the normative measure role in the POIS.

14.12 The chords and the natural measures of the angles, directions and midpoints

It may appear that the angles, directions and midpoints gave us natural
measures of the chords but the appearance is deceptive. The measures were
salient, yes, but it must be remembered that none of those were measures

of the chords. Indeed, it was this very fact that gave rise to most of the frailties of Bertrand's procedure and forced us to prove and use the theorems of Chapter 2, in places making use of quite esoteric mathematics such as quotient spaces, to produce satisfactory versions of Bertrand's cases.

What was done before was merely to treat those natural measures on $\mathbb{R}$ or $\mathbb{R}^2$ *as if* they were measures of the chords. That was a kind of blunder, because the base space of the chords is not a subset of either of those but of the power set of $\mathbb{R}^2$. Measures on $\mathbb{R}$ or $\mathbb{R}^2$, however salient, *cannot* be measures on the chords just because their domains are subsets of $P(\mathbb{R})$ and $P(\mathbb{R}^2)$ respectively, whereas a measure on the chords is a function whose domain is a subset of $P(P(\mathbb{R}^2))$. Thus did we draw some early lessons from our Best Theory premiss before that premiss was made explicit: we must attend to the definitions of measure theory and we are not allowed to substitute the measure of something else as if it were the measure of the chords.

The Theorem of Induced Measure got us out of these difficulties. Instead of treating these salient measures as if they were measures on the chords we used an equal ignorance function from the chords to their measure space to induce measures on the chords. In so doing, it remained the case that none of these salient measures is a measure on the chords and none of the induced measures on the chords is one of these salient measures: they can't be just because they have distinct domains.

14.13 TIM is Strongly Unavoidable

Our method of fixing the frailties was, therefore, *not* to discover a natural measure on the chords to be the normative measure in POIS. Now we know why: there isn't one to discover. So we didn't merely happen to use the TIM, we *had* to. That is to say, to provide a normative measure for POIS we had to use the Theorem of Induced Measure *in a stronger sense* than is given by the TIM is Unavoidable Theorem 23 and the Equal Ignorance Function Is Unavoidable Theorem 28. It wasn't simply that any probability measure on the chords *can* be got via TIM, it was that without it we had *no place to start*, no normative measure candidate. In other words, for the chord paradox *TIM is Strongly Unavoidable*, as defined by

> *TIM Is Strongly Unavoidable* for a set of events iff it has no natural measure and so the Theorem of Induced Measure *must* be used to provide the normative measure for the Principle of Indifference for Sets.

14.14 The paucity paradoxes

Paucity paradoxes produce paradoxical probabilities because, having no natural measures, TIM is Strongly Unavoidable. This means the Deluge

Theorem 24 always threatens to drown them, that for any event space we can induce a measure that gives any event any probability in [0,1]. Of course, some of the measure spaces used in TIM to beget a normative measure on the events may be obviously irrelevant or arbitrary, which is why I then went on to develop the concept of an equal ignorance function. When TIM is Is Strongly Unavoidable, it also means that an equal ignorance function is strongly unavoidable, because that is what is needed if TIM is to produce a measure that can not merely play the normative measure role but also fulfil it. As we saw in the rigorous construction of Bertrand's cases, using functions from the chords to various properties of the chords, the properties whose measures were the seemingly natural measures of the chords in Bertrand's treatment, gave us equal ignorance functions which induced distinct measures on the chords. Being distinct measures, when they were used as normative measures in the POIS to beget probability measures they gave us paradoxical probabilities.

14.15 The plenitude paradoxes

Plenitude paradoxes produce paradoxical probabilities because the event space has more than one natural measurement and therefore more than one natural measure. What produces the paradox is none being more natural than any of the others and at least two being non-linearly related, which, by Corollary 13, means the probability measures begotten by those natural measures in POIS will be paradoxical.

Our example is the square paradox. The first thing to say is that as Bertrand expounds it, it isn't a very good paradox. Asking the chance that a number chosen at random between 0 and 100 is greater than 50 and offering a uniform distribution over the squares of those numbers as if it were as good as over the numbers is obviously wrong headed. This is no doubt why van Fraassen adapted it to being a question about cubes (van Fraassen 1989:303). I have done the same, considering the probability of getting a random square of area 1 when chosen at random from squares with areas up to 4. Equal ignorance over the side lengths gives the answer ½ but over the areas gives answer ¼.

In Bertrand's original version about numbers the difference between the minimum and maximum is the prominent natural measurement and the difference between the squares of the minimum and maximum looks arbitrary, even though that difference bears a necessary relation to the numbers. In my version, the area is an obvious natural measurement, but it is no longer obvious that the alternative is wrong headed because the side length is also a natural measurement. Whether the area is more natural than the side length is disputable: it is this that makes the paradox cogent. In van Fraassen's version, the volume is the obvious natural measurement but the side length and side area are also natural measurements, which make his paradox cogent.

In the square paradox, for square S, we have the measure of the squares got from the side lengths with Lebesgue $\mathbb{R}$ measure λ and the measure of the areas with Lebesgue $\mathbb{R}^2$ measure α, which when used as the normative measures in the POIS produce inconsistent probability measures. Letting Δ be a σ–algebra generated by the squares in S, the function mapping one measure to the other is the squaring function, giving us the commuting diagram (Figure 27) between the measures:

Applying POIS and Theorem 7 gives for $\delta \in \Delta$, $\delta \neq S$

$$P_\alpha(\delta) = \frac{\alpha(\delta)}{\alpha(S)} = \frac{\lambda(\delta)^2}{\lambda(S)^2} < \frac{\lambda(\delta)}{\lambda(S)} = P_\lambda(\delta) \text{ since } \frac{\lambda(\delta)}{\lambda(S)} < 1$$

So the way the square paradox works is by having more than one salient natural measure on the set of events to play the normative measure role, and recalling Corollary **13**, the damage is then done by the map between those measures being the squaring function, which is non-linear.

Evidently, the equations just given would apply for any set of similar shapes whatsoever, however irregular they be. λ could be any natural one-dimensional measure and α any natural two dimensional measure of whatever shape and they would have the same squaring function, up to a multiplicative constant, between them. Thus by Corollary **13** they would produce paradoxical probabilities and give an extended version of the square paradox. Likewise, higher dimensional examples, such as van Fraassen's cubes, produce the same results for any set of similar shaped regions.

It is perhaps worth noting that Corollary **13** can also be exploited rather slackly, but for that reason less convincingly. Fisher, for example, objects to the principle of indifference on the basis of the plenitude of arbitrary parameterizations usually possible. He contrast a parameterization by x with one by θ defined by $\sin \theta = 2x - 1$ without actually giving a case in

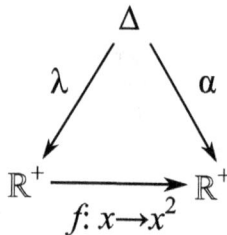

Figure 27 Commuting diagram of functions.

which such these might be equally natural (Fisher 1922:325). Told baldly in that way, it is weak in the same way the Bertrand's original square paradox is weak. Zabell notes this kind of (slack) critique of the principle appearing earlier than Bertrand when noting the attribution by 'Hald 1998, 269–72, who discusses the little-known critique of the Danish actuary Frederik Moritz Bing in 1879' (Zabell 2016a:135).

14.16 Naturalness solution strategy difference between plenitude and paucity paradoxes

Now we are clear on the difference in their mechanisms, we can see that the plenitude paradoxes have a resource for solution that the paucity paradoxes lack entirely. The mechanism of plenitude paradoxes is that they have more than one salient measure available to play the normative measure in POIS. So, in principle at least, there is a strategy of solution: to defend one such measure as the most natural, or the only natural one, which is therefore uniquely available to fulfil the normative measure role in POIS. This would be an instance of the Well-posing strategy suitably adapted to apply to plenitude paradoxes.[169]

14.17 Naturalness Well-posing applied to the square paradox

Some plenitude paradoxes have at least a good candidate for the most natural measurement, and therefore for *the* natural measure. For the square paradox it is the area. It can be argued to be more natural than the side because it is fundamentally constitutive of the square in a way that the side is not: they may both be necessary but the difference between being a square and being a side is the difference in dimensionality, which difference makes the area more fundamental and therefore a more natural measurement than the side length. On being more fundamental, one might say that the area grounds the side length but the side length does not ground the area. Similarly, for the cube paradox it can be argued that the volume is more natural than the side.

I have a further couple of arguments for area being more fundamentally constitutive of a square than lines making use of my best theory premiss and I expect more could be found. We can redistribute constituent areas of the square in space and their total area remains the same as the square. However, we can arrange the vertical lines that are constituents of the square, and whose union is the square, by translating them by a bijection

---

169 For the approach to solving the square paradox based on invariance over units for measuring side and area see Joyce 2005:169–70. For a critique of this approach, in particular to its application to the water/wine paradox, see Milne 1983.

that maps their endpoints to a Cantor set on [0,1]. The measure of the lines that are constituents of the square when arranged in that way can be zero or any value in [0,1) you like (Bruckner *et al.* 2008:theorem 1.23).

Second argument: measures of the plane using a one-dimensional pre-measure needn't agree with Lebesgue measure or even give any measure to squares. As already mentioned, Bruckner *et al.* 2008 exhibit a measure defined by using the diameters of squares as a premeasure on the semialgebra of squares and the measure this produces cannot satisfy even finite additivity for small squares constituting a larger square and consequently no square in the semialgebra of the premeasure is measurable. This example can be given with a premeasure being the side length, which I interpolate in this quotation:

Example 3.5

Take $X = \mathbb{R}^2$, let $\mathscr{T}$ be the family of open squares in X, and choose as a premeasure $\tau(T)$ to be the [side length] of T. We apply Method I to obtain an outer measure μ^* and then a measure space $(\mathbb{R}^2, \mathscr{M}, \mu)$. What would we expect about the measurability of sets in T? Since [side length] is essentially a one-dimensional concept, while T consists of two-dimensional sets, perhaps we expect that every nonempty T has infinite measure.

Let $T_0 \in \mathscr{T}$ have side length 3, and let T_1, T_2, T_3 and T_4 be in $\mathscr{T}$, each with side length 1, [they are the four corner squares of the nine such squares making up T_0]. Then $\tau(T_0) = 3$, while $\tau(T_i) = 1$ for $i = 1, 2, 3, 4$. It is easy to verify that, for all $T_0 \in \mathscr{T}$, $\mu^*(T) = \tau(T)$ and that

$$\mu^*(T_1 \cup T_2 \cup T_3 \cup T_4) \le \mu^*(T_0) = 3 < 4$$
$$= \mu^*(T_1) + \mu^*(T_2) + \mu^*(T_3) + \mu^*(T_4)$$

It follows that none of the sets T_i, $i = 1, 2, 3, 4$, is measurable. A moment's reflection shows that no nonempty member of T can be measurable.

(Bruckner *et al.* 2008:110)

So dimensionality matters. First, measuring constituent lines of an area fails to preserve measure under rearrangement. Perhaps more importantly, measuring squares by measuring side lengths results in a measure under which squares are unmeasurable. What underlies this difficulty is that in wanting to use side length to produce a measure of $\mathbb{R}^2$ we are essentially trying to use a metric structure of $\mathbb{R}^2$, that is, distances, to measure areas. Of course, one can evade this difficulty by bluntly using the Lebesgue measure on side lengths as if it were a measure on squares. But this difficulty shows that measuring squares by measuring side lengths is in general less natural than the area measure and hence the evasion of the difficulty is done at the cost of using a less natural measure.

14.18 Naturalness Well-posing in the literature

This approach to plenitude paradoxes is the one that Huemer takes to Fumerton's car trip paradox (Fumerton 1995:215):

> Sue has taken a trip of 100 miles in her car. The trip took between 1 and 2 hours, and thus, Sue's average speed was between 50 and 100 mph. Given only this information, what is the probability that the trip took between 1... and 1½ hours?.... [if] we assign a flat probability density over the ...possible durations of the trip....the probability... is ½... [if] we assign a flat probability density over the... average velocities.... [for a]... journey... between 1... and 1½ hours... her velocity was between 66 2/3 mph... and 100 mph.... the probability... is 2/3.
>
> (2009:349)

Huemer's proposed solution is that 'The duration of Sue's trip is causally explained by the speed at which she drove, not vice versa' (2009:356) so the second answer is correct. The solution is an instance of applying his

> Explanatory Priority Proviso to the Principle of Indifference... in applying the Principle of Indifference, one ought to assign equal probabilities (or a uniform probability density) at the most explanatorily basic level.
>
> (2009:354–5)

Huemer has earlier listed 5 kinds of explanatory priority between facts or propositions (2009:352–3), casual, temporal, part-whole, in-virtue-of relation, supervenience and he remarks that

> The sort of priority invoked in the Explanatory Priority Proviso is metaphysical rather than conceptual. What matters is that Sue's velocity is metaphysically prior to the duration of her trip.
>
> (2009:357)

My appeal to naturalness is also about metaphysical priority and is therefore in line with his proviso. Furthermore, there is a broader notion of naturalness than that which I used that could be the metaphysical basis of explanatory priority as used by Huemer, although here is not the place to go into that. The point for us here is that Huemer's proposal is an instance of the strategy of claiming there is a most natural measurement to be used in the principle of indifference. Recalling Climenhaga's explanationism from Chapter 2.10, it is evident that Climenhaga could treat Fumerton's case similarly.

This is also the approach that Mikkelson takes to the water/wine paradox. Having shown Bertrand to be the originator of Feldman's and van
Fraassen's kind of plenitude paradoxes, I will take a moment to show that
he is also an originator of the water/wine paradox. Deakin notes that

> The origins of the [water/wine] paradox are somewhat obscure.
> Mikkelson attributes it to von Mises, but gives no details. However, von
> Mises himself claims to follow Poincaré in attributing it to Bertrand.
>
> (Deakin 2006:201)

Von Mises' presentation is found in (Von Mises 1951/1981:77–78).
Poincaré is right in this sense: the water/wine paradox is about events that
are dimensionless ratios (as we are about to see) and therefore is a simplified version of Bertrand's celestial sphere paradox, which makes use of
angle versus solid angle. Angles are dimensionless ratios, in the first case
the ratio of arc length to the radius and in the second the ratio of the area
of the cap to the area of the sphere (see Figure 3). Von Kries is also a close
originator when he points out that the dimensioned ratios of specific gravity and specific volume are inverses of one another. Measures of events in
terms of one will be non-linearly related to the other, thus producing paradoxical probabilities (by Corollary **13**):

> the probability that a *specific weight* is between 1.0 and 2.0 would ap-
> pear to be the same as the probability that it is between. 10.0 and
> 11.0... [for the *specific volume*, however] the range 1.0 to 0.5 would
> correspond to the former, 0.1 to 0.09 correspond to the latter, and the
> former would then appear about 50 times more probable than the
> latter.
>
> (1886:31, my translation and emphasis)

Anyway, onto the water/wine paradox: In a mixture of water and wine,
the ratio of water to wine and wine to water are inverses of one another so
the function mapping the event space measure of water:wine to the event
space measure of wine:water is a non-linear function, whence, by Corollary
13, the probability measures begotten from the measures by POIS are
inconsistent.

For example, suppose we know only that one of these ratios is at most
2:1. Let σ-algebra $\mathcal{M}$ be $[\frac{1}{2},2] \cap \mathcal{L}$ and let μ and ν be Lebesgue measure
$\lambda|_{[\frac{1}{2},2]}$. Let the event measure space for wine:water be $([\frac{1}{2},2], \mathcal{M}, \mu)$ and for
water:wine be $([\frac{1}{2},2], \mathcal{M}, \nu)$. Since the ratios are inverse, the event, e, of
wine:water $\in [x, y]$ is the same event as water:wine $\in [1/y,1/x,]$ and the
function between the measures is the non-linear $\nu(e) = \mu(e)/xy$ and hence
$P_\mu(e) = P_\nu(e)/xy$. The event of wine:water $\in [\frac{1}{2},1]$ is the same event as

water:wine$\in$[1,2]. Using μ and ν as the normative measures in POIS and Corollary **9** gives

$$P_\mu\left(\text{wine}:\text{water}\in\left[\frac{1}{2},1\right]\right)=\frac{\mu\left(\left[\frac{1}{2},1\right]\right)}{\mu\left(\left[\frac{1}{2},2\right]\right)}=\frac{1}{3}$$

$$P_\nu\left(\text{water}:\text{wine}\in\left[1,2\right]\right)=\frac{\nu\left(\left[1,2\right]\right)}{\nu\left(\left[\frac{1}{2},2\right]\right)}=\frac{2}{3}$$

Mikkelson shows that if the problem is formulated in terms of dimensioned quantities (volumes) instead of ratios, consistent probabilities are obtained (2004). He justifies the dimensioned approach as the right one on the ground of distinguishing primary facts about quantities from derivative facts about ratios. In other words, he is claiming that the natural measurement of volumes is more natural than the dimensionless measurement of ratios.

Of course, this whole strategy can be controverted:

> The side length of the square is not explanatorily prior to its area, and the area of the square is not explanatorily prior to its side length. We simply have a choice of how to parameterise the problem, and no one way of doing so seems more "basic" than the other.
>
> (Lando 2021:345)

What is more natural can also be disputed: Norton's symmetric description treatment of the water/wine paradox (Norton 2008:49–51) takes the dimensionless formulation as fundamental, whereas Mikkelson took the dimensioned formulation to be fundamental.

Although this naturalness strategy merits further investigation, for my purposes here I do not need to claim that the strategy succeeds. The point of importance for us is that this is a strategy that is not available for the paucity paradoxes. No such defence could even be mooted for the chord paradox precisely because the chords lack any natural measures to attempt to justify as the most natural.

14.19 The sources of paradoxical probabilities

It might be thought that a source of the paradoxical probabilities was what we noted in Chapter 2, that we Need-a-Measure-to-Get-a-Measure.

Certainly that sets up the possibility, yet it was also simply a necessity of the structure of the principle of indifference, so that cannot really be a source. There is a tendency to point to non-linearity as a source of paradoxical probabilities (for early instances see Borel 1909:85, Keynes 1921/1973:51). However, recall Corollary 13, that the measures begetting the probability measures are non-linearly related if and only if probability measures produce paradoxical probabilities. Non-linearity is therefore not diagnostic of the source of the paradoxicality. What is diagnostic is how non-linearity ends up being admitted. We have now seen that the mechanisms of plenitude and paucity paradoxes are distinct in this respect.

In the case of the plenitude paradoxes, the non-linearity is admitted immediately because there are too many natural measures of the event space. For paucity paradoxes the non-linearity is much further down stream because there are no natural measures at the source.

The difference between the two kinds then manifests in their relation to the Theorem of Induced Measure. Because the event space of the square already had salient natural measures to play the role of the normative measure in POIS, we did not have to use the Theorem of Induced Measure to get started but were able to apply the POIS directly to derive the probabilities. This is true for plenitude paradoxes in general. Although TIM is Unavoidable (Theorem 23) for the square paradox and for other plenitude paradoxes, TIM is not *Strongly* Unavoidable for them.

By contrast, TIM *is* Strongly Unavoidable for the Paucity paradoxes and so before we can get to using POIS we must find an equal ignorance function for the events. In the chords, there are too many equal ignorance functions based on properties of the chords and at least two of which, those for angles and direction, when used in TIM, beget non-linearly related measures on the chords which then produce the paradoxical probabilities.

I have lost count of the number of discussions in the literature which speak of nothing more than a plenitude paradox, usually some variation on the square paradox (e.g. Lando 2021:342, Hájek 2019:15–16) and in so doing take it that they have addressed Bertrand's paradoxes *in toto*. It is this last fact, of the downstream non-linearity, which has led to this tendency to treat the chord paradox as if it were essentially the same as the plenitude paradoxes and so to cover one of the latter is to have covered one and all. That, we now see, is an error, since the mere presence of non-linearity is trivial and what distinguishes them is the plenitude or paucity of natural measures and the consequence this has for their relation to the TIM is Unavoidable theorem, which is unavoidable for both but only strongly unavoidable for one.

There is therefore no such thing as *the* Bertrand's paradox, if by that one means not the chord paradox alone but one groups the chord paradox with the square paradox and all the other plenitude paradoxes as being essentially the same. Such an equation is a failure to understand the difference between Bertrand's square and chord paradoxes, more generally, between plenitude and paucity paradoxes, which last pose an entirely different challenge to the principle of indifference than that posed by the plenitude paradoxes.

14.20 Objections to naturalness

I have drawn the difference between plenitude and paucity paradoxes in terms of naturalness and pointed out that it can be drawn in rather broader terms than those in which it was originally introduced, namely, in terms of the metaphysical underpinnings of Huemer's explanatory priority proviso. There are objections to the objectivity of naturalness.

Huemer remarks that

> In some cases we may have two ways of characterizing the possibilities, neither or which is intuitively more natural than the other, and neither of which classifies the alternatives in terms of explanatorily prior propositions.
>
> (Huemer 2009:357–8)

This is, of course, the distinguishing feature of the plenitude paradoxes but I do not think that Huemer intends this remark to undermine the objectivity of naturalness, or at least, it would do so only if its objectivity implies there is a most natural measure. If that were true then on the one hand it would support the power of the naturalness Well-posing strategy. On the other hand, if there were cases where not only were we unable to determine the most natural measure, which might be a merely epistemic inability, but where the best explanation of the inability was that there was none, they would undermine its objectivity. Whether there is such an implication is not clear and I don't think I need to settle that matter one way or the other. In this case, at least, I think Huemer accepts the objectivity of naturalness without endorsing the implication.

Lando objects to Huemer that 'we've been given no reason to privilege partitions that classify the alternatives in terms of explanatorily basic propositions' (Lando 2021:345). This sounds like unmotivated burden shifting. The reason to privilege them is that they are explanatorily basic. One could discuss the general significance of such basicity, such as its relevance to prediction and so on, but it is not clear why it is needed. Lando amplifies:

> What is the thought behind the Explanatory Proviso? Is the thought supposed to be that our reasons themselves are really only symmetric

with respect to the cells of the privileged partition—the one that divides the space into the explanatorily most basic alternatives? Or is the thought that in spite of the fact that our reasons are symmetric with respect to cells of different partitions, we should still only adopt credences that are uniform over one of those partitions? On the first horn, we posit asymmetries in our reasons where there seemingly are none; on the second we abandon the clear and compelling thought behind POI— namely, that symmetries in reasons require sameness of doxastic response. Huemer is not clear about which of these two ways of understanding the Proviso he endorses.

(Lando 2021:345)

There seems to be an obvious way to go between the horns. For each partition, considered in isolation, we may have no reason to discriminate between its cells. But we *do* have reason to discriminate between partitions that are more or less explanatorily basic. What is that reason? Just whatever the reasons are that mean they *are* more or less explanatorily basic. Considering the most basic partition, being the *most* explanatorily basic entails, among other things, that the symmetry of its reasons are not affected by the reasons concerning the other partitions but its reasons may affect the symmetry of the reasons of the partitions that are less explanatorily basic. Consequently, when we take into account the reasons of greater or lesser basicity, it is only the most basic partition that is guaranteed to retain the equal epistemic status for each of its cells. The epistemic status of the cells of the less basic partitions are then influenced by their relation to the reasons of the most basic cells.

Lando continues

Huemer is not clear about which of these two ways of understanding the Proviso he endorses. Instead, as far as I can tell, his justification for adopting it seems to be that it delivers the "intuitively correct" distributions in a number of particular examples. But for many of us there simply is nothing more intuitively correct about those distributions. Assigning equal weight to "red" "blue" and "green" is not intuitively preferable over assigning equal weight to "on" and "off".

(Lando 2021:345)

The case he is discussing at the end is one in which whether a lamp is on or off is causally determined by the colour of ball selected at random. I don't myself find there to be any equality in this clash of intuitions. Furthermore, Huemer is not merely *announcing* the consistency with intuition. The proviso also *explains* why those intuitions are truth conducive: it is because they are responding to *how things are and how things go*. At this point

I think Lando is revealing not that he has an argument against the objectivity of naturalness and explanatory priority, but only that this is how things strike subjectivists.

Having distinguished his distinct versions of Bertrand's question, Marinoff says

> To dispute which of these questions... "best" represents Bertrand's generic question Q is to relinquish geometry for aesthetics.... It seems unlikely that a majority of mathematicians or philosophers would support the claim that *any particular demand is the most "natural" one to impose on the problem....* the choice... remains grounded in subjectivity.... To paraphrase the Bard: "There is nothing paradoxical, but thinking makes it so".
>
> (1994:21–2 my emphasis)

Recall that Marinoff's questions specify instances of random processes for choosing the chords, which specifications then make a particular equal ignorance function the correct one to use in TIM. I am not sure whether Marinoff thinks this is an argument against the objectivity of naturalness, although his final remark tends to suggest that he does. Yes, one explanation would be that naturalness is subjective. But there is a better explanation of why, in the chord paradox, the choice between Marinoff's questions might be a matter of irrelevant aesthetics, why no one would support the claim of a most natural one: this is exactly what we would predict if *there is no natural measure of the chords.* Indeed, paucity paradoxes are powerful precisely because the arbitrary aesthetics of assigning a normative measure is obvious.

Klyve agrees with Marinoff:

> Philosophically, there is not much difference between choosing a number at random, and choosing a chord at random. Some of Bertrand's methods are biased toward choosing more short chords, and some to choosing more long chords. A mathematician may feel that one method is the 'most natural' (I have such a disposition myself towards Bertrand's third method), but this is primarily an aesthetic judgment.
>
> (Klyve 2013:369)

It is not clear exactly what work is supposed to be done by the word 'philosophically'. In one sense, there is not much difference between choosing among *any* different things randomly, but that hardly suffices even to support the claim of it being an aesthetic judgement.

The aforementioned literature is typical on this question. I haven't been able to find an actual discussion of the objectivity of naturalness in the literature on Bertrand's paradox and the principle of indifference. The two sides have tended to do as we have seen: to articulate positions compatible with or grounded in an assumption of objectivity or subjectivity and left it at that. There are cases that tend to support a side and other cases that tend to undermine it. That being said, in my response to Marinoff I think I have shown that the tendency to deploy the chord paradox as a support of subjectivity is mistaken: the cited phenomena are at least as well, and I would say better, explained by the objectivity of naturalness and the absence of any natural measure of the chords.

Nevertheless, there is at least a question over to what extent the distinction I have drawn between plenitude and paucity paradoxes depends on a realist metaphysics. Anti-realists may regard naturalness as subjective: constructive empiricists, for example, may say it is merely a way of registering the convenience of our theoretical instruments. That being said, and although I am personally of the view that naturalness is ontologically objective, I don't think the distinction I have drawn stand or falls with its ontological objectivity. It may stand or fall with its epistemic objectivity. That, however, would not trouble me, since for reasons mentioned in the earlier chapter on permissivism, subjectivists about epistemic probability have no problems with any of our paradoxes.

Again, I think my Best Theory premiss has a role here. The difference between the plenitude and paucity of natural measures has aspects that are unavoidable, whether or not they are advanced *in terms of naturalness*. Lines, squares and cubes just do have lengths, areas and volumes by which to beget a measure and sets of chords just don't have a measurement by which to so beget; that is as much true for mathematical nominalists and intuitionists as it is for Platonists. The measures of properties of chords just are not measures of the chords and so cannot be used as if they were in order to provide a normative measure for the POIS. TIM just is strongly unavoidable for the chords and not strongly unavoidable for the squares. Plenitude paradoxes just do have the strategy of defending certain salient measures as being more basic than others and paucity paradoxes just don't.

These are unavoidable differences between the plenitude and paucity paradoxes, and we can speak of them in terms of naturalness and in terms of explanatory priority. Those terms are most straightforward for realists but the differences remain for anti-realists, whether or not the differences are best formulated by them in terms of their concept of naturalness. So for these reasons, having started by drawing the distinction in the natural way of naturalness, and now thinned it out in a way that shows it to be unavoidable even for anti-realists, I shall continue to articulate it in terms of the presence or absence of *natural* measures. I leave it to the anti-realists to articulate the unavoidable distinction between the plenitude and paucity paradoxes in whatever terms they prefer.

15 Bertrand's Temptations

15.1 Introduction

The purpose of this chapter is to outline temptations that mislead us in our struggle with the paradox. Three originate in legitimate methods of scientific and mathematical enquiry, methods which can be misapplied when addressing the philosophical problem of Bertrand's paradox. I call them substitution, convention and distraction temptations. I examine these by drawing together instances we have seen in earlier chapters. The fourth is philosophical and I examine it by a case study of the response of objective Bayesians to the paradox. First, I set this in a needed context of a distinction between philosophical and methodological theories of probability.

15.2 Methodological *versus* philosophical theories of probability

Probabilisms such as the classical, logical, Bayesian, frequentist and propensity interpretations of probability are not heuristic or methodological theories of probability: they are philosophical theories. By contrast, mathematicians and scientists are mostly concerned with an heuristic or methodological view of probability. They are concerned with developing the mathematical theory and its application, concerned with what appears to work in advancing scientific enquiry, concerned with what methods their peers are willing to accept as helping get at the truth. Often the philosophical problems are regarded with impatience as the obstruction of windbags (sometimes true, no doubt) when what they want to do is to get on with doing the maths in order to do the science. They care neither about what probability is nor about the epistemological grounding of method.

This is fair enough so far as it goes but it cannot be allowed as a means of evading the philosophical problems. Rather, it must be understood as a retreat into methodology, a retreat that can legitimately appeal to useful tricks and heuristics for those purposes, but which tricks and heuristics

DOI: 10.4324/9781003456308-15

must not be substituted for the philosophical theory or assumed to be a philosophical theory without proper philosophical grounding. To be clear about this, for all statisticians talk about being Bayesians, what is really going on for some of them, without them necessarily being aware of it, is that Bayesian methodology is regarded as a more reliable method to get at propensities or frequencies than is Fisherian statistics. So despite calling themselves Bayesians, Bayesianism is for them a merely heuristic or methodological theory of probability, not a philosophical theory.

This distinction between philosophical theory and methodology also appears in the way 'probabilism' can be equivocal in philosophy. As a theory within epistemology it is the view that the normativity of belief and knowledge is importantly related to probabilities. Proponents of the principle of indifference are committed to this view. Some avowed probabilists, philosophical Bayesians such as de Finnetti and Ramsey, philosophical logicists such as Carnap, and more recently Jeffrey, Stalnaker and Maher, even reject the ideology of reasons for belief and justification. Instead, the normativity of belief is held to be entirely the matter of rational degrees of belief, which are identical to or determined by probability. Varieties of probabilism, in this sense, and without necessarily rejecting the aforementioned ideology, are philosophical theories of probability because they are epistemological theories.

By contrast, 'probabilism' also refers to the use of the mathematical theory to model various issues in epistemology, such as confirmation, coherence, aggregation of evidence, aggregation of opinion and so on. Such probabilism, valuable as it undoubtedly is, does not *on its own* constitute a philosophical theory *of probability*. Rather, it presupposes that we have such a theory and then uses the formal methodology to represent and theorize such issues by giving *probabilistic* theories of *confirmation, coherence, aggregation of evidence*, and so on. Bayesianism used in this way is a methodological theory of probability, not a philosophical theory.

I am not here suggesting a simple cleaving apart of methodological and philosophical theories, but rather clarifying the demands of the latter. A methodological theory as described may be part of a philosophical theory in virtue of the philosophical embedding it is given. However, a merely methodological theory cannot be a philosophical theory. It cannot because of its nature it eschews philosophical questions about probability: at best it presupposes a certain philosophical background. As illustrated earlier, it may even presuppose a philosophical background (frequentism or propensity in the example) that is at odds with the philosophical theory going by the same name as the methodology (Bayesianism in the example). Rather than any claim to establish some part of normative epistemology on the basis of the theory, at best the methods are found to be instrumentally rational at the service of whatever philosophical background is being presupposed. At worst they are philosophically naïve methodology, sometimes accompanied by explicit and dogmatic dismissals of philosophical issues as mere matters of convention that can be set aside by whatever bag of heuristic tricks are convenient.

A philosophical theory, by contrast, must attempt philosophical completeness. It must locate probability within a range of epistemological and metaphysical questions and ensure consistency with respect to them. It will include methodology but not substitute methodology for philosophy. It may not evade philosophical difficulties by wielding ad hoc method or retreating from philosophical commitments necessary to distinguish it philosophically from other theories.

15.3 Scientific temptations

With the distinction between philosophical and methodological theories of probability in place, we can define the main scientific temptation: it is to reach for more or different or better methodology with which to tackle Bertrand's paradox, to reach for the various tricks and heuristics that scientists use, in which anything goes if it advances the enquiry, and to assume that doing so can suffice to solve the philosophical problem. This is a lure in van Fraassen's trap. It also appears to be the reason scientists seem not to see some of the objections in earlier chapters which, once we had hacked our way through the mathematical thicket surrounding their underlying philosophical assumptions, confuted them.[170]

Falling for this temptation is a characteristic mistake made by physicists and mathematicians when addressing the philosophical problems posed by Bertrand's paradox. The temptation is especially tempting because it usually amounts to using a standard tactic in the face of an awkward mathematical problem or scientific question. When you can't prove what you want to prove, you can add hypotheses to your theorem until you get something you can prove. When addressing a question you can't answer, you substitute one that you can.

These are all perfectly legitimate tactics, provided that when you use them you don't then pretend that you've achieved the original objective. Nearly all the examples we have seen are, in the end, instances of this pretence hidden from their authors by motivated reasoning. I shall call this the *substitution temptation*.

A pair of temptations for scientists originates in their sometimes warranted impatience with philosophical argumentation. Whether a putative epistemic norm has an appropriate philosophical ground is a philosophical question that requires a substantive answer. An opposing view is that this

---

170 See also Fine, T. 1973:242 on physicists' naivety about probability.

question is merely terminological or taxonomical or a matter of convention. Rather, we need only consider whether the norm is generally plausible and can defend it solely on the ground of its successes and failures in solving substantive problems. I call this the *success programme for epistemic norms* and it originates in the same legitimate drive to advance the enquiry.

I believe the success programme is mistaken. First, it is possibly contradictory, since the appeal to general plausibility is an appeal to intuitions of its epistemic grounds. The rejection of any concerns with grounding is therefore not tenable. Second, of course such successes and failures count, but they are not the only things that count. Third, the question of appropriate ground is neither merely terminological nor merely taxonomical.

When we come to the matter of convention, we enter some admittedly deep waters, since there are anti-realists whose philosophy of epistemic norms is that they *are* a matter of convention.[171] This philosophy faces a dilemma: either it is itself a matter of convention or it is not. If it is, we must accept a social relativism for epistemic norms. This is epistemic subjectivism and for reasons I gave in the chapter on permissivism, we are concerned only with epistemic objectivism.

If it is not, then the problem is that if it isn't a matter of convention, in other words, if the answer to the question of why epistemic norms are a matter of convention is not something we are free to adopt by convention, then at least the epistemic norms that apply to answering this very question are not matters of convention but are objective requirements of rational belief. And if that is the case, why should there be *any* epistemic norms, properly so called, that *are* matters of convention? Rather, what are being *called* epistemic norms of convention are, in fact, no such thing. To the extent that they are adopted conventions they are merely rules of thumb, heuristics or methodologies that we have found instrumentally helpful in certain regions of enquiry. It is exactly this confusion between heuristics and objective epistemic norms that I am accusing scientists of being inclined to make.

I am not going to take this particular question further and will happily accept as commitments of my enquiry whatever assumptions about the objectivity of epistemic norms are needed to rule out their conventionality.

All that being said, philosophical questions may indeed require enquiry into questions of terminology, taxonomy or convention, without being any less important for the necessity. Nor need the answers to those included questions be philosophically trivial. Having the contrary view on their

---

171 Notoriously, the pragmatist Rorty is reported to have said 'Truth is what your contemporaries let you get away with saying' (Sartwell 2016).

significance, however, tempts one to think that philosophical difficulty can be evaded by what are in the end essentially verbal means: that some different words, a convention or distinction one may freely adopt or define, and asseveration, can replace addressing the difficulty. I shall call this (for want of a better word) the *convention temptation*.

Commitment to the success programme combined with legitimate skill and interest in heuristics and methodology leads to directing attention away from philosophical concerns and towards mathematics without showing whether the mathematics fulfils the philosophical needs. I shall call this the *distraction temptation*.

In practise, one may find all three of these temptations manifested in a single attempt to solve Bertrand's paradox. I am going to illustrate the three with what I hope are obvious examples from our earlier chapters. Before doing that, I should perhaps say that philosophers are quite as capable of falling for these temptations as mathematicians and scientists!

15.4 Substitution temptation

Most of the frailties amount to making substitutions, and given the minimal attention they received prior to this book, virtually invisible substitutions. We have seen instances of substitution in both the Distinction strategy and Well-posing strategy. Recall the diagnosis of the Distinction strategy—Bertrand's question is unanswerable *as it stands*, being a clumsy way of raising whilst confounding distinct well-posed questions with unique answers; the strategy is an exact instance of substituting a problem you can't solve for one you can. Hence we had Marinoff and the 12 others I quoted all falling into van Fraassen's trap by accusing Bertrand's question of being 'vaguely posed', 'a single fuzzy question' and so on. There is a tendency for physicists in particular to interpret the paradox in terms of tossing ideal sticks, pebbles and other things, onto ideal circles, as we saw Jaynes, Porto *et al.* and Aerts and Sassoli de Bianchi doing. Although no doubt relevant to considering specific empirical situations, the move to treat them separately ends in making the Distinction strategy substitution, as does taking Bertrand's 'on trace'—one draws—literally.

The philosophical problem they have all evaded is the proof that Bertrand's question is unanswerable in a sense that allows it to be rejected altogether, thereby justifying the Distinction strategy. When we did the needed philosophical work, first in Chapter 3 of analysing determinate and indeterminate problems and separating that question from a determinate problem having a unique solution, and second in Chapter 5 of distinguishing a generic and a general question, that entire strategy collapses.

This temptation manifests in the Well-posing strategy by the application of the principle needed to restrict the domain of probability measures that ends up substituting a covert restriction of Bertrand's question, giving a solution that falls into the Well-posing Danger. Jaynes' proposed principle is the requirement to respect symmetry. We saw Jaynes' use of the symmetries of the Euclidean plane relied on imposing a coordination of regions across many circles, a coordination that is no part of the original question and therefore that amounts to a restriction of the question. Wang and Jackson thought that the homogeneity of random lines solved the paradox but it ended up being a substitution of a different question for Bertrand's questions.

Rizza's principle is a demand that our mathematics must be adequate to represent the intuitive content of any problem it is going to be used to solve. He claimed the difficulty had with the paradox was because measure theory was inadequate and that a supplementation using Sergeyev numbers would succeed. His restriction was the assumption that only his way of using Sergeyev numbers, naming endpoints on the circle, gave an approximate name space for the chords and a counting Sergeyev measure to be used in POIS. I showed that if this restriction was legitimate, we could just as well have solved the paradox with standard measure theory and the name space we used for the angle case in §3.11. Furthermore, Sergeyev numbers can be used just as well to name the angles of diameters and points on diameters, giving us an approximate naming the chords by naming the angle of the diameter they intersected at their midpoint and the point of intersection on that diameter. So Rizza fell to the Well-posing Danger and this alternative naming gave a different answer.

The final substitution we have met is the assumption that the chord paradox poses the same problem as the plenitude paradoxes. For example, Klyve assumes that

> there is not much difference between choosing a number at random, and choosing a chord at random.
>
> (Klyve 2013:369)

Bangu's treatment and Gyenis and Rédei's General Bertrand's paradox both make this substitution, although I did attempt to find a way to formulate their treatment as applicable to the chord paradox. Lando's introduction is a typical way of making the substitution

> The original paradox was given in Bertrand [1907], and had to do with the probability that a chord chosen at random in a circle is longer than a side of the inscribed equilateral triangle. Simpler examples were given in recent years; for our purposes, the following modification in White [2009] of an example in van Fraassen [1989] will suffice: Mystery Square.
>
> (2021:341)

Another way of being led to substitution is assuming that Bertrand's procedures give measures on the chords:

> A modern probabilist, however, will take comfort in remembering that the set of diameters forms a set of measure zero in the in the [sic] set of all chords of a circle.
>
> (Klyve 2013:368)

Klyve fails to notice that the measure of which he speaks is not a measure of the chords at all, but of their midpoints in the disk.

In §14.19 we saw how cogency of this assumption rests on Corollary **13**, but that the inevitability of that corollary being fulfilled whenever and for whatever reason we encounter paradoxical probabilities refuted the assumption. The explanation in terms of the relation of paucity and plenitude paradoxes to the Theory of Induced Measure shows you cannot substitute a plenitude paradox for the chord paradox.

15.5 Convention temptation

The first manifestation of this was Bangu's attempted to deny the transmission of randomness across event entailment. We were offered an argument that achieved nothing: the attempt to distinguish the randomness of the entailing event from non-randomness of the entailed event was left as mere asseveration. We saw a multiplication of distinctions that failed to do the job it was supposed to do. This was Gyenis and Rédei's distinction between relabelling invariance and labelling irrelevance, where the claim was that it was only the latter that mattered for the consistency of the principle of indifference. On this basis, the necessary violation of the former, being a corollary to van Douwen's theorem, was supposed to be irrelevant to the truth of the principle. We showed that the latter implied the former and hence the distinction could not save the principle. In short, both of these attempts at the Irrelevance strategy failed because they made distinctions without differences, and in so doing they fell for the convention temptation.

We saw an attempt to define uniqueness as a criterion of identity for the principle of indifference. There was no reason given for this, only an insistence that if the principle were well defined it *would* satisfy uniqueness. This, I think, is a characteristic error that a mathematician might make, precisely for the reason that this insistence is often quite correct in mathematics. If your definitions and putative axioms end up producing contradictions something is wrong with them and you are entitled to fix them and move onto happily proving theorems for the objects of which the axioms are true. But the principle of indifference is not up for theoretical correction and redefinition in this way, as if we hadn't yet got our definitions and axioms in order. Uniqueness is a demand originating in epistemic objectivity and for this reason is a criterion of success logically prior to the principle of indifference. It therefore cannot be defined to be a criterion of identity of the principle that justifies redefining the principle just because we are ending up with paradoxical probabilities from the principle as standardly given.

A typical error for a mathematician is to think a metaphysical question is a matter of aesthetics, where the latter is additionally thought of as a matter of arbitrary convention. We didn't see anyone attempt an Irrelevance strategy on this basis, but the thought is frequently found in the literature. In our discussion of naturalness we found Klyve challenging its objective significance because 'this is primarily an aesthetic judgment' (Klyve 2013:369). There is, especially in mathematics, an important question of whether aesthetic properties such as simplicity and beauty are indicative of truth. Here is not the place to go into it. Many mathematicians and scientists would say the answer is yes and if so this is no challenge. In addition, being an aesthetic judgement doesn't challenge objective significance absent a subjective theory of aesthetic properties. Marinoff raised it in these terms and joins Klyve in the doubt of there being a most natural among Bertrand's cases but we saw this reason for this doubt is better explained by the absence of any natural measure for the chords. Admittedly, if a judgement of naturalness is an aesthetic judgement independent of truth conducivity, it poses a different question about the significance of naturalness. Yet as I showed, the formal features of the distinction I drew in terms of naturalness cannot be avoided.

15.6 Distraction temptation

The necessity of careful mathematical thinking can easily distract us from thinking about how exactly the offered mathematics satisfies the philosophical demands of solving the paradox. Working through Jaynes' clever paper leaves us failing to notice the dependence of his solution on the restriction I identified. Similarly, the interest and novelty of applying Sergeyev numbers distracts us from thinking about whether it is surprising that we get consistent probabilities by interpreting all of Bertrand's cases into the same approximate name space, let along thinking of other ways of naming the chords using Sergeyev numbers. Similarly, the interest of the Haar measure distracts us from the complete absence of any attempt to apply it to the chords.

Aerts and Sassoli de Bianchi promised us a meta-indifferent average: a philosophically significant form of meta-indifference over all the probability measures in play. They immediately backed away from actually doing that because it is too hard:

To be able to define a uniform average over all possible uncountable $\rho(x)$, without being confronted with insurmountable technical problems related to the foundations of mathematics.

(2014:7)

In doing this they are applying a standard heuristic for physicists to avoid mathematical difficulties: to make the mathematics tractable by restricting the domain and simply ignoring the awkward cases. This is legitimate when there is an empirical check since if the check goes against you the methodology is rejected for the case. The danger is that the restriction of the domain makes the claimed solution covertly ad hoc and thereby amounts to a retreat from a solution adequate for philosophical theory into mere methodology.

In place of what we were promised, we were given Aerts and Sassoli de Bianchi's universal average defined by the limiting process we saw. When we examined what this amounted to, their universal average turned out to be nothing more than a limit of a constant sequence of measures, each member of which is the uniform measure on [0,2]. We examined the sense in which this could be justified as a meta-indifferent average and showed there was no good grounding for it. In the end, the philosophical work was done only by the verbalism of calling what we were given by a name that sounded like it was what we wanted. Its inadequacy was hidden, perhaps from the authors themselves, by the distraction of several different kinds of proof going on. First, there were the proofs that the step functions were, in their special sense, dense in set of probability densities. Second, there was the complexity of the proof extending over two and a half pages which gave no clue to the fundamental triviality of why their universal average is the uniform measure on [0,2].

15.7 Philosophical temptation

The philosophical temptation is to find some way to evade the commitment to uniqueness, or at least, to find some way to evade embarrassment by the failure to meet that commitment which is forced on one by Bertrand's paradox. Permissivism, of course, denies any such commitment. Subjective permissivism has no problem with the paradox anyway: it is only a problem for epistemically objective theories. I argued that on the only good ground for objective permissivism, we do not end up, as we were supposed to, in an Irrelevance strategy solution of the paradox, but instead in giving up the principle of indifference altogether. And if you do that, you need not worry about any problem that the paradox poses the principle anyway. If, however, you want to keep the principle without being an epistemic subjectivist, you are in for some difficulties.

We now turn to a case study on what happens when proponents of an epistemically objective theory of probability are tempted by evasion. Classical probabilists, logical probabilists and objective Bayesians have been the main proponents of objective epistemic probability. Classical probabilism and logical probabilism were both held to be refuted by Bertrand's paradox showing that they failed uniqueness. Recently, objective Bayesians appear to have decided to deny uniqueness. Is that tolerable? I now argue here that it is not. Among those things that distinguish a philosophical theory of probability from philosophically naïve methodology is their provision of a principled satisfaction of singleness. Objective Bayesianism cannot therefore be given a pass if it too, fails uniqueness. Rather, it must resolve any problems it has with uniqueness if it is to be accepted as the advance on logical probabilism it claims to be whilst remaining philosophically (as opposed to merely methodologically) distinct from subjective Bayesianism.

15.8 Objective Bayesianism requires uniqueness

Objective Bayesianism is supposed to be an advance on classical and logical probabilism and distinct from subjective Bayesianism.[172] What distinguishes it from them? It is more advanced methodologically, in some ways, but on the other hand, nothing of its methodology could not be adopted by the other three. As mentioned before, one could even be a propensity or frequency theorist and regard objective Bayesian methodology as better for discovering chances.

Both objective and subjective Bayesians hold probability to be ontically subjective, being identical to rational degrees of belief. Where objective and subjective Bayesianism apparently part company is over epistemic normativity: subjective Bayesianism allows probability to be epistemically subjective whilst objective Bayesianism purports to be epistemically objective. The word 'objective' has been used by objective Bayesians to signal the ambition to solve, rather than submit to, the problem of arbitrary priors. In so doing, they mark a commitment to uniqueness in opposition to the permissivism of subjective Bayesianism. Galavotti identifies Jeffreys and perhaps also Reichenbach as the earliest objective Bayesians on the basis of their combination of Bayesian method with a commitment to uniqueness (Galavotti 2016:139&136 respectively).

A tool that has been offered to solve that problem is the Maximum Entropy principle, which principle Jaynes offered as a generalizing supercession of the principle of indifference. In adopting that tool, objective

---

172 See Rowbottom 2008 for an argument that logical probabilism and objective Bayesianism are not clearly distinct.

Bayesians give up on updating credences by conditioning (because it is proved that the latter is incompatible with updating using the Maximum Entropy principle).

As a consequence, the word 'objective' has also been used to signal the *methodological* commitment to the Maximum Entropy principle as opposed to the subjectivist's commitment to conditioning. If at this point objective Bayesians give up on uniqueness (and we shall see shortly that some appear to do so) then we slip unnoticed from the philosophical opposition between objective and subjective Bayesians (with its consequent philosophical opposition of uniqueness versus permissivism) to a merely methodological opposition. Conversely, there are Bayesians (such as White 2005 and Christensen 2007) who remain committed to updating by conditioning and who reject permissivism without calling themselves objective Bayesians.

So there is a danger of the terminology of 'objective Bayesianism' drifting into marking a merely methodological difference at the expense of sustaining a philosophical difference. Methodology is, of course, a perfectly proper focus for the mathematicians and scientists who have contributed to the development of objective Bayesianism. Yet such a focus brings with it a certain danger, the danger of a theory that is no more than philosophically naïve methodology. If we are to treat objective Bayesianism as a serious philosophical programme with a history and a line of development, we cannot treat it as a bag of methodological tricks that can discard awkward philosophical problems as mere matters of convention. We must discern just what that philosophical programme is, what distinguishes it from others in the region and then we must judge to what extent it furnishes a satisfactorily philosophical theory of probability.

Consequently, insofar as the objective Bayesian is a philosophical theorist, that is to say, unless the objective Bayesian means nothing more by 'objective' than a methodological disagreement with subjective Bayesianism, then he is committed to epistemic normativity being objective, if not on matters of taste at least on general matters of fact. They are also committed to the principle of indifference, since it is a restriction of the Maximum Entropy principle. Consequently, they cannot be permissivists, since as we saw previously, objective permissivists abandon the principle of indifference.

So objective Bayesianism, insofar as it is intended to be understood as a philosophical theory rather than a collection of methods or heuristics, and insofar as it is intended to be an advance on classical or logical probabilism and distinct from subjective Bayesianism, is committed to uniqueness. Uniqueness for objective Bayesians is a polite word for consistency. If they cannot fulfil it, they cannot maintain consistency without relativising probabilities to persons, and that amounts to abandoning what is supposed to distinguish them philosophically from subjective Bayesians.

15.9 Objective Bayesianism aims at uniqueness

Objective Bayesians such as Jaynes (2003), Rosenkrantz (1977), Paris (1999) and Williamson (2010) use the Maximum Entropy principle to address the ignorance requirement of §1.2. Jaynes' own justifications of the principle are that 'the probability [function]... which has maximum entropy subject to whatever is known, provides the most unbiased representation of our knowledge' (Jaynes 1957b:171) and that it is the 'maximally non-committal with regard to missing information' (Jaynes 1957a:620). Both of these remarks are claims that the maximum entropy function is least presumptuous, in the sense we introduced in §1.2. A recent justification from Paris (1999) is based on a number of the principles that Paris calls common sense, whose satisfaction he proves constrains us to the Maximum Entropy principle. Some of these principles are identified ways of not being presumptuous[173] and the others are formal constraints.[174] Jaynes and Williamson are the only avowed objective Bayesians who say much about our concerns here so I shall review relevant material from them.[175]

15.10 Jaynes

Jaynes' proposal (Jaynes 1957a, b, 1968, 1985) is that the Maximum Entropy principle can play the role played by the principle of indifference in logical probabilism. He intended that the Maximum Entropy principle should result in epistemic objectivity, that, for example, 'prior probabilities can be made fully ... "objective"' (Jaynes 1968:228) even in the face of complete ignorance (more below). Epistemic objectivity will be obtained because

> we interpret the principle in the broadest sense which gives it the widest range of applicability, i.e., whether or not any random experiment is involved, the maximum-entropy distribution still represents the most "honest" description of our state of knowledge.
>
> (Jaynes 1968:235)

So the Maximum Entropy principle will measure the rational base probabilities whilst being non-presumptuous. Jaynes is here acknowledging the fully general problem the Maximum Entropy principle is supposed to be solving.

---

173 Irrelevant information, obstinacy, relativization, equivalence, continuity, weak independence (Paris 1999:79–82).
174 Symmetry and renaming (Paris 1999:78–79).
175 Rosenkrantz also, but he adds nothing to Jaynes.

the maximum-entropy principle will lead to a definite, parameter-independent method for setting up prior distributions. (Jaynes 1968:236)

Hence the Maximum Entropy principle will solve the problem of non-unique priors giving non-unique rational probabilities and instead result in uniqueness. So Jaynes' intended objective Bayesianism should achieve epistemic objectivity and uniqueness.

It is sometimes thought that Jaynes may have weakened his ambition by the time of his final book. In some ways that may be true. He expresses some impatience with philosophical concerns not motivated by empirical applications. In other ways, his ambition is more philosophically comprehensive.

> Our topic is the normative principles of logic... to direct attention to constructive things and away from controversial irrelevancies, we shall invent an imaginary being... designed by us, so... it reasons according to certain definite rules... [to]... be deduced from simple desiderata which... would be desirable in human[s]; i.e.... that a rational person, should he discover that he was violating one of these desiderata, would wish to revise his thinking.
>
> (Jaynes 2003:8)

His use of 'normative principles' and 'rational person' proves that he is concerned with philosophical theory, not only methodology. Jaynes states three basic desiderata to be met by an ideal plausible reasoner, desiderata that he claims 'uniquely determine the rules by which our robot must reason' (2003:19).[176] In Chapter 2 he then derives on the basis of his desiderata the standard probability axioms and calculus in an approach originating in Cox's proof of Cox's theorem (Cox 1946). The calculus itself does not supply numerical probabilities beyond the extremes and hence the 'poorly informed robot' (2003: 274) can make some headway with the principle of indifference and must eventually use the Maximum Entropy principle (2003:chapters 11 and 12).

Jaynes' first and third desiderata include 'degrees of plausibility are represented by real numbers' (2003:17) and 'if a conclusion can be reasoned out in more than one way, then every possible way must lead to the same result' (2003:18). So it is quite clear that his commitment to a unique degree of rational belief is at the very foundations of his defence of probability as the rational ideal. Clearly, then, Jaynes was not presenting a theory

---

176 These desiderata are similar to Cox's requirements of divisibility and comparability, common sense and consistency (Cox 1946, 1961), whose work inspired Jaynes's approach here (and see also Arnborg and Sjödin 2000).

of probability that is no more than a narrow piece of methodology. Jaynes had the ambition to advance objective Bayesianism as a philosophical theory of probability grounding epistemic objectivity and therefore uniqueness. His known dislike of philosophy suggests he would not own up to the ambition under that description: this does not change the fact that his ambition falls under that description as herein defined.

15.11 Williamson

Williamson has on at least one occasion committed objective Bayesianism to epistemic objectivity and uniqueness:

> The aim of objective Bayesianism is to constrain degree of belief... [so]... that only one value for $p(v)$... [is]... rational... [given]... an agent's background knowledge.... probability varies as background knowledge varies but two agents with the same background knowledge must adopt the same probabilities as their rational degrees of belief.
>
> (Williamson 2009:502)

He has also committed it to constraint by the person's state of ignorance:

> An agent's degrees of belief should also be fixed by her lack of knowledge of the world.
>
> (Williamson 2009:505)

Put these together and we have objective Bayesianism committed to a method satisfying

> The *objective ignorance requirement*: for any state of ignorance over a domain of propositions, there exists a single probability measure over that domain determined by the method for anyone.
>
> (§1.2)

On the other hand, towards the end of his book Williamson gives the appearance of continuing in this ambition whilst also to some extent backing off from it. We will discuss this next.

15.12 Objective Bayesianism fails uniqueness

As mentioned in §8.2, the Maximum Entropy principle has been applied to some of the paradoxes that trouble the principle of indifference and it has been claimed to resolve some of them. We have seen above that it doesn't solve Bertrand's paradox. The paradox confronts the principle with such an extent and depth of difficulties, difficulties that occur for any event space sharing certain features with the set of chords, that no rescue is possible. The Maximum Entropy principle therefore fails uniqueness without prospect of amelioration. Since this is proven, the problem of non-uniqueness cannot be ignored by proponents of the Maximum Entropy principle.

15.13 Retreating into methodology

Modern proponents of objective Bayesianism have been willing to lower their ambitions. They have backed off from claiming epistemic objectivity in general. They are, rather, concerned to develop the theory in detail and to explore the extent it provides useful methodology and models. In some cases that physicists care about they can parameterize a space in a way that is empirically adequate and for which, because of that parameterization, the Maximum Entropy principle works well. Similarly, as Williamson (2010) has ably shown, relativised to a language, the Maximum Entropy principle can produce useful models of issues such as coherence, confirmation, testimony, judgement aggregation and so on.

This modern, narrower, project is without doubt valuable, so far as it goes. Nevertheless, as I argued earlier, on its own this project is insufficient for a philosophical rather than methodological theory of probability. Insofar as objective Bayesianism is supposed to be an *advance* on logical probabilism as a philosophical theory of probability (and distinct from subjective Bayesianism), it cannot simply back away from epistemic objectivity and uniqueness. After all, if we were allowed to narrow the concerns of logical probabilism to exclude the areas where it cannot furnish uniqueness, it *too* could offer useful methodology and models.

Jaynes certainly did not wish to take any such retreat. Jaynes argues for the necessity to meet the ignorance requirement when he says

> how do we find the prior representing "complete ignorance?"... To reject the question... on the grounds that the state of complete ignorance does not "exist" would be... absurd.... In the study of inductive inference, the notion of complete ignorance intrudes itself into the theory just as naturally and inevitably as the concept of zero in arithmetic. If one rejects the consideration of complete ignorance on the grounds that the notion is vague and ill-defined, the reply is that the notion cannot be evaded in any full theory of inference.
>
> (1968:236)

Again, his unwillingness to abandon this shows his commitment to philosophical theory and his rejection of the philosophically dogmatic methodologism of those he is criticising here.

Insofar as Williamson has addressed the problem of non-uniqueness, we must look to Chapter 9 of his book. We find therein neither a satisfactory variety of epistemic objectivity nor a principled ground for non-unique precise probabilities. In fairness to Williamson, and despite his discussion of non-uniqueness and epistemic objectivity, it is not clear that he is claiming to have achieved them. The chapter may instead be intended merely to outline the aspiration, to explain how his account of objective Bayesianism, at that point in the book defined only for propositional and predicate languages, may be extended to 'richer languages' including 'the usual mathematical language for probability theory' (2010:148), and to locate clearly his account with respect to some wider issues.

In §9.3.2 Williamson offers something he later calls epistemic objectivity, which is not epistemic objectivity relativised to evidence alone or to epistemic reasons, but is also relativised to a language (2010:158). This is defended by the thought that 'there is a sense in which an agent's language constitutes empirical evidence' (2010:156). Although an interesting suggestion, a suggestion with which I am in sympathy, this defence is quite undeveloped. Williamson leaves unexplained whether and how this countenancing of evidence fits his calibration norm, the norm that his official position advances as the way objective Bayesianism countenances evidence. Furthermore, language relativity entails non-uniqueness and has been regarded as a failure of the principle of indifference for logical probabilism, so absent some very good reasons it ought to be regarded as a failure of objective Bayesianism. Williamson briefly outlines an example of a possible virtue of language dependence: handling the well known difference in statistics that apply to bosons and fermions (2010:157) by language relativity. Setting aside whether language dependence is in fact *required* to handle this, one example hardly suffices to make the case.

In §§9.1.3, 9.2 and 9.3.1 Williamson offers some remarks on non-uniqueness for his system originating in paradoxical cases. These are very brief and for the most part he merely notes ways in which non-uniqueness arises in paradoxical cases. He mentions Jaynes' treatment of Bertrand's paradox but does not address its refutation in Shackel 2007. He makes no attempt to apply the Maximum Entropy principle to Bertrand's paradox nor does he make any statement on whether doing so would result in unique probabilities. Instead he moves on immediately to indicate that he is untroubled by non-uniqueness in general. There is one remark apparently intended to deflect worries about non-uniqueness:

the role of the infinite is just to facilitate our reasons about the large but finite and discrete universe in which we dwell (Hilbert 1925). In which case, the chief constraint on the behaviour of the infinite is that it conserves the behaviour of the finite.

(2010:152)

This is not widely agreed but depends on highly contentious theoretical and empirical propositions. If any physical quantities are continua, this will never do. Furthermore, it denies the extremely important role that probability plays in domains that are essentially infinite, such as its use in proofs in number theory, most especially in analytic number theory (in which the infinite continua of analysis are used to obtain results concerning the discrete infinity of natural numbers). No explicit case is made out of this remark for why non-uniqueness is not failure: instead we are shown how his system would countenance the infinite as a mere ideal to aid in handling the finite.

He then goes on to explain at some length the technicalities of how his system gives rise to and formally handles non-uniqueness for continuous languages, concluding that

> the language of the mathematical theory of probability is expressive, but... it offers little in the way of guidance concerning the implementation of the Equivocation norm.[177] One needs to appeal to other considerations in order to determine an appropriate set of... potential equivocators and to determine respects in which closeness to these potential equivocators is desired.
>
> (2010:155)

This amounts to nothing more than conceding non-uniqueness and offers nothing towards justifying why lack of uniqueness in Bertrand's paradox is not a failure.

Williamson concludes the chapter with a section discussing objectivity.

> An agent's evidence and language may not uniquely determine a belief function.... an agent is rationally entitled to set her degrees of belief according to any one of these belief functions.... This choice is one of the features that distinguishes objective Bayesianism from the logical interpretation of probability.
>
> (2010:158)

The claim of rational entitlement is mere asseveration: no argument for the normative claim is given. The distinction amounts to no more than announcing acceptance of non-uniqueness. Yet as I proved earlier, for those committed to epistemic objectivity and the principle of indifference, non-uniqueness is not an acceptable permissivism but just a euphemism for inconsistency. If accepting inconsistency is what distinguishes objective Bayesianism from logical probabilism, it does the former no credit at all.

---

177 Williamson's version of the principle of indifference.

Williamson then once more notes ways in which non-uniqueness arises in his system and suggests this is not a problem because 'although objective Bayesianism often yields objectivity, it can hardly be blamed where none is to be found' (2010:158). Whether it can be blamed, however, depends on why none is to be found.

This perhaps has some plausibility in some special cases. Earlier in the book he discussed the case of a known biased coin, for which on his analysis there is no admissible function with maximum entropy (because the function with maximum entropy is the limit of the admissible functions but is not itself in the set of admissible functions) and therefore non-uniqueness arises. It fits with his pragmatic concerns to slacken the condition of his Equivocation norm:

> If the agent can—or need—only distinguish probability values to a fixed finite number of decimal places then any probability measure that is indistinguishable from any other [closer to the maximum entropy function] will be sufficiently close.

> (2010:30)

This can be challenged, but one takes the point: because of the specifics of an empirical situation we can make some sense of a narrow band of functions near the inadmissible maximum entropy function, between which we are unable to distinguish, and so, for empirical reasons, epistemic objectivity eludes us.

In the case of Bertrand's paradox, however, this has things exactly backwards. The mere appeal to a system that allows non-unique precise probabilities in a mathematically systematic way is not going to help. Here, epistemic objectivity eludes us, not for a principled reason based in the case, but because non-uniqueness means that objective Bayesianism is failing the objective ignorance requirement. Retreating here to empirical cases where more is known, in the hope of similar moves to the just discussed coin case being made, is just another way of falling into van Fraassen's trap.

Williamson then turns to discussing something he calls 'full objectivity.... probabilities uniquely determined, independently of evidence and language' (2010:159). Since we have not been given it before, here is where we might expect finally to meet an explicit attempt at either a satisfactory grounding of epistemic objectivity or a principled grounding of non-uniqueness in cases such as Bertrand's paradox (if to do one or other were an object of the book). That, however, is exactly what we don't get. His concern instead is to show that this 'ultimate belief notion of probability' could be countenanced in his system 'by considering some ultimate evidence... and ultimate language' (2010:159) and he then discusses how this might be related to objective chances.

So whilst Williamson may be understood as announcing his acceptance of non-uniqueness, to the extent he is doing that he is abandoning epistemic objectivity. He is doing it without a principled defence and the retreat counts either as a retreat from a philosophical theory of probability altogether, or if the intention is to be proffering a philosophical theory, it is an epistemically subjective theory and therefore not distinct philosophically from subjective Bayesianism.

15.14 Conclusion of the case study

It is the claim that objective Bayesianism is a philosophical theory that is the subject here and to the extent there are objective Bayesians who abjure such an ambition they are irrelevant, being merely methodological Bayesians who do not matter philosophically.

The absence of uniqueness manifested by Bertrand's paradox was regarded as refuting for classical and logical probabilism and objective Bayesianism should not be given a pass that was denied classical and logical probabilism. Insofar as contemporary objective Bayesians have allowed non-unique probabilities, the thought that this amounts to resolving rather than falling to Bertrand's paradox is manifestly ad hoc. The challenge is to justify why, *here*, is lack of uniqueness not a failure. The challenge therefore remains, and remains stark: objective Bayesians must either solve the problem of epistemic objectivity posed by Bertrand's paradox or give principled reasons for the acceptability of non-uniqueness.

Their commitment to the Maximum Entropy principle means they do not have a principled way of avoiding non-uniqueness unless they give up their commitment to epistemic objectivity. Giving up the latter is to give up exactly the *philosophical* point on which they sought to distinguish themselves from subjective Bayesians.

The current insouciance on the part of objective Bayesians over non-uniqueness is therefore whistling past the graveyard. They should come clean and admit that, philosophically, objective Bayesianism is no advance on logical probabilism because it has fallen to the very same issue that was taken to refute logical probabilism. Jaynes was right and recent objective Bayesians who follow Williamson are wrong. An objective Bayesianism worthy of being a distinct probabilism cannot abandon epistemic objectivity and therefore can neither abandon the objective ignorance requirement nor uniqueness.

We seek a philosophy of probability that either satisfies epistemic objectivity for graded belief or grounds its absence in a principled manner.

Epistemic objectivity requires uniqueness. The Maximum Entropy principle fails this by falling to Bertrand's chord paradox. If falling to Bertrand's paradox stands against classical and logical probabilisms, it stands equally against objective Bayesianism.

A fall back by logical probabilists to non-unique rational probabilities and a restricted principle of indifference has been regarded as a failure. It neither grounds a satisfactorily comprehensive epistemic objectivity nor grounds its absence in a principled manner. Objective Bayesians have fallen back similarly. I have given reasons why this is unsatisfactory and also contrasted such a diminution with Jaynes' original ambitions and Williamson's commitments. Furthermore, considering the extent and depth of difficulties with non-uniqueness demonstrated herein, difficulties not confined to Bertrand's paradox alone but that occur for many event spaces, it is very difficult to see how such a fall back would be a principled grounding of the absence of epistemic objectivity. The current attitude of insouciance to non-uniqueness is, rather, indifference to the philosophical demands of epistemic objectivity. Consequently, if objective Bayesianism is supposed to be philosophically as opposed to only methodologically distinct from subjective Bayesianism, it is required to give a principled account of its failure at uniqueness rather than staying with evasion.

Of course, the temptation to evade is entirely understandable, and perhaps even more so in the light of this book. For we have seen no way out of Bertrand's paradox for epistemically objective theories of probability. But it is a temptation that must be resisted, for as we have seen in this case study, the evasion amounts to a retreat from offering a distinctive objective philosophical theory of probability into one that is at best only methodologically distinct from a subjective theory, and at worst is philosophically naïve methodology.

16 Rational Strength

16.1 Introduction

No strategy of solving Bertrand's paradox has worked. In this chapter, for completeness I briefly explain two responses to Bertrand's Paradox that are available if one gives up certain views of probability: Bertrand's original aim, finitism, and the position of ontologically objective theories of probability. I discuss whether our ultimate understanding should be that the paradox proves the principle false and I defend the continuing paradoxicality of the paradox. I then propose my own solution, an instance of the Entirely Unanswerable strategy, justifiable both by the failures of the other strategies and by the unearthed root. We would like to know where the proposal leaves the principle of indifference and in particular how it fits with its central commitment to epistemic reason. The rest of the chapter sketches this in. I show that the proposal does not vindicate a renewed Distinction strategy and we look at some literature compatible with it. From our proportioning ground I derive a principle more general than the principle of indifference, the Principle of Rational Strength. Making some further assumptions about the nature of epistemic reasons proves the Principle of Rational Strength true and gives us a principled restriction of the domain of the principle of indifference. I then articulate the shape of my Entirely Unanswerable solution to Bertrand's paradox as a whole. The chapter concludes with a summary of the book.

16.2 Finitism

It is possible to avoid the chord paradox whilst retaining the principle of indifference by allowing finite additivity but denying countable additivity for probabilities. Bertrand himself takes the view that his question is Entirely Unanswerable because the principle of indifference cannot be applied to infinite state spaces, and therefore finitism and giving up countable additivity for probabilities must be true.

DOI: 10.4324/9781003456308-16

Finitism can be motivated independently of Bertrand's paradox. De Finetti held that 'no-one has given a real justification of countable additivity' (1970:119) and that it is a defect of measure theory that we end up with unmeasurable sets in continua, which corresponds to events without probabilities. Kolmogorov regarded his sixth axiom, the axiom of continuity, (which is equivalent to countable additivity) as needed only for 'idealised models of real random processes' (Kolmogorov 1933/1956:15). Nevertheless, it is needed for differentiating probability measures, which can give us probability density functions, and for proofs of limits of probability measures, such as the central limit theorem.

It is true that finitism for probability might, in the end, be a position we have to accept. However, finitism is a severe restriction and may amount to an unacceptably impoverished or hobbled theory of probability.

> The conventional wisdom is that finite additivity is harder than countable additivity, and does not lead to such a satisfactory theory.
>
> (Bingham 2010:5)

Edwards is more forthright: 'Finitely additive measures can exhibit behaviour that is almost barbaric' (Edwards 1995:213). Furthermore, it is well known that

> the Stone representation and the Drewnowski techniques allow one to reduce the finitely additive case to the countably additive case. They both show that a finitely additive measure is just a countably additive measure that was unfortunate enough to have been cheated on its domain.
>
> (Uhl 1984)

See Bingham 2010:6 ff. for the nature of the hobbling, in particular, for the inability of finite additivity to satisfy de Finetti's motivating criterion of being defined on all subsets (de Finetti 1970) for spaces with dimensions greater than 2. For the dispute see, for example, Howson 2014a. Furthermore, some philosophers (e.g. Williamson 1999) have been willing to argue contra de Finetti that subjectivists must accept countable additivity. For good reason, then, we have been unwilling to give in without a fight, and have continued to try to solve the paradox whilst retaining countable additivity.

16.3 Ontologically objective theories of probability

The second response can be advanced on the basis of ontologically objective theories of probability such as frequentist or propensity theory.

Defining probability in terms of frequency, or regarding frequencies as evidence of propensities, and distinguishing reference classes in terms of specifics of empirical situations (for example, a circular flower bed and chords defined by entrance and exit points of overflying birds, a circular container of gas and chords defined by successive collisions with the wall by particles) could well result in determinate solutions for such empirical situations. Such solutions might be regarded as examples of the Distinction strategy.

However, the original point and the continuing importance of the paradox is the challenge it poses to the principle of indifference, and hence to theories of probability that have some reliance on that principle. Frequentist and propensity theories can reject that principle altogether, or accept it as a mere heuristic or (since they answer the Ascertainability criterion in terms of empirical detection of ontologically objective probabilities) even advance it as a reversed principle that defines equal epistemic status by equal probability. Consequently, such solutions to Bertrand's paradox are somewhat beside the point of the paradox.

16.4 Real and apparent paradoxes

We call something a paradox if it strikes us as peculiar in a certain way, if it strikes us as something that is not simply nonsense, and yet it poses some difficulty in seeing how it could be sense.[178] When we examine paradoxes more closely we find that for some the peculiarity is relieved and for others it intensifies. Some are peculiar because they jar with how we expect things to go, but the jarring is to do with imprecision and misunderstandings in our thought or failures to appreciate the breadth of possibility consistent with our beliefs. Other paradoxes, however, pose deep problems. Closer examination does not explain them away. Instead, they challenge the coherence of certain conceptual resources and hence challenge the significance of beliefs which deploy those resources. For brevity, I call the former kind *apparent* paradoxes and the latter *real* paradoxes. Whether a particular paradox is apparent or real is sometimes a matter of controversy—sometimes it has been realized that what was thought real is in fact apparent, and vice versa—but the distinction between the two kinds is generally worth drawing.

The pressure of paradox has often been a spur to intellectual endeavour. Apparent paradoxes have on occasion led us to greater clarity and

---

178 On paradoxes in general see also introductory discussions of Salmon 1970; Sorensen 2003; Clark 2007; Cook 2013. Sainsbury's definition—'an apparently unacceptable conclusion derived by apparently acceptable reasoning from apparently acceptable premises' (Sainsbury 1995:1)—is compatible with but narrower than mine.

precision in our thought. Real paradoxes have on occasion led us to radical conceptual innovation, indeed, have been the basis of entire research programmes. Such programmes often bifurcate. On the one hand, various means of evading the paradox are instituted, such as conceptual refinement, restriction or substitution. On the other hand, we continue to think about the paradox and think about what status should be accorded the means that avoids the paradox. One way for a real paradox to be resolved is for the means of evasion to be shown to be adequate to the issues raised by the paradox. For example, it is at least arguable (but also contested) that the mathematical resources developed by nineteenth century mathematicians are adequate to the conceptual problems in understanding time and space that Zeno's paradoxes raised.

Probability is especially rich in apparent paradoxes, since (it turns out) we are not good probabilistic thinkers but are rather prone to probabilistic fallacy. Nor need apparent paradoxes be easy to set at rest. Bertrand is also an originator here, with his coin paradox being the origin of the three prisoners paradox (Gardener 1959) and the Monty Hall problem (Selvin 1975).

> Three boxes are identical in appearance. Each has two drawers, each drawer contains a medal. The medals in the first box are in gold; those of the second box, in silver; the third box contains a gold medal and a silver medal.
>
> We choose a box; what is the probability of finding, in its drawers, a gold coin and a silver coin?
>
> Three cases are possible and are also equally possible since the three boxes are identical in appearance.
>
> Only one case is favorable. The probability is 1/3.
>
> The box is chosen. We open a drawer. Whatever medal is found there, only two cases remain possible. The closed drawer may contain a medal whose metal differs or not from that of the first. Of these two cases, one alone is favourable for a box whose coins are different. The probability of getting your hands on this box is therefore 1/2.
>
> How to believe, however, that it is enough to open a drawer to change the probability and from 1/3 to raise it to 1/2?
>
> (Bertrand 1889:2)

In opening a drawer you have new information and probabilities can change with new information. Nevertheless, ½ is incorrect.

The coin paradox[179] relies on our failure to distinguish two kinds of identity, qualitative identity and numerical identity. If we say that Jack and

---

179 Bertrand uses both médaille d'or and piece d'or but the paradox has in English always been called the coin paradox.

John are the same age we mean that there is a property that they have in common, their age. This is qualitative identity, because it is a matter of the identity of a property rather than of an object. But if we say that Jack and John are the same person we don't mean that Jack and John are distinct objects who share the property of personhood (if we meant that we'd say that they are *both* persons), but that Jack and John are one and the same person. This is numerical identity and it is a matter of the identity of an object.

Returning to the coins, supposing the first coin is gold, it is true that so far as qualitative identity goes, there are only two distinguishable options for the other coin, namely, gold or silver. But the possibilities we must distinguish are distinguished in terms of numerical identity. There are in fact *three* different gold coins that you might have revealed on opening the drawer. Only one of those coins is paired with a silver coin, the other two are paired with each other. So the chance is ⅓.

Greater care in our thought resolves the coin paradox. And yet despite being apparent, its further developments in the following paradoxes can still leave us struggling to find a definite error in our way of thinking about them that tells us which way to go.

I have twins and one of them, a boy, wanders in. I ask you 'What is the chance the other is a girl?' Determined not to be caught out in the way you were by Bertrand's box, you reason similarly that it must be ⅓—since either both are boys or one is a girl and one a boy, so there are three boys you might be seeing, only one of which is paired with a girl. It then occurs to you that the son you see is either the first or the second born. You might be seeing the eldest, who might be paired with a younger brother or sister, or you might be seeing the youngest, who might be paired likewise. So there are four possibilities, two of which are cases in which the sibling is a girl, so the chance is ½.

My friend also has twins and I tell you that one is a girl. What is the chance that the other is a boy? Wise to my tricks, you see your way to reasoning to either of ⅓ and ½ again. It then strikes you that the possibilities consistent with my statement are that the first child is a boy and the second a girl, or vice versa, or both are girls, and these possibilities are equally likely so the chance the other is a boy is ²/₃!

My other friend also has twins and I tell you that his first child is a boy. What is the chance that the other is a girl? Relieved, you conclude that, in at least this case, by telling you which child is a boy I have knocked out one of the three possibilities that is consistent with me telling you that *a* child is a boy, the possibility in which the first child is a boy and the second a girl. So there are only two left, either BG or BB, so the chance is ½. Or is it?

And we are not finished. We return to Bertrand's boxes, only now we have not gold and silver coins but two kinds of bosons, calls them yellow and blue. You open a compartment and see a yellow. What is the chance that the other particle in the box is a blue? Our earlier reasoning would imply that the answer is ⅓, but astonishingly, both physical theory and empirical investigation show it to be ½! Why is that? The physicists say bosons (and fermions) are *indistinguishable* particles, by which they seem to mean that they lack numerical identity. If that is the case, the earlier argument we gave based on numerical identity lapses and instead we can reason only on the basis of qualitative identity and distinction. Since the other particle is either yellow or blue the chance is ½.

The idea of particles lacking numerical identity is very difficult to understand. One way of interpreting the indistinguishability of particles would be similarly to the twins case when I tell you that one child is a girl. It might be that seeing a yellow is merely acquiring the information that *a* particle is yellow, so the possibilities consistent with that information are both yellow or one yellow and one blue, hence the chance is ½. However, that can't be the whole story. In the case of the children, they are numerically distinct, there is a fact about whether this child is the same child as that child, and it is what we know that fails to distinguish them. In the case of indistinguishable particles, whilst they are countable, the claim is not the epistemological claim that what we know fails to distinguish them but the metaphysical claim that there is no fact about whether this particle is the same particle as that particle. That is a deeply puzzling claim, but we shall leave its further investigation to the philosophers of physics.

16.5 Bertrand's paradox is still a real paradox

The question arises, is Bertrand's paradox an apparent or a real paradox? No strategy of solving Bertrand's paradox has worked. Our classification of solution strategies, our examination of all extant attempts at those strategies, the wide variety of thoughts and methods deployed in those attempts, and the exhibition of temptations that underlie those deployments, leaves very little reason to think there ever could be a solution. The soundness of the Rigorous argument remains unimpugned and its conclusion is that the principle of indifference is false. It may therefore appear that Bertrand's chord paradox is only an apparent paradox, a paradox that is in fact a settled refutation of the principle of indifference.

And yet can the negation of the principle really be true?

Negation of the Principle of Indifference: it is possible that probabilities of possibilities between which you lack any reason to discriminate epistemically are not equal.[180]

---

180 To make the logical form absolutely clear as before, negating the proposition in footnote 9 gives: Possibly, there exist possibilities, x, y, such that we have no reason to discriminate between x and y and for which Probability$(x) \neq$ Probability(y).

My only argument for its truth is the Rigorous argument. But examining the statement itself, and bearing in mind that we are concerned only with epistemic probability, it seems obviously and flatly false. If probabilities are to respect epistemic reasons, and surely they must do so, I do not see how having unequal probabilities in this way can be rational. Even granted that in such a case we are satisfying the ignorance requirement of Chapter 1, so we have a unique probability measure, if that measure gives distinct probabilities for possibilities of equal epistemic status, that measure is not respecting the state of ignorance but treating it arbitrarily in some way. In other words, we may be satisfying the ignorance requirement but we are failing the non-presumptuous requirement.

On the assumption that epistemic probability exists, it seems that the principle of indifference must be true, a view supported by the argument I gave and whose cogency is vindicated by the sustained support for and defence of the principle that continues to appear in the literature. And yet I have given the Rigorous argument for its falsity, have refuted all extant attempts to prove it unsound, and addressed the final hope from symmetry and shown it to be forlorn. It has proven impossible to solve the paradox.

The paradox poses a challenge to epistemic probability itself. By producing paradoxical probabilities it threatens our confidence in the coherence and comprehensiveness of probability and contradicts the principle of indifference, a principle which we have independent reasons to think true and which, moreover, appears self evidently true. On these grounds I believe that Bertrand's chord paradox remains a real paradox.

16.6 Bertrand's question is Entirely Unanswerable

The Rigorous argument works by deriving a contradiction and applying reductio ad absurdum to reject the premiss that the principle of indifference is true. There remains, however, a further assumption of the argument: that all events have probabilities. Consequently an alternative possibility for the conclusion of the argument is to reject that assumption: not all possibilities have probabilities. Instead of proving the principle false, the argument proves that the chords have no probabilities. I cannot see any other solution. It is the only alternative left.

The root we uncovered suffices for a way to defend this alternative. To get a measure you need a measure. This manifested for the Full Principle of Indifference for Sets by the need for a logically prior normative measure as an input, whose output is then equal probability for events with equal normative measure, leading then to the probability measure itself via Theorem 7 and its Corollaries 8 and 9. But there is no natural measure of the chords to fulfil the normative measure role. So there are no probabilities for the chords. This alternative is therefore based on the root of the paradox, the absence of natural measures for the chords.

This proposal does at least vindicate something of Bertrand's underlying intuition, since it is an instance of the Entirely Unanswerable strategy. If the paradox proves there are no probabilities for the chords, then Bertrand's question has an irreparable false presupposition, the error of supposing all events—*a forteriori* the chords—have probabilities. His question is therefore entirely unanswerable. So I am denying Premiss 3, that his question poses a determinate probability problem, because it poses no probability problem at all.

Rejecting the assumption that all events have probabilities also amounts to denying Premiss 2 of the Rigorous argument because it also presupposes the falsehood. If we add the premiss that all events have probabilities, however, the point of Premiss 2 is retained if we replace it with the conditional that *if* an event space has probabilities, then there is a single epistemically rational probability measure for that event space. We would then no longer be denying that premiss.

We have a reason why the chords lack probabilities and therefore why Bertrand's question is entirely unanswerable. We would like to know where the proposal leaves the principle of indifference and in particular how it fits with its central commitment to epistemic reason. The rest of the chapter sketches in the shape of the solution as a whole.

16.7 A covert Distinction strategy?

An immediate worry is whether this proposal doesn't resurrect the Distinction strategy. The difference between it and the Entirely Unanswerable strategy is only that the former allows the question to be resolvable. If, however, we have now proved that there are no probabilities for the chords, why shouldn't the Distinction strategist give up that part of his diagnosis? Instead of saying Bertrand's question is a clumsy way of raising whilst confounding distinct well-posed questions with unique answers, he could simply accept that it makes no such error. He can rest instead on the point that the question is unanswerable as it stands just because it presupposes the falsehood that the chords have probabilities. Then, instead of offering his various and distinct empirical realizations as *disambiguations* of Bertrand's question, he can simply offer them as empirical circumstances of choosing chords randomly. Hence, whilst we may have been correct in principle in our earlier critiques of the Distinction strategy, the absence of probabilities for the chords allows its resurrection in a modified form.

The first thing to say is that van Fraassen's trap immediately opens its ugly maw—but if in fact the chords have no probabilities, may not its bark now be worse than its bite? No! The feature of the earlier disambiguations that defeated them was not that they were disambiguations but that they work by specifying the random processes by which the chords are chosen.

Doing that is no solution because it replaces equal ignorance with knowledge of the random process and so amounts to a covert abandonment of the principle of indifference.

Absent the additional information about random processes, this modified Distinction strategy may still end up with paradoxical probabilities. Suppose for a moment that we have circumstances which justify equal ignorance over the directions of chords but not over the other properties of the chords. In §6.13, whilst discussing Rizza's solution, I exhibited what I called a new source of Bertrand's paradox in which not even a single variety of equal ignorance, namely equal ignorance over a single property, could avoid paradoxical probabilities. We demonstrated uniform parallel chord cross sections in two different namespaces, in the original one $[-R, R] \times [0,\pi)$ and in $[0,2\pi] \times [0,2\pi]/\sim$, where the former gives the probability of longer of ½ and the latter gives ⅓. Of course, if we define $[-R, R] \times [0,\pi)$ to be the correct namespace, that is define the naming function $\rho: C \rightarrow [-R, R] \times [0,\pi)$ from §3.12 to be *the* equal ignorance function, we can avoid this, but now we have fallen into van Fraassen's trap with bark and bite intact.

We therefore cannot assume we can block paradoxical probabilities for the distinct probability problems that the modified Distinction strategist is now going to offer us. Once again, the only way out is that there are no probabilities for the chords, even under the proffered distinct more informationally rich questions that the Distinction strategist claims are well-posed.

This conclusion is just one more paradox, since not only have we now got no probabilities for the chords in general, but none even for cases that were supposed to be constrained adequately. The physicist may be unworried by this, since they may say they only cared about real empirical situations. That, however, is the retreat into methodology, where now the principle of indifference is a mere heuristic whose results are to be corrected by observed frequencies. That is being corrected in the wrong kind of way to preserve the philosophical theory of them being epistemic probabilities derived from the principle of indifference.

16.8 No probabilities for continua?

If the reasoning of the preceding section is correct, the conclusion is worrying. Now we have lost entirely our grip on probabilities for the continua of chords, what other continua may slip through our fingers? The only consolation I can offer is the distinction between paucity and plenitude paradoxes. We have lost our grip on continua without natural measures but we still have the continua with them and can defend in some cases there being most natural measures, which are therefore the measures that should be used for the normative measures in POIS.

16.9 Giving up on probabilities

I now turn to literature on the principle of indifference that is not about probabilities. Comparative probabilities[181] that preserve linearity (all probabilities are comparable) and transitivity are the first step in giving up quantitative probabilities. It is provable that this much cannot be represented numerically (Fine 1973:18). Surprisingly little has to be added, however, to guarantee such representation, namely, that the event σ-algebra, taken as a topology, has a countable base. The representation is not necessarily unique but can be restricted to give numerical probabilities (Fine 1973:19). So it is unclear that this kind of comparative probability can avoid the paradox. In the following, we will see comparisons of rational belief that at least give up linearity and that may do so for that reason.

The position that not all events have quantitative probabilities has a respectable history. Boole remarks

> While extending the real power of the theory of probabilities, the method tends in some cases to diminish the apparent value of its results. For all problems in which the data admit of logical expression can be solved by it; but the resulting solutions, founded upon the bare data, may be of an indeterminate character.
>
> (Boole 1862:251)

von Kries held that

> there are relationships of the logical connection between certain things regarded as certain and others, which relation constitutes a greater or lesser degree of probability without a numerical measure existing for them.
>
> (von Kries 1886:26)

Keynes argues in Chapter 3 of his book that, contrary to 'generally accepted opinion' (1921/1973:21), not all probabilities are numerical. Moreover, 'comparisons of more and less are not always possible' (1921/1973:38) and so even if 'every probability lies on a path between impossibility and certainty' (1921/1973:41) they need not be totally ordered, let alone numerical quantities. In addition to probability, he introduced something he called the weight of arguments, which 'turns upon a balance, not between the favourable and the unfavourable evidence, but between the *absolute* amounts of relevant knowledge and relevant ignorance' (1921/1973:77). He acknowledges his views have been influenced by 'certain German logicians' (1921/1973:84) including von Kries. Fioretti (2001) offers a valuable articulation of their view of non-numerical probabilities and traces the history from von Kries to Keynes.

---

181 See axioms in Fine, T. 1973, 2016; Hawthorne 2016.

A recent response to the paradoxes, a kind of halfway house, is to give up on the precision of probabilities, which is not exactly giving up on numerical probabilities but it is using them in a way that loses being a total order. Joyce diagnoses the vulnerability to paradox in this way:

> It is wrong-headed to try to capture states of ambiguous or incomplete evidence using a single credence function.... Proponents of the Principle of Insufficient Reason were right to think that good epistemology requires us to treat hypotheses for which we have symmetrical evidence in the same way; they went wrong in thinking that equal treatment requires equal investments of confidence.
>
> (Joyce 2005:170–1)

Weatherson offers an account of imprecise probabilities intended to model 'the underlying motivations of Keynes' theory' (Weatherson 2002:50) and later he offers a development of the model as adequate for equal ignorance (Weatherson 2007).

In this halfway house, imprecise probabilities for an event space are sets of probability measures. Equal reason can be held to give the maximally imprecise probability for events, the interval [0,1]. I have not found anyone actually demonstrating the application of imprecise probabilities to the chords. The Deluge theorem **24** proves that for any random event of the chords and for each value in [0,1] there is a probability measure giving that value for the probability of that event. So the set of all those probability measures would be the kind of thing to be the imprecise probability for the event space for the chords. This might be regarded as an instance of the Well-posing strategy. Nevertheless, I have found Rinard's rebuttal, based on the fundamental premiss that 'it is never rational to have an interval-valued credence with an extreme endpoint' (Rinard 2014:111, defended in Rinard 2013), convincing and therefore do not regard imprecise probabilities as a solution to the paradox.

Recent authors have also proposed entirely non-numerical versions of the principle of indifference in the context of partial orders for rational belief. Norton defines his

> Principle of Indifference (PI). If we are indifferent among several outcomes, that is, if we have no grounds for preferring one over any other, then we should assign equal belief to each.
>
> (2008:47)

He then argues that the failure apparently manifested by his principle in the face of the paradoxes is in fact due to assuming that rational epistemic states are probability distributions. He concludes that 'the epistemic state of complete ignorance is not a probability distribution' (2008:45). He then offers a 'structure hospitable to nonprobabilistic belief states' (2008:59) which is defined as a partial order of 'nonnumerical degrees of confirmation' (2008:47). In the case of complete ignorance this gives 'a single ignorance degree of belief, "*I*", assigned to all contingent propositions' (2008:59), thereby preserving his principle.

For Novack

> The Principle of Indifference (Poi).... says that for any agent *x*, if *x* has no more reason to believe *A* than *B* and vice versa, then *x* ought not have a belief state leaning towards *A* over *B* and vice versa. It has an antecedent about evidence, and a normative consequent about the directionality (or lack thereof) of belief states.
>
> (2010:656)

He rejects

> assuming that degrees of belief ought rationally to conform to the probability calculus ("belief-probabilism"), that degrees of belief are the only interesting kinds of belief, or even that degrees of belief exist at all (Poi is consistent with the view that all belief is of the all-or-nothing type).
>
> (2010:657)

Instead, he uses 'an intuitive notion of a belief state that's tilted [asymmetric]' and defines 'to be credentially symmetric regarding *A* and *B* is to have a belief-state that isn't tilted' (2010:657). On this basis, numerical probability plays no role in his principle. He then defends the principle against various strengthenings of the square paradox by denying successively transitivity for symmetry and monotonicity for asymmetry (2010:660) and finally (and minimally) what he calls heredity (2010:665).

Eva defends the principle of indifference by using

> comparative confidence judgments about propositions drawn from some fixed Boolean algebra $\mathcal{B}$ of equivalence classes of logically equivalent sentences of a fixed language *L*.... They can be strictly more confident in the truth of *p* than they are in the truth of *q* [*p* > *q*]. Alternatively, *A* can be equally confident in the truth of *p* and *q* [*p* ~ *q*].
>
> (2019:394)

Taking comparative confidence to give a weak partial order ≥ (by the relation of being equal or more confident), he uses $p \odot q$ for when *p* and *q* are incomparable, that is when neither $p \geq q$ nor $q \geq p$. He assumes that the partial order obeys

(A1) T≥⊥
(A2) For any $p, q \in \mathcal{B}$, if $p \vdash q$ then $q \geq p$.
(A4) If $p \geq q$ then $\neg q \geq \neg p$

(2019:395 & 401)

In so assuming, it is clear that the weak partial order is the order of rational comparative confidence judgements that is not a matter of numerical probabilities. Eva illustrates how such an order might originate in or correspond with imprecise probabilities but stresses that this is only

> to aid the reader's intuitive understanding.... this is not indicative of any substantive philosophical commitments on my part.
>
> (2019:402)

The final version of the principle offered by Eva is

> Comparative Principle of Indifference 2 (CPI2): Let $X = \{x_1, x_2, ..., x_n\}$ be a partition of the set W of possible worlds into n mutually exclusive and jointly exhaustive possibilities. In the absence of any relevant evidence pertaining to which cell of the partition is the true one, a rational agent should not make any comparative confidence judgments regarding the pairs (x_i, x_j) for $i, j \leq n$. Equivalently, for all $i, j \leq n$, $x_i \odot x_j$ should hold in the agent's confidence ordering.
>
> (2019:402)

He then proves

> Theorem. Let $\mathcal{B}$ be the Boolean algebra encoding the logical structure of a language L. If an agent does not have access to any evidence pertaining to any non-trivial partitions of $\mathcal{B}$, then the only way for them to satisfy both CPI2 and A4 is for their confidence ordering over $\mathcal{B}$ to be isomorphic to $\mathcal{B}$ itself.[182]

Finally, he gives a convincing demonstration of how this treats the square paradox consistently, provided we have confined ourselves to finite partitions of the possibilities.

16.10 The Principle of Rational Strength

A virtue of offering the proportioning ground (or something similar) for the principle of indifference is that it gives us the possibility of a strategic

---

182 The paper switches between speaking of partitions of possible worlds and partitions of Boolean algebras of sentences of a language without defining their relation. The proof relies on identifying these without further explanation.

retreat. Recall the principle of indifference was derived from the Principle of Proportionate Reason on the assumption that credal commitments are necessarily rationally related to probabilities. If, however, there are possibilities without probabilities, then for those possibilities we must give up that assumption and make it conditional on possibilities that have probabilities. We still have the superordinate principle, however.

At this point I must introduce something I'm going to call *rational strength*, for want of a better term. Rational strength is a more general feature that we need to introduce if we are going to give up on probabilities. The literature we discussed previously is using concepts that may be subsumed under rational strength and its ordering. Rational strengths also bear some relation to Keynes' proposal that not all probabilities are measurable. Certainly rational strengths may fit with well with what Keynes has to say about his non-numerical probabilities, since for him 'probability... compris[es] that part of logic which deals with arguments that are rational but not conclusive' (1921/1973:241). It may also fit with the views of von Kries and other 19th-century German logicians who influenced Keynes on this point. Perhaps his terminology would in the end be better, especially since our common talk of chance and chances already allows likely and unlikely to be ordered but non-quantitative and not necessarily always comparable. Nevertheless, for the time being it avoids confusion to leave probabilities as the quantitative properties I have defined them to be and to find a different term for the more general feature. I shall not attempt to resolve the question of its relation to Keynes' non-numerical probabilities.

When we take credal commitments to be necessarily rationally related to rational strength, the Principle of Proportionate Reason issues in

> *The Principle of Rational Strength*: necessarily, the rational strengths of possibilities between which you lack any reason to discriminate epistemically are equal.

There is an obvious step we can now take. Recall that possibilities have equal reason just in case the balance of reasons for each of them is equal. That balance is the resultant of reasons for and against. Earlier we allowed that we might have no more from this balance than of it being equal or not equal. However, that reasons can be more or less strong is entirely defensible and that the strength of reasons should be at least partially ordered is a viable position.[183]

---

183 There is a very large literature on the strength and weight of reasons and here is not the place to review it. For an overview see Cullity 2018 and for recent work see Lord and Maguire 2016.

If reasons themselves have strengths, then the balance of reasons is the resultant strength of reasons for and against. If we define the rational strength of a possibility to be the resultant strength of reasons for and against, the Principle of Rational Strength is certainly true! It might be thought thereby to be trivial. Nevertheless, rational strength can do work.

Should rational strength for some domain of events be totally ordered and scalar, giving us rational weights (say), then they themselves give a route to probabilities. The route, however, is not straightforward. First, possessed reasons can be fine grained enough to distinguish intensional contexts and modes of presentation, whereas propositions may be more coarse grained, and so rational weights may be finer grained than probabilities. Second, the balance of reasons can be against a proposition which would make the weight negative. Third, as I mentioned before, if there are no reasons for a possibility that possibility is in fact (epistemically) impossible and so isn't in the event space, or alternatively, it is the empty event. Consequently, rational weight should obey some version of regularity, although the next point shows that this can't be defined by being non-zero as it is for probability. Finally, rational weights are arguably unbounded, at least for continuum sized event spaces, E. The absolute certainty of E has a greater rational weight than any event that is E-less-countably-many-atomic-events and the absolute impossibility of the empty event has a lesser rational weight than any atomic event. So we'll either need a signed measure to the extended reals (Halmos 1974:118) or define a hyper-real measure to an interval of the reals extended with infinitesimals. To get probabilities from rational weights in the latter case we would take the real part of the hyper-real measure and in the former use one of the standard functions that maps the extended reals to an interval, such as arctan or a logistic (composed with an appropriate translation and linear scaling).

All that being said, even if rational strengths are sometimes totally ordered and scalar, the weight of reasons literature shows we don't know how to measure them. Nevertheless, what we have done here is shifted a number of problems that arise for probabilities onto rational weight. My suggestion is that that is the correct place to place them. They are problems for probabilities just because they are problems for rational weight and if we could solve them for rational weights we'd have solved them for probabilities.

16.11 Restricting the domain of the principle of indifference

The line of argument I have been developing implies that, contrary to my original statement, the domain of the principle of indifference is restricted because not all events have probabilities. The problem that has been had generally in defending the principle by use of domain restriction is that attempts to do so are methodological and heuristic rather than philosophical and principled. Adopting rational strength gives us a principled distinction by which to characterise the restricted domain of the principle of indifference.

> *The Domain of the Principle of Indifference*: The principle applies to possibilities iff the rational strengths for those possibilities are measurable.

We get this from a virtue of the principle of indifference that we noted before. It allows us to set aside the difficulties of weighing rational strengths because it does not rely on rational strength being discriminable beyond being equal or unequal. With equal reason as an input it returns to us a total quantitative order on the events. Bearing in mind the remarks about the relation of rational weights on the extended reals to probability, we can map the probabilities to a signed measure of rational strengths.

Here, of course, we are appealing to the traditional view of probability as being intimately related to a measure of rational strength. Conversely, for events for which the rational strengths have weights, then those weights imply probabilities. Despite our inability to measure those weights, if they have weights then equal and unequal reason is defined for those events and hence a full epistemic state over the event space is defined. Hence FPOIS applies and the FPOIS begotten probability measure is the probability measure implied by the rational weights.

I concede that this domain restriction is of little practical use. Its virtue is that it is principled on the assumption that epistemic reasons have strengths and some such strengths can have weights in the ways described. It is accompanied by a criterion for whether rational strengths are measurable and therefore weighable. If we can derive paradoxical probabilities from the principle of indifference, that is a proof that the rational strengths of those events are not weighable.

16.12 The Entirely Unanswerable solution to Bertrand's paradox

I can now articulate the shape of my Entirely Unanswerable solution to Bertrand's paradox. The negation of the principle of indifference means that events between which we have no epistemic reason to discriminate can have distinct epistemic probabilities, but that is impossible. And yet the paradox still stands, and the firmness with which it stands is shown both by the failures of the other strategies and by the unearthed root. Consequently, there is only one answer left: that the paradox is a proof that not all events have probabilities.

To sustain the solution we need a principled rather than ad hoc domain restriction for the principle of indifference. To do this I rely on distinguishing rational strength from probabilities. Credal commitments are rationally related to probabilities for all and only possibilities that have probabilities. In those cases, the Principle of Proportionate Reason implies the Principle of Indifference. Where there are no probabilities, possibilities yet have rational strength determined by the balance of reasons for and against them. When we take credal commitments to be necessarily rationally related to rational strength, the Principle of Proportionate Reason entails the Principle of Rational Strength.

Admittedly, rational strength is so far only thinly defined. Nevertheless, it is compatible with Keynes', von Kries' and the earlier German logicians' ideas of non-numerical probabilities and also with the ideas of more recent authors defending a principle of indifference that is about a non-numerical rational relation. The natural next step is to use the strength of reasons and identify rational strength with the strength of the balance of reasons defined by the strength of the reasons. This proves the Principle of Rational Strength true. If the strength of reasons gives us no more than balances being equal or unequal, we can still ascertain probabilities where they exist with the Principle of Indifference. If additionally the strength of reasons is weighable, equal weights of balances can be used to define the normative measure in the Principle of Indifference. Finally, wherever we can use the Principle of Indifference we can weigh the rational strengths, thereby vindicating the traditional idea that probability measures the strength of epistemic reasons.

Altogether, this gives us the principled Domain of the Principle of Indifference, whereby the principle applies only to those possibilities having measurable rational strengths. Instead of paradoxical probabilities proving the Principle of Indifference false, they are a criterion for proving that we are outside the domain. Therefore Bertrand's paradoxical probabilities prove not that the Principle of Indifference is false but that the chords are outside its domain and don't have probabilities. Hence Bertrand's question is Entirely Unanswerable, exactly as Bertrand himself said it was, and the classical probabilism defined in §13.11 may be revived.

16.13 Conclusion

The principle of indifference supplies a satisfaction of Salmon's Ascertainability criterion for philosophical theories of probability that take an epistemic view of probability. I examined how the requirement to proportion our belief to our possessed epistemic reasons brought into view the normative ground presupposed by the principle. This led to a comprehensive definition of the principle including a definition of its inputs and outputs, both in its most general form and in its generation of subordinate principles. I then defined the subordinate Principle of Indifference for Sets and articulated a vital distinction between the two roles played by measures in the principle, the logically prior role that I call the normative measure role and the logically posterior role defining probabilities.

I expounded Bertrand's four paradoxes for infinite state spaces. Bertrand's derivations of probabilities for the chord paradox have been followed without query for more than a century. Nevertheless, his procedures suffer from six frailties, each of which may suffice to reject the paradox. I developed a range of analytic tools and used them to replace Bertrand's procedure with mathematically rigorous treatments. My analysis showed that Bertrand's angle and direction cases can be entirely reformulated with neither reliance on Bertrand's own flawed procedures nor vulnerability to any of his frailties. Bertrand's midpoint case, however, cannot, and I also gave a measure-theoretic argument to reject it. An analysis of the distinction between determinate and indeterminate questions put me in a position to give the Rigorous argument from the paradox to the falsity of the principle of indifference. I distinguished the four kinds of strategies for solving the paradox that are mutually exclusive and jointly exhaustive. For each strategy I gave a characteristic diagnosis, including whether Bertrand's question is a good or bad question, and a response to the Rigorous argument in terms of premises rejected.

I showed that Marinoff's attempted solution using the Distinction strategy falls into what I called van Fraassen's trap. Falling into van Fraassen's trap is not specific to Marinoff's solution but applies to the Distinction strategy as such, and I identified the same error in other literature pursuing this strategy.

Piecemeal attempts at the Well-posing strategy won't succeed. What is required is an appeal to general principle. But there is a Well-posing Danger that proposed principles, whilst general in themselves, impose in their use a covert restriction on Bertrand's question, thereby substituting a restriction of the problem for the general problem rather than comprehending the general problem. Jaynes, Wang and Jackson and Rizza all backed their solutions with principles but fell variously for the danger.

I examined the two attempts at the Irrelevance strategy in the literature from Bangu and from Gyenis and Rédei. Both relied on drawing distinctions that I showed to be without differences, and their solutions failed for that reason.

Applying the Maximum Entropy Principle counts as an instance of the Well-posing strategy pursued by use of meta-indifference. In the chapter appendix I solved the substantial difficulties that lie in the way of even applying the principle and in the chapter I showed that the paradox recurs in various ways, including in an original revenge paradox of the chords that is unique to the Maximum Entropy Principle. Aerts and Sassoli de Bianchi offer a measure theoretic meta-indifference they call the universal average. My analysis showed that their universal average is meta-indifferent only in some very etiolated sense and that even in that sense, the paradox still recurs. I concluded the examination of Well-posing by meta-indifference by addressing meta-indifference in the general terms of the two evident kinds. There isn't a general argument to be given against the first kind so I noted some general dangers that emerged in our chapter on the Maximum Entropy principle and left it as defeated so far as any known contenders go. I addressed the second kind by defining a Principle of Meta-Indifference for Sets and showed that this principle fails due to a recurrence of the paradox.

According to permissivism a particular pair of the premises in the Rigorous argument are contraries, the pair that are denied by the Irrelevance strategy. I argued that epistemic objectivity requires uniqueness and showed that objective permissivism amounts to the denial of the principle of indifference. Consequently theories of epistemic probability that subscribe to both the principle of indifference and epistemic objectivity are committed to uniqueness. Two papers in the literature take uniqueness to be a criterion of identity for the principle of indifference, giving thereby an instance of the Irrelevance strategy. However, because uniqueness arises independently of the principle, being a requirement of epistemic objectivity, uniqueness is not a criterion of the principle's identity but of its success and so this solution cannot work.

The hope from symmetry requires a Justified Guarantee that the principle of indifference will not produce paradoxical probabilities. Gyenis and Rédei's Justified Guarantee for their compact topological group representation of symmetry is false. I used group theory to analyse the hope from symmetry in full generality, noting that if equal reason is an equivalence relation then this analysis of the hope from symmetry covers all hopes. I gave a mathematical proof that the obvious candidate for a Justified Guarantee fails to guarantee consistency. Finally, I proved that no condition can be a Justified Guarantee and so the hope from symmetry is forlorn.

Returning once again to measure theory allowed us to unearth the root of the failed attempts at solutions: some state spaces have a natural measure of their own and some don't. Apart from my earlier work, prior discussions of Bertrand's paradox have failed to attend to this distinction. It explains why the chord paradox is rightly renownedly known as Bertrand's paradox, even if not so known for this reason. The chord paradox has no natural measure, no measure of its own, and this is why no one has solved it.

Three temptations that mislead us in our struggle with Bertrand's paradox originate in legitimate methods of scientific and mathematical enquiry, methods which can be misapplied when addressing the philosophical problem of Bertrand's paradox and which I showed being misapplied during the book. A fourth temptation is philosophical, the evasion of epistemic objectivity's commitment to uniqueness, and I examined it with a case study of the response of objective Bayesians to the paradox.

Finally, I defended the continuing paradoxicality of the paradox and proposed my own solution, an instance of the Entirely Unanswerable strategy. The Rigorous argument has a premiss other than the principle of indifference to reject in the face of the contradiction, namely, that all events have probabilities, giving a view that is compatible with some literature on partially ordered comparative probability. From our proportioning ground I derived a principle more general than the principle of indifference, The Principle of Rational Strength.

Making some further assumptions about the nature of epistemic reasons proves the Principle of Rational Strength true and gave us a solution to Bertrand's paradox using a principled rather than ad hoc restriction of the domain of the principle of indifference. The rational strengths of the events of choosing chords are not measurable because applying the principle of indifference to Bertrand's paradox produces paradoxical probabilities. Consequently Bertrand's question is Entirely Unanswerable because the chords do not have probabilities.

Bibliography

Aerts, D. & Sassoli de Bianchi, M. 2014. Solving the Hard Problem of Bertrand's Paradox. *Journal of Mathematical Physics*, 55, 083503.

Aerts, D. & Sassoli de Bianchi, M. 2015a. The Unreasonable Success of Quantum Probability I: Quantum Measurements as Uniform Fluctuations. *Journal of Mathematical Psychology*, 67, 51–75.

Aerts, D. & Sassoli de Bianchi, M. 2015b. The Unreasonable Success of Quantum Probability II: Quantum Measurements as Universal Measurements. *Journal of Mathematical Psychology*, 67, 76–90.

Andersen, K. 1985. Cavalieri's Method of Indivisibles. *Archive for History of Exact Sciences*, 31, 291–367.

Ardakani, M. & Wulff, S. S. 2014. An Extended Problem to Bertrand's Paradox. *Discussiones Mathematicae Probability and Statistics*, 34, 23–34.

Arnborg, S. & Sjödin, G. 2000. On the Foundations of Bayesianism. Preprint submitted to Elsevier, http://www.stats.org.uk/bayesian/ArnborgSjodin2000.pdf

Banach, S. & Tarski, A. 1924. Sur la Décomposition des Ensembles de Points en Parties Respectivement Congruentes. *Fundamenta Mathematicae*, 6, 244–77.

Bangu, S. 2010. On Bertrand's Paradox. *Analysis*, 70, 30–5.

Bartha, P. & Johns, R. 2001. Probability and Symmetry. *Philosophy of Science*, 68, S109–S22.

Basano, L. & Ottonello, P. 1996. The Ambiguity of Random Choices: Probability Paradoxes in Some Physical Processes. *American Journal of Physics*, 64, 34–9.

Beck, C. & Schlögl, F. 1993. *Thermodynamics of Chaotic Systems: An Introduction.* Cambridge: Cambridge University Press.

Bernoulli, J. 1713. *Ars Conjectandi Opus Posthumum.* Basileae: Impensis Thurnisiorum, Fratum.

Bertrand, J. L. F. 1889. *Calcul des Probabilités.* Paris: Gauthier-Villars et Fils.

Bingham, N. 2010. Finite Additivity Versus Countable Additivity. *Electronic Journal for History of Probability and Statistics*, 6, 1–35.

Boole, G. 1862. On the Theory of Probabilities. *Philosophical Transactions of the Royal Society of London*, 152, 225–52.

Borel, E. 1909. *Eléments de la Théorie Des Probabilités.* Paris: Librairie Scientifique J. Hermann.

Bovens, L. & Hartmann, S. 2003. *Bayesian Epistemology.* Oxford: Clarendon Press.

Brentano, F. 1915. Gedankengang Beim Beweise Für Das Dasein Gottes. *Theologie*, 24. Graz: Franz Brentano Archive.

Brentano, F. 1917. Zum Bertrand'schen Problem. *Erkenntnislehre und Logik, 22.* Graz: Franz Brentano Archive.

Bruckner, A. M., Bruckner, J. B. & Thomson, B. S. 2008. *Real Analysis.* 2nd ed. ClassicalRealAnalysis.com.

Burock, M. 2005. Indifference, Sample Space, and the Wine/Water Paradox. PhilSci Archive.

Butler, J. 1736. *The Analogy of Religion, Natural and Revealed, to the Constitution and Course of Nature: To Which Are Added Two Brief Dissertations: I. Of Personal Identity. II. Of the Nature of Virtue.* London: James, John and Paul Knapton, at the Crown in Ludgate Street.

Cantor, G. 1955. *Contributions to the Founding of the Theory of Transfinite Numbers.* New York: Dover Publications.

Capinski, M. & Kopp, P. E. 2004. *Measure, Integral, and Probability.* 2nd ed. London: Springer.

Carathéodory, C. 1914. Über Das Lineare Mass Von Punktmengen – Eine Verallgemeinerung Des Längenbegriffs. In *Gesammelte Mathematische Schriften: Hrsg. Im Auftrag Und Mit Untersttzung Der Bayerischen Akademie Der Wissenschaften.* Vol. 4. Munchen: Beck. 249–75.

Carathéodory, C. 1963. *Algebraic Theory of Measure and Integration.* New York: Chelsea Pub. Co.

Carnap, R. 1950. *Logical Foundations of Probability.* Vol 48. Chicago: University of Chicago Press.

Carnap, R. 1955. Statistical and Inductive Probability. In *Philosophy of Probability.* Ed. Eagle, A. Abingdon: Routledge. 317–26.

Carnap, R. 1971. A Basic System of Inductive Logic, Part I. In *Studies in Inductive Logic and Probability.* Vol. 1. Eds. Jeffrey, R. & Carnap, R. University of California Press: Los Angeles. 34–165.

Carroll, L. 1962. *Through the Looking Glass.* London: Folio Society.

Castell, P. 1998. A Consistent Restriction of the Principle of Indifference. *The British Journal for the Philosophy of Science*, 49, 387–95.

Chiu, S. S. & Larson, R. C. 2009. Bertrand's Paradox Revisited: More Lessons about That Ambiguous Word, Random. *Journal of Industrial and Systems Engineering*, 3, 1–26.

Christensen, D. 2007. Epistemology of Disagreement: The Good News. *Philosophical Review*, 116, 187–217.

Clark, M. 2007. *Paradoxes from A to Z.* 2nd ed. London: Routledge.

Climenhaga, N. 2020. The Structure of Epistemic Probabilities. *Philosophical Studies*, 177, 3213–42.

Cook, R. T. 2013. *Paradoxes.* Cambridge: Polity Press.

Cox, R. T. 1946. Probability, Frequency and Reasonable Expectation. *American Journal of Physics*, 14, 1–13.

Cox, R. T. 1961. *The Algebra of Probable Inference.* Baltimore: Johns Hopkins Press.

Cozman, F. G. 2016. Imprecise and Indeterminate Probabilities. In *The Oxford Handbook of Probability and Philosophy.* Eds. Hájek, A. & Hitchcock, C. Oxford: Oxford University Press. 296–311.

Crofton, M. W. 1885. Probability. *Encyclopaedia Britannica.* Edinburgh: Adam and Charles Black. 768–88.

Crossley, M. D. 2010. *Essential Topology*. London: Springer.

Cullity, G. 2018. Weighing Reasons. In *The Oxford Handbook of Reasons and Normativity*. Ed. Star, D. 423–42.

Dancy, J. 2004. *Ethics without Principles*. Oxford: Clarendon Press.

Dancy, J. 2018. *Practical Shape: A Theory of Practical Reasoning*. Oxford: Oxford University Press.

de Cristofaro, R. 2008. A New Formulation of the Principle of Indifference. *Synthese*, 163, 329–39.

de Finetti, B. 1970. *Theory of Probability*. New York: Wiley.

Deakin, M. 2006. The Wine/Water Paradox: Background, Provenance and Proposed Resolutions. *The Australian Mathematical Society Gazette*, 33, 200–5.

Dogramaci, S. & Horowitz, S. 2016. An Argument for Uniqueness about Evidential Support. *Philosophical Issues*, 26, 130–47.

Donkin, W. F. 1851. On Certain Questions Relating to the Theory of Probabilities. *The London, Edinburgh, and Dublin Philosophical Magazine and Journal of Science*, 1, 353–68.

Drory, A. 2015. Failure and Uses of Jaynes' Principle of Transformation Groups. *Foundations of Physics*, 45, 439–60.

Dynkin, E. B. 1965. *Markov Processes*. Vol 2. Berlin: Springer-Verlag.

Eagle, A. 2016. Probability and Randomness. In *The Oxford Handbook of Probability and Philosophy*. Eds. Hájek, A. & Hitchcock, C. Oxford University Press. 440–59.

Edwards, R. E. 1995. *Functional Analysis*. New York: Dover Publications.

Eva, B. 2019. Principles of Indifference. *Journal of Philosophy*, 116, 390–411.

Feldman, F. 2007. Reasonable Religious Disagreement. In *Philosophers without God: Meditations on Atheism and the Secular Life*. Ed. Antony, L. Oxford: Oxford University Press. 194–214.

Fine, K. 1994. Essence and Modality. *Philosophical Perspectives*, 8, 1–16.

Fine, T. 1973. *Theories of Probability: An Examination of Foundations*. New York; London: Academic.

Fine, T. 2016. Mathematical Alternatives to Standard Probability That Provide Selectable Degrees of Precision. In *The Oxford Handbook of Probability and Philosophy*. Eds. Hájek, A. & Hitchcock, C. Oxford: Oxford University Press. 203–47.

Fioretti, G. 2001. Von Kries and the Other 'German Logicians': Non-Numerical Probabilities before Keynes. *Economics and Philosophy*, 17, 245–73.

Fishburn, P. C. 1986. The Axioms of Subjective Probability. *Statistical Science*, 1, 335–45.

Fisher, R. A. 1922. On the Mathematical Foundations of Theoretical Statistics. *Philosophical Transactions of the Royal Society of London. Series A, Containing Papers of a Mathematical or Physical Character*, 222, 309–68.

Folland, G. B. 2009. *A Guide to Advanced Real Analysis*. Washington, D.C.: Mathematical Association of America.

Friedman, K. S. 1975. A Problem Posed. *Foundations of Physics*, 5, 89–91.

Fumerton, R. A. 1995. *Metaepistemology and Skepticism*. Lanham, MD: Rowman & Littlefield.

Funkenbusch, W. W. 1962. A New Probability Model for Bertrand's Paradox. *Mathematics Magazine*, 35, 144.

Galavotti, M. C. 2016. The Origins of Probabilistic Epistemology: Some Leading 20th-Century Philosophers of Probability. In *The Oxford Handbook of Probability and Philosophy*. Eds. Hájek, A. & Hitchcock, C. Oxford: Oxford University Press.

Gardener, M. 1959. Mathematical Games: Problems Involving Questions of Probability and Ambiguity. *Scientific American*, 201, 174–82.

Garwood, F. & Holroyd, E. M. 2016. The Distance of a "Random Chord" of a Circle from the Centre. *The Mathematical Gazette*, 50, 283–6.

Gillies, D. A. 2000. *Philosophical Theories of Probability*. London; New York: Routledge.

Greco, D. & Hedden, B. 2016. Uniqueness and Metaepistemology. *Journal of Philosophy*, 113, 365–95.

Grice, H. P. 1989. Logic and Conversation. In *Studies in the Way of Words*. Cambridge, MA: Harvard University Press. 22–40.

Gyenis, Z. & Rédei, M. 2015a. Defusing Bertrand's Paradox. *The British Journal for the Philosophy of Science*, 66, 349–73.

Gyenis, Z. & Rédei, M. 2015b. Why Bertrand's Paradox Is Not Paradoxical but Is Felt So. In *Recent Developments in the Philosophy of Science*. Eds. Maki, U., Ruphy, S. & Schurz, G. London: Springer.

Gyenis, Z. & Rédei, M. 2016. Measure Theoretic Analysis of Consistency of the Principal Principle. *Philosophy of Science*, 83, 972–87.

Haar, A. 1933. Der Massbegriff in der Theorie der Kontinuierlichen Gruppen. *Annals of Mathematics*, 2, 147–69.

Hájek, A. 2003. What Conditional Probability Could Not Be. *Synthese*, 137, 273–323.

Hájek, A. 2019. Interpretations of Probability. In *The Stanford Encyclopedia of Philosophy*. Ed. Zalta, E. N., https://plato.stanford.edu/archives/fall2019/entries/probability-interpret

Hájek, A. & Hitchcock, C. 2016. *The Oxford Handbook of Probability and Philosophy*. Oxford: Oxford University Press.

Halmos, P. R. 1974. *Measure Theory*. New York: Springer-Verlag.

Hawkins, T. 1970. *Lebesgue's Theory of Integration: Its Origins and Development*. Madison: University of Wisconsin Press.

Hawthorne, J. 2016. Logic of Comparative Support: Qualitative Conditional Probability Relations Representable by Popper Functions. In *The Oxford Handbook of Probability and Philosophy*. Eds. Hájek, A. & Hitchcock, C. Oxford: Oxford University Press.

Hawthorne, J. et al 2017. The Principal Principle Implies the Principle of Indifference. *British Journal for the Philosophy of Science*, 68, 123–31.

Holbrook, J. & Kim, S. S. 2000. Bertrand's Paradox Revisited. *The Mathematical Intelligencer*, 22, 16–19.

Howson, C. 2014a. Finite Additivity, Another Lottery Paradox and Conditionalisation. *Synthese*, 191, 1–24.

Howson, C. 2014b. What Probability Probably Isn't. *Analysis*, 75, 53–9.

Howson, C. & Urbach, P. 2006. *Scientific Reasoning: The Bayesian Approach*. 3rd ed. Chicago: Open Court.

Huemer, M. 2009. Explanationist Aid for the Theory of Inductive Logic. *British Journal for the Philosophy of Science*, 60, 345–75.

Hume, D. 1739/1978. *A Treatise of Human Nature*. 2nd ed. Eds. Selby-Bigge, L. A. & Nidditch, P. H. Oxford: Oxford University Press.

Jackson, C. S. 1903. The Following Paradox May Serve to Illustrate M. Bertrand's Remark "L'infini N'est Pas Un Nombre" (Calcul Des Probabilités, P. 4). *The Mathematical Gazette*, 2, 360.

Jacobson, N. 2009. *Basic Algebra*. Unabridged 2nd ed. New York: Dover.

Jaynes, E. T. 1957a. Information Theory and Statistical Mechanics. *Physical Review*, 106, 620–30.

Jaynes, E. T. 1957b. Information Theory and Statistical Mechanics II. *Physical Review*, 108, 171–90.

Jaynes, E. T. 1963. Information Theory and Statistical Mechanics. In *Statistical Physics*. Vol. 3. Ed. Ford, K. W. New York: Benjamin Inc. 182–218.

Jaynes, E. T. 1968. Prior Probabilities. *IEEE Transactions on Systems Science and Cybernetics*, 4, 227–41.

Jaynes, E. T. 1973. The Well Posed Problem. *Foundations of Physics*, 4, 477–92.

Jaynes, E. T. 1985. Some Random Observations. *Synthese*, 63, 115–38.

Jaynes, E. T. 2003. *Probability Theory: The Logic of Science*. Cambridge: Cambridge University Press.

Jeffreys, H. 1998. *Theory of Probability*. 3rd ed. Oxford: Clarendon Press.

Jost, J. 2015. *Mathematical Concepts*. London: Springer.

Joyce, J. M. 2005. How Degrees of Belief Reflect Evidence. *Philosophical Perspectives*, 19, 153–79.

Joyce, J. M. 2009. Accuracy and Coherence: Prospects for an Alethic Epistemology of Partial Belief. In *Degrees of Belief*. Vol. 342. Eds. Huber, F. & Schmidt-Petri, C. London: Springer. 263–97.

Joyce, J. M. 2015. The Value of Truth: A Reply to Howson. *Analysis*, 75, 413–24.

Keynes, J. M. 1921/1973. *A Treatise on Probability*. Royal Economic Society ed. London: Macmillan.

Khinchin, A. I. 1957. *Mathematical Foundations of Information Theory*. New York: Dover Publications.

Klyve, D. 2013. In Defense of Bertrand: The Non-Restrictiveness of Reasoning by Example. *Philosophia Mathematica*, 21, 365–70.

Knight, F. H. 1921. *Risk, Uncertainty and Profit*. Boston: Houghton Mifflin.

Kolmogorov, A. N. 1933/1956. *Foundations of the Theory of Probability*. 2nd English ed. New York: Chelsea Pub. Co.

Kyburg, H. E. 1970. *Probability and Inductive Logic*. London: Collier Macmillan.

Kyburg, H. E. & Teng, C. M. 2001. *Uncertain Inference*. Cambridge: Cambridge University Press.

Lakatos, I. 1976. *Proofs and Refutations: The Logic of Mathematical Discovery*. Cambridge: Cambridge University Press.

Landes, J., Wallmann, C. & Williamson, J. 2021. The Principal Principle, Admissibility, and Normal Informal Standards of What Is Reasonable. *European Journal for Philosophy of Science*, 11, 1–15.

Lando, T. 2021. Evidence, Ignorance, and Symmetry. *Wiley: Philosophical Perspectives*, 35, 340–58.

Laplace, P. S. 1812/1886. *Théorie Analytique Des Probabilités*. Paris: Gautheir-Villars.

Laplace, P. S. 1814/1995. *A Philosophical Essay on Probabilities*. Translated by F. L. Emory and F. W. Truscott. New York: Dover.

Lax, P. D. & Terrell, M. S. 2017. *Multivariable Calculus with Applications.*, London: Springer.

Loeb, P. A. 1979. An Introduction to Nonstandard Analysis and Hyperfinite Probability Theory. In *Probabilistic Analysis and Related Topics*. Ed. Bharucha-Reid, A. T. London: Academic Press. 105–42.

Lord, E. & Maguire, B. 2016. *Weighing Reasons*. New York: Oxford University Press USA.

Luce, R. D. & Raiffa, H. 1957/1985. *Games and Decisions: Introduction and Critical Survey*. New York: Dover Publications.

Lyon, A. 2016. Kolmogorov's Axiomatization and Its Discontents. In *The Oxford Handbook of Probability and Philosophy*. Eds. Hájek, A. & Hitchcock, C. Oxford University Press. 155–66.

Marinoff, L. 1994. A Resolution of Bertrand's Paradox. *Philosophy of Science*, 61, 1–24.

Mellor, D. H. 2005. *Probability: A Philosophical Introduction*. London: Routledge.

Mikkelson, J. M. 2004. Dissolving the Wine/Water Paradox. *British Journal for the Philosophy of Science*, 55, 137–45.

Milne, P. 1983. A Note on Scale Invariance. *British Journal for the Philosophy of Science*, 34, 49–55.

Moore, G. E. 1903. *Principia Ethica*. Cambridge: Cambridge University Press.

Nathan, A. 1984. The Fallacy of Intrinsic Distributions. *Philosophy of Science*, 51, 677–84.

Nickerson, R. 2005. Bertrand's Chord, Buffon's Needle, and the Concept of Randomness. *Thinking and Reasoning*, 11, 67–96.

Norton, J. D. 2008. Ignorance and Indifference. *Philosophy of Science*, 75, 45–68.

Novack, G. 2010. A Defense of the Principle of Indifference. *Journal of Philosophical Logic*, 39, 655–78.

November, D. D. 2019a. The Indifference Principle, Its Paradoxes and Kolmogorov's Probability Space. *PhiSciArchive*.

November, D. D. 2019b. Interpretive Implications of the Sample Space. *PhiSciArchive*.

Nualart, D. 2004. Kolmogorov and Probability Theory. *Arbor*, 178, 607–19.

Parfit, D. 2011a. *On What Matters*. Vol. 1. Oxford: Oxford University Press.

Parfit, D. 2011b. *On What Matters*. Vol. 2. Oxford: Oxford University Press.

Paris, J. 1999. Common Sense and Maximum Entropy. *Synthese*, 117, 75–93.

Paris, J. 2014. What You See Is What You Get. *Entropy*, 16, 6186–94.

Paris, J. & Vencovská, A. 1997. In Defense of the Maximum Entropy Inference Process. *International Journal of Approximate Reasoning*, 17, 77–103.

Pearl, J. 1988. *Probabilistic Reasoning in Intelligent Systems: Networks of Plausible Inference*. San Francisco: Morgan Kaufmann.

Pettigrew, R. 2016a. *Accuracy and the Laws of Credence*. Oxford: Oxford University Press.

Pettigrew, R. 2016b. Accuracy, Risk, and the Principle of Indifference. *Philosophy and Phenomenological Research*, 92, 35–59.

Pettigrew, R.2020. The Principal Principle Does Not Imply the Principle of Indifference. *British Journal for the Philosophy of Science*, 71, 605–19.

Poincaré, H. 1912. *Calcul Des Probabilités*. 2nd ed. Paris: Gauthier-Villars.

Popper, K. R. 2002. *The Logic of Scientific Discovery*. London: Routledge Classics.

Porto, P. D. et al 2011. Bertrand's Paradox: A Physical Way out Along the Lines of Buffon's Needle Throwing Experiment. *European Journal of Physics*, 32, 819–25.

Pressley, A. 2012. *Elementary Differential Geometry*. 2nd ed. London: Springer.

Qin, X. & Chen, M. 2023. Bertrand's Problem in Making a Face Mask. *The Mathematical Intelligencer*.45, 71–2.

Rényi, A. 1955. On a New Axiomatic Theory of Probability. *Acta Mathematica Academiae Scientiarum Hungarica*, 6, 285–335.

Rinard, S. 2013. Against Radical Credal Imprecision. *Thought: A Journal of Philosophy*, 2, 157–65.

Rinard, S. 2014. The Principle of Indifference and Imprecise Probability. *Thought: A Journal of Philosophy*, 3, 110–4.

Rizza, D. 2018. A Study of Mathematical Determination through Bertrand's Paradox. *Philosophia Mathematica*, 26, 375–95.

Rosenkrantz, R. D. 1977. *Inference, Method and Decision: Towards a Bayesian Philosophy of Science*. Dordrecht: Reidel.

Rowbottom, D. P. 2008. On the Proximity of the Logical and 'Objective Bayesian' Interpretations of Probability. *Erkenntnis*, 69, 335–49.

Rowbottom, D. P. 2013. Bertrand's Paradox Revisited: Why Bertrand's 'Solutions' Are All Inapplicable. *Philosophia Mathematica*, 21, 110–4.

Rudin, W. 1987. *Real and Complex Analysis*. 3rd ed. New York: McGraw-Hill.

Rudin, W. 1993. Autohomeomorphisms of Compact Groups. *Topology and its Applications*, 52, 69–70.

Rynne, B. P. & Youngson, M. A. 2008. *Linear Functional Analysis*. 2nd ed. London: Springer.

Sainsbury, R. M. 1995. *Paradoxes*. 2nd ed. Cambridge: Cambridge University Press.

Salmon, W. C. 1967. *The Foundations of Scientific Inference*. Pittsburgh, PA: University of Pittsburgh Press.

Salmon, W. C. 1970. *Zeno's Paradoxes*. Indianapolis, IN: Bobbs-Merrill.

Sartwell, C. 2016. Richard Rorty: An Intellectual Memoir. *Splice Today*.

Selvin, S. 1975. Letters to the Editor. *The American Statistician*, 29, 67.

Sergeyev, Y. 2009. Numerical Computations and Mathematical Modelling with Infinite and Infinitesimal Numbers. *Journal of Applied Mathematics and Computing*, 29, 177–95.

Shackel, N. 2005. The Vacuity of Postmodernist Methodology. *Metaphilosophy*, 36, 295–320.

Shackel, N. 2007. Bertrand's Paradox and the Principle of Indifference. *Philosophy of Science*, 74, 150–75.

Shackel, N. 2008. Paradoxes of Probability. In *Handbook of Probability Theory with Applications*. Ed. Rudas, T. Thousand Oaks, CA: Sage. 49–66.

Shackel, N. 2018. Scope or Focus? Normative Focus and the Metaphysics of Normative Relations. *Journal of Philosophy*, 115, 281–312.

Shackel, N. 2021. Constructing a Moorean 'Open Question' Argument: The Real Thought Move and the Real Objective. *Grazer Philosophische Studien*, 98, 463–88.

Shackel, N. 2022. Uncertainty Phobia and Epistemic Forbearance in a Pandemic. In *Values and Virtues for a Challenging World*. Vol. 92. Eds. Jefferson, A. et al Cambridge: Cambridge University Press. *Royal Institute of Philosophy Supplement*. 271–91.

Shackel, N. n.d.-a. Enkrasia and the Fixed Point Thesis Are Equivalent. Manuscript.

Shackel, N. n.d.-b. Theoretical Reason, Modifiers and Probability. Manuscript.

Shackel, N. n.d.-c. Wild Uncertainty. Manuscript.

Shackel, N. & Rowbottom, D. P. 2020. Bertrand's Paradox and the Maximum Entropy Principle. *Philosophy and Phenomenological Research*, 101, 505–23.

Shafarevich, I. R. 1990. *Basic Notions of Algebra*. Translated by M. Reid. Berlin: Springer-Verlag.

Shakespeare, W. 1750. *Romeo and Juliet*. London: Printed for J. and R. Tonson and S. Draper.

Shannon, C. E. 1948. The Mathematical Theory of Communication. *The Bell System Technical Journal*, 379–423, 623–56. Pages numbers in text refer to the reprint online.

Shimony, A. 1985. The Status of the Principle of Maximum Entropy. *Synthese*, 63, 35–53.

Soranzo, A. & Volčič, A. 1998. On the Bertrand Paradox. *Rendiconti del Circolo Matematico di Palermo*, 47, 503–9.

Sorensen, R. A. 2003. *A Brief History of the Paradox: Philosophy and the Labyrinths of the Mind*. Oxford; New York: Oxford University Press.

Spohn, W. 1986. The Representation of Popper Measures. *Topoi*, 5, 69–74.

Strevens, M. 1998. Inferring Probabilities from Symmetries. *Noûs*, 32, 231–46.

Suppes, P. & Zinnes, J. 1963. Basic Measurement Theory. In *Handbook of Mathematical Psychology, Volume I*. Eds. Luce, D. & Bush, R. London: John Wiley & Sons. 1–76.

Tao, T. 2011. *An Introduction to Measure Theory*. Providence: American Mathematical Society.

Tissier, P. E. 1984. Bertrand's Paradox. *The Mathematical Gazette*, 68, 15–19.

Titelbaum, M. G. & Hart, C. 2020. The Principal Principle Does Not Imply the Principle of Indifference, Because Conditioning on Biconditionals Is Counterintuitive. *British Journal for the Philosophy of Science*, 71, 621–32.

Uhl, J. J. 1984. Book Reviews. *Bulletin of the London Mathematical Society*, 16, 431–2.

van Douwen, E. K. 1984. A Compact Space with a Measure That Knows Which Sets Are Homeomorphic. *Advances in Mathematics*, 52, 1–33.

van Fraassen, B. C. 1989. *Laws and Symmetry*. Oxford: Clarendon Press.

Vidovic, Z. 2021. Random Chord in a Circle and Bertrand's Paradox: New Generation Method, Extreme Behaviour and Length Moments. *Bulletin of the Korean Mathematical Society*, 58, 433–44.

Vidovic, Z. 2022. Bertrand's Paradox: New Probabilistic Models. *Kragujevac Journal of Mathematics*, 49, 61–4.

von Kries, J. 1886. *Die Principien Der Wahrscheinlichkeitsrechnung, Eine Logische Untersuchung*. Tubingen: Mohr.

Von Mises, R. 1951/1981. *Probability, Statistics, and Truth*. 2d rev. English ed. New York: Dover Publications.

Walley, P. 1991. *Statistical Reasoning with Imprecise Probabilities*. London: Chapman and Hall.

Wang, J. & Jackson, R. 2011. Resolving Bertrand's Probability Paradox. *International Journal of Open Problems in Computer Science and Mathematics*, 3, 2–103.

Weatherson, B. 2002. Keynes, Uncertainty and Interest Rates. *Cambridge Journal of Economics*, 26, 47–62.

Weatherson, B. 2007. The Bayesian and the Dogmatist. *Proceedings of the Aristotelian Society*, 107, 169–85.

Weir, A. 1973. *Lebesgue Integration and Measure*. Cambridge: Cambridge University Press.

White, R. 2005. Epistemic Permissiveness. *Philosophical Perspectives*, 19, 445–59.

White, R. 2009. Evidential Symmetry and Mushy Credence. In *Oxford Studies in Epistemology*. Vol. 3. Eds. Szabo Gendler, T. & Hawthorne, J. Oxford: Oxford University Press. 161–86.

Willard, S. 2004. *General Topology*. Mineola, NY: Dover.

Williams, D. 1991. *Probability with Martingales*. Cambridge: Cambridge University Press.

Williamson, J. 1999. Countable Additivity and Subjective Probability. *British Journal for the Philosophy of Science*, 50, 401–16.

Williamson, J. 2009. Philosophies of Probability: Objective Bayesianism and Its Challenges. In *Handbook of the Philosophy of Mathematics*. Ed. Irvine, A. Amsterdam: Elsevier. 493–534.

Williamson, J. 2010. *In Defence of Objective Bayesianism*. Oxford: Oxford University Press.

Williamson, J. 2018. Justifying the Principle of Indifference. *European Journal for Philosophy of Science*, 8, 559–86.

Zabell, S. L. 2005. *Symmetry and Its Discontents: Essays on the History of Inductive Probability*. Cambridge: Cambridge University Press.

Zabell, S. L. 2016a. Johannes Von Kries's "Principien": A Brief Guide for the Perplexed. *Journal for General Philosophy of Science/Zeitschrift für allgemeine Wissenschaftstheorie*, 47, 131–50.

Zabell, S. L. 2016b. Symmetry Arguments in Probability. In *The Oxford Handbook of Probability and Philosophy*. Eds. Hájek, A. & Hitchcock, C. Oxford: Oxford University Press. 315–40.

Index

For Product Safety Concerns and Information please contact our EU
representative GPSR@taylorandfrancis.com
Taylor & Francis Verlag GmbH, Kaufingerstraße 24, 80331 München, Germany